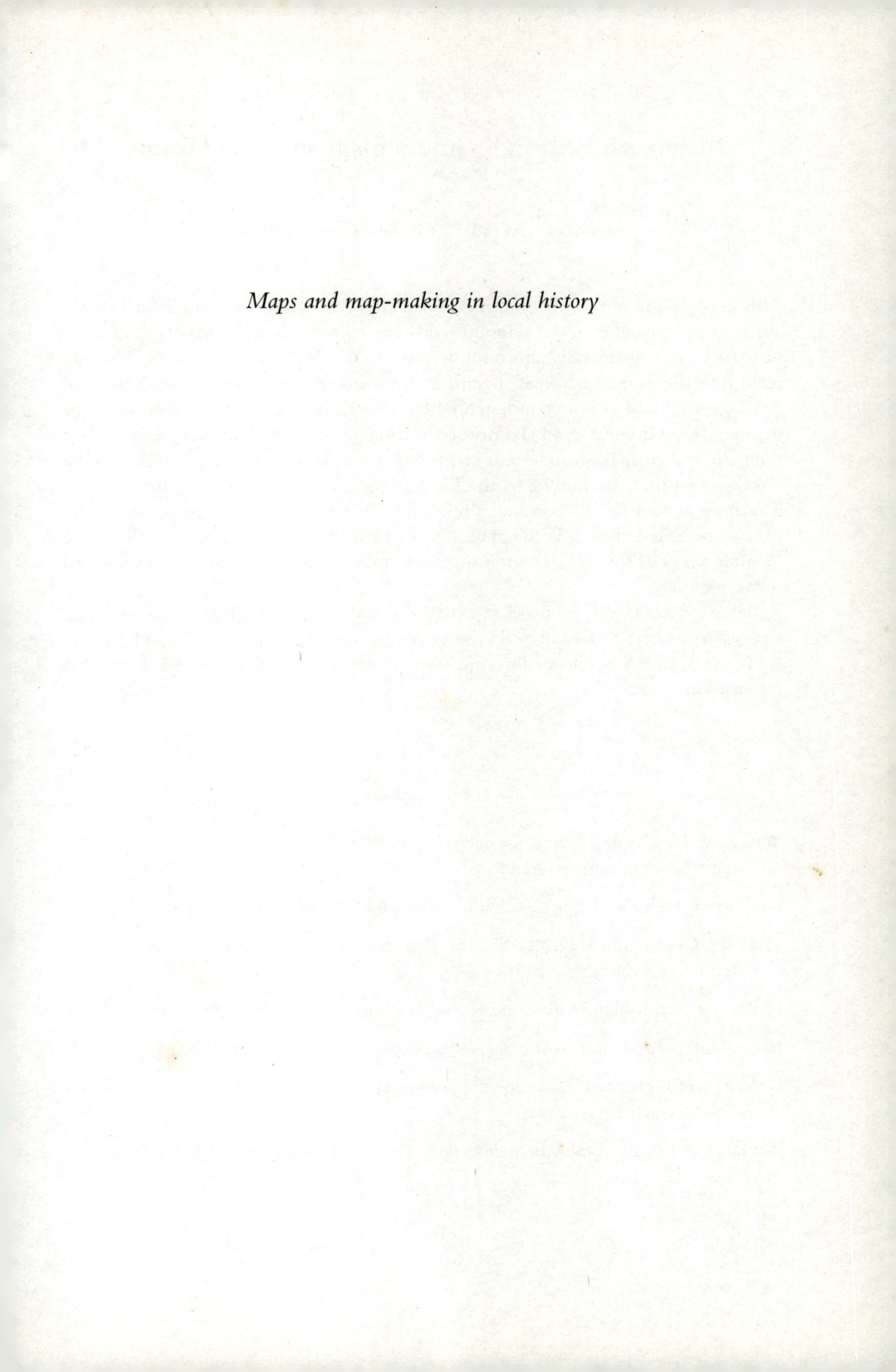

Maps and map-making in local history

Maynooth Research Guides for Irish Local History

GENERAL EDITOR Mary Ann Lyons

This book is one of the Maynooth Research Guides for Irish Local History series. Written by specialists in the relevant fields, these volumes are designed to provide historians, and specifically those interested in local history, with practical advice regarding the consultation of specific collections of historical material, thereby enabling them to conduct independent research in a competent and thorough manner. In each volume, a brief history of the relevant institutions is provided and the principal primary sources are identified and critically evaluated, with specific reference to their usefulness to the local historian. Readers receive step by step guidance as to how to conduct their research and are alerted to some of the problems which they might encounter in working with particular collections. Possible avenues for research are suggested and relevant secondary works are also recommended.

The General Editor acknowledges the assistance of both Dr Raymond Gillespie, Co-ordinator of the MA in Local History programme, NUI Maynooth and Dr James Kelly, St Patrick's College, Drumcondra, in the preparation of this book for publication.

IN THIS SERIES

Terence A.M. Dooley, *Sources for the history of landed estates in Ireland* (Irish Academic Press, 2000)

Raymond Refaussé, *Church of Ireland records* (Irish Academic Press, 2000)

Patrick J. Corish and David C. Sheehy, *Records of the Irish Catholic Church* (Irish Academic Press, 2001)

Philomena Connolly, *Medieval record sources* (Four Courts Press, 2002)

Brian Gurrin, *Pre-census sources for Irish demography* (Four Courts Press, 2002)

E. Margaret Crawford, *Counting the people: a survey of the Irish censuses, 1813–1911* (Four Courts Press, 2003)

Jacinta Prunty, *Maps and map-making in local history* (Four Courts Press, 2004)

Maynooth Research Guides for Irish Local History: Number 7

Maps and map-making in local history

Jacinta Prunty

FOUR COURTS PRESS

Set in 10.5 pt on 12.5 pt Bembo by
Carrigboy Typesetting Services for
FOUR COURTS PRESS
7 Malpas Street, Dublin 8
e-mail: info@four-courts-press.ie
and in North America by
FOUR COURTS PRESS
c/o ISBS, 920 N.E. 58th Avenue, Suite 300, Portland, OR 97213

A catalogue record for this title
is available from the British Library.

ISBN 1–85182–870–2 hbk
ISBN 1–85182–699–8 pbk

SPECIAL ACKNOWLEDGEMENT

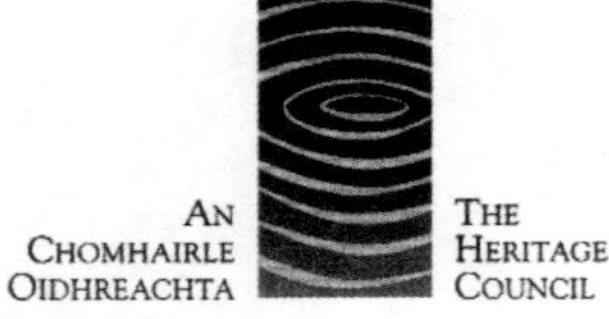

This publication was grant-aided by the Heritage Council
under the 2004 Publication Grant Scheme.

Jacinta Prunty is a Member of the
National Institute for Regional and Spatial Analysis

Printed and bound in Great Britain by
Antony Rowe Ltd, Chippenham, Wilts.

Contents

Illustrations

Foreword

There are certain subjects whose history cannot be studied without the help of maps. One, not the main subject of this book, is the early development of map-making itself, together with cognate branches of science, technology and business. Another is the landscape, defined in a broad geographical sense that includes not just material objects but many invisible features such as placenames and territorial boundaries. Finally, each map is itself an event in the history of a particular place and of the people whose lives are related to that place: whatever the cartographer's choice of subject-matter, he is ultimately concerned with man's attempts at exercising power – sometimes over nature, more often over other men. As Jacinta Prunty makes clear in the following pages, this last theme holds a special relevance for Irish readers.

For five centuries Ireland ranked among the most frequently and variously mapped countries of the world. Most of its maps were made for the benefit of foreigners who had money to spend on correcting their geographical ignorance. When the foreigners withdrew, much of their cartographic legacy remained, but many years went by before an independent Ireland could afford to house, classify and catalogue its historic maps as carefully as they deserved, or adequately to facilitate their use by students. In the last few decades, this situation has been transformed – to such an extent that the achievements of present-day Irish map curators now urgently need to be digested and summarised in a single convenient work of reference. It is a task for which no one can be better qualified than Dr Prunty, a distinguished historical geographer who has devoted much of her career to teaching the subject of this book at both undergraduate and graduate level in the National University of Ireland, Maynooth.

Part of her agenda is to answer, for Ireland, the kind of question that has already been addressed in Britain by J.B. Harley (*Maps for the local historian: a guide to the British sources*, 1972), David Smith (*Maps and plans for the local historian and collector*, 1988) and Paul Hindle (*Maps for historians*, 1998). The most important of these questions is: what can the historian learn from maps that cannot be learnt from other sources? Before replying, we must understand how the early cartographer's 'language' (including its silences) differed from that of his modern successors. We must then ask whether, in any given case, that language happens to be telling us the truth. Some cartographic scholars are reluctant to pose this question: some can hardly bring themselves to write of truth and accuracy at all except under the protection of inverted commas. Local historians cannot afford to be so squeamish: they must somehow persuade the early map-maker to communicate what we want to know as well as what he wanted to say. It is a difficult and perhaps impossible task, not to be undertaken without closely studying the aims, methods, sources and professional standards that lie behind the map under consideration. Such studies lead on to the categorisation of maps as illustrated in Dr Prunty's chapter 1. They also reveal how easily a map (with so much space to fill) can defy our classifications by stepping out of the character we have assigned to it. From which there follows a maxim that not every student will be grateful for: look at all the maps that cover your chosen territory, regardless of their scale and their ostensible subject-matter. From an avowed map enthusiast, such a counsel of perfection comes easily enough. Whether readers have the time and energy to act accordingly in real life is a question that not even Dr Prunty can answer on anyone else's behalf: what she can do, and has done, is supply the wherewithal for making a decision.

This book also provides two useful services not to be found in the British manuals mentioned above. First, its advice on the actual tracking-down of Irish maps and cartographic study-aids is commendably specific and down-to-earth: does one need an appointment, is there a fee to be paid, where are the catalogues kept, what is on open access – no practical detail escapes the author's vigilance. Secondly, and with equal thoroughness, she gives valuable lessons in cartographic design and execution to local historians who wish to map their own results. How to find maps, how to read maps, how to draw maps: put these themes together and we have a book that every serious student of Irish history must keep within easy reach.

J.H. ANDREWS

For Anngret Simms

Acknowledgements

My first debt of gratitude is to the series editor Mary Ann Lyons who proposed this text, and watched over its preparation with great patience and endless encouragement. I would like to acknowledge the kind assistance of colleagues in NUI Maynooth, UCD and TCD, and in a variety of repositories in Ireland and in the UK: Vincent Comerford, Raymond Gillespie, Terence Dooley and David Worthington (Modern History, NUIM); Patrick J. Duffy, Jim Walsh, Sheelagh Waddington, Jim Keenan and Brídín Feeney (Geography Department, NUIM); Anngret Simms, William Nolan, Joe Brady, Arnold Horner and Stephen Hannon (Geography, UCD), Howard Clarke (Medieval History, UCD), and Paul Ferguson (Glucksman Map Library, TCD), who made available his own typescript bibliography and guide to secondary work on the history of cartography in Ireland. Interdisciplinary support has come through NIRSA at NUI Maynooth; thanks to Rob Kitchin, current director, to Mary O'Brien for technical support, and to all colleagues who make up that research community. Special thanks are due to John H. Andrews who advised on numerous matters and gave very generous support throughout. Thanks are extended to John Bradley (Modern History, NUIM) and Paul Walsh and Chris Corlett (Dúchas) who assisted with the section on archaeological mapping, and to Susan Murphy of the Northern Ireland Office (sites and monuments). I am indebted to Niall Dardis of Dublin Port archives, to Brian Smyth and Michael Corcoran of Dublin Corporation Drainage Division, and to Georgina Laragy and Kieran Swords at the County Library, Tallaght for introductions to the maps in their care. John Kenny, Offaly Historical Society, kindly located maps of Killurin and Kenmare for use in this study. Student assistance came via Lorcan Egan, Bríd Heslin Miriam Moffitt, Maeve Mulryan Moloney, and Frank Cullen (Modern History, NUI Maynooth). At the National Archives of Ireland the assistance of Frances McGee, who is responsible for the OS archive, was generously given and much appreciated; I am also indebted to Aideen Ireland, Caitríona Crowe, Gregory O'Connor and the reading room staff who graciously retrieved so many documents. Julia Barrett, Sheila Astbury and Jane Nolan (Architecture Library UCD); Niall McKeith (National Science Museum Maynooth / Experimental Physics, NUIM), Raymond Refaussé and Susan Hood (RCBL), Bernadette Cunningham and Dymphna Moore (RIA Library), Dympna Kelledy (CIE), Ann McVeigh and Ian Montgomery (PRONI), Colum O'Riordan (IAA), Matthew Parkes (GSI), Máire Kennedy (Dublin City Libraries), David Simms (Mathematics, TCD), Kevin Bright (RDS), Olive Cullen (SPCM), Colette Ellison (RSAI) and also Eoghan O'Regan, formerly of the OS and later of Léaráid maps, have all assisted in various ways. At the National Library of Ireland, Joanna Finegan, Elizabeth Kirwan and Colette O'Flaherty also provided practical advice. Kevin

Mooney and Paddy Prendergast, DIT Bolton Street, helped advance my computer mapping skills through their evening courses, while Dennis Pringle (Geography Department, NUIM) introduced me to GIS in the most user-friendly way. Sarah Gearty and Angela Murphy of the *Irish Historic Towns Atlas* project (RIA) gave practical assistance and unfailing encouragement. Library colleagues at Maynooth provided every assistance: at the John Paul II Library (Agnes Neligan, Helen Fallon, Susan Durack, Mary Kearney) and at the Russell Library (Penny Woods and Celia Kehoe). Map library staff outside Ireland were also unfailingly helpful; I would like to extend thanks to Annie Pinder, Robert Harrison and Mari Takayanagi at the House of Lords Record Office, Westminster; to Fred Musto at the Sterling Map Library, Yale University; to Edward Redmond at the Geography and Map Division, Library of Congress; to Tony Campbell and Peter Barber at the British Library Map Room and to Lorraine Quinn, BL bookshop; to Neil Cobbett and Rose Mitchel at the Public Record Office (now the National Archives) Kew; and to staff at the Caird Library, National Maritime Museum Greenwich; the National Library of Scotland, Edinburgh, and at Cambridge University Library.

This book grew out of the third year 'Maps in history' special topic BA course at NUI Maynooth, the MA (Local History) and BA (Local and community studies) programme in Maynooth and Kilkenny, and the NUI Certificate in Local History offered at various centres throughout the country. These are the students with whom and for whom I have developed most of this material. Their eagerness to locate maps which will help unwrap the history of their home area, and their readiness to engage with maps as complex historical documents is the inspiration behind this guide. The keen desire of so many local history students to make their own maps is behind the final section.

Thanks must be extended to the Holy Faith communities at Celbridge and Killester where this work was carried out, to my parents, Agnes and Joe, my family and friends, especially Kieran Prunty and Bernie O'Hanlon, and to Edel and Paul Fitzgerald with Vivienne (and now Will), who provided me once more with a home from home when researching in London.

Thanks are extended for permission to reproduce the following illustrations: to the Royal Irish Academy (cover illustration, figs. 3a, 8, 43, 59); the National Library of Ireland (fig. 2); the Ordnance Survey of Ireland (figs. 1, 21, 32, 33, 34, 36, 47, 48, 50, 51, 53, 56, 57, 58); NUI Maynooth (fig. 15), the British Library (fig. 12), the National Archives of Ireland (figs. 16, 17, 18, 23, 24, 25, 26, 28, 29, 31, 60); the Irish Land Commission (fig. 19), the Geological Survey of Ireland (fig. 42), the Russell Library (figs. 27, 40), the Ordnance Survey of Northern Ireland (fig. 37), the Department of the Environment, Heritage and Local Government (fig. 41); the Central Statistics Office and Department of Geography NUI Maynooth (fig. 68); to Arnold Horner and the *Irish Historic Towns Atlas*, RIA (fig. 59), James Keenan and Jim Walsh (Geography, NUI Maynooth), and the Central Statistics Office (fig. 68), Martina O'Donnell (figure 61) and Anna Byrne (fig. 62). Permission to reproduce OS extracts is granted under permits number APL0000104 to APL0000404.

Abbreviations

ACE	Address Centred Extract
BL	British Library
CSO	Central Statistics Office
CUL	Cambridge University Library
DDA	Dublin Diocesan Archives
DED	District Electoral Division
DS	Down Survey
ESB	Electricity Supply Board
GIS	Geographic Information System
GSGS	Geographical Section, General Staff
GSI	Geological Survey of Ireland
GSNI	Geological Survey of Northern Ireland
LC	Library of Congress, Washington
MHTI	Mining Heritage Trust of Ireland
NAI	National Archives of Ireland
NLI	National Library of Ireland
NLS	National Library of Scotland, Edinburgh
NMM	National Maritime Museum, Greenwich
NUI Maynooth	National University of Ireland, Maynooth
OD	Ordnance Datum
OS	Ordnance Survey
OS*i*	Ordnance Survey of Ireland
OSNI	Ordnance Survey of Northern Ireland
PLACE	Planning, Legal, Agricultural, Construction, Engineering
PRO	Public Record Office, Kew
PRONI	Public Record Office of Northern Ireland
QRO	Quit Rent Office
RIA Proc.	*Proceedings of the Royal Academy of Ireland*
RC	Roman Catholic
RCBL	Representative Church Body Library
RD	Rural District
RIA	Royal Irish Academy
RIC	Royal Irish Constabulary
RPDCD	*Reports and printed documents of the Corporation of Dublin*
TCD	Trinity College Dublin
TNA	The National Archives, Kew
UCD	University College Dublin
VO	Valuation Office
WD	War Department
WO	War Office

Introduction

WHAT ARE MAPS?

> Maps are graphic representations that facilitate a spatial understanding of things, concepts, conditions, processes or events in the human world.[1]

In this definition, the editors of the monumental *History of cartography* series (1987–) try to answer the question: what, in essence, is a map? It is a drawing or graphic, though the materials with which it is produced can be anything available to human ingenuity, from baked clay, rock walls, desert sand, shells and sticks to metal, paper and electronic media. The 'spatial understanding' can include all efforts to make sense of relationships in space, and to communicate them to others. Tracing routeways to hunting grounds or pilgrimage sites, planning the drainage of bogland, determining battle formations (real or imagined, before or after the event), describing the location of a distant land, predicting the movements of stars and planets – all of these, and much more, may result in maps.

Cartography has been defined as 'the study of maps, their historical evolution, methods of cartographic presentation, and map use'.[2] This definition provides key entry points to the responsible and fruitful use of maps in local studies. Several individuals and groups are likely to have been involved in the creation of the map: those who conducted the survey or otherwise provided information; those who sponsored or commissioned the product; cartographers, technicians and artists; printers, publishers and distributors. Who were the map users? What did they contribute to the process? Taking account of even one of these parties, and considering their formative milieu, previous experiences, circle of contacts, political, social and economic pressures, access to previous output in the field, contemporary fashions in science, art and religion, instrumentation available to them, and their personal purpose at the time of the map-making, one embarks upon a more thoughtful – and informed – study.

Maps have popularly been regarded as a type of 'snapshot in time', a necessarily simplified but nonetheless 'true' representation of reality at the time of their production. As the cartographic historian Denis Wood, so ably demonstrates in *The power of maps* (London, 1992), that perception is false, under several headings. Most obviously, the map rarely represents a single moment in time. In the very process of map-making, there is always some time-lag, however brief, between surveying, map-making and publication. Maps are regularly used to represent an area in more

1 J.B. Harley and David Woodward (eds), 'Introduction', *The history of cartography*, i, *Cartography in prehistoric, ancient and medieval Europe and the Mediterranean* (Chicago and London, 1987), p. xvi. **2** *Cartographic Journal*, i (1964), p. 17.

than one period, as new survey material is superimposed onto an existing framework, as in the reworking of the Ordnance Survey maps for Valuation Office purposes. But more importantly, each map is a human construct serving selected interests. Those interests may be readily discerned, as in the case of maps produced for colonial land redistribution which present an image of territory devoid of native settlement. Alternatively, the interest behind the creation of a map may be much more subtle, as is true of general topographic maps which purport to be neutral, dispassionate, 'true'. Every map serves the interests of its creators, and the challenge to both historian and geographer is to enter into the world within which it was produced, to question not alone the motives of those who worked on the project, but also the philosophical basis upon which they operated. Questions must be asked regarding the sort of knowledge that was respected, promoted and advanced at the time, and the reasons why this was so. In the case of overseas expansion, such as in the late sixteenth century, the map was an active tool, in Ireland as in other colonies. The very act of mapping was part of the appropriation of land, in psychological as well as practical terms. Cultural imperialism blatantly promoted links between literacy, civilisation and scientific knowledge; the map as both a scientific and artistic achievement was an ideal vehicle to convey this idea. Seventeenth-century surveyors such as Sir William Petty advertised that they were at the cutting edge of technological advances by incorporating drawings of impressive instruments into their border decoration. Eighteenth-century geographers, surveyors and map publishers, including Herman Moll, John Rocque and Bernard Scalé, emphasised the role their products played in what they viewed as the great task of bringing civilisation to 'savage' foreign peoples, including the Irish. Elaborate titles and dedications, with their royal insignia, studied displays of classical learning and beckoning horizons were all part of the iconography of British imperialism. In the promotion of nationalism and Eurocentrism, nineteenth-century governments made widespread use of historical atlases produced to very specific criteria, that is, skewed towards the period of their own domination.[3] And in the twenty-first century, the making of maps in the context of boundary disputes, plans for 'resettlement', controlling public assembly and so on continues to exert potent influence, made perhaps all the more so by the way in which the map, as both scientific and artistic achievement, is still accepted by so many as 'value free'.[4]

J.B. Harley, the British map historian, provides a model for analysis in the effort to 'unwrap' the map at this deeper level.[5] He suggests firstly 'deconstructing' the map in a manner already common in literary criticism. Here the 'language' of the map is explored, with attention to code, context, and content. Delving further one can pursue questions about the changing readership for maps, levels of carto-literacy, conditions of authorship, aspects of secrecy and censorship, and also the nature of the political statements made by maps. Harley's second theoretical vantage point, iconology, is the identification of the symbolic layers in maps, the layers at

3 Jeremy Black, *Maps and history: constructing images of the past* (New Haven and London, 1997). **4** Mark S. Monmonier, *How to lie with maps* (Chicago, 1991). **5** J.B. Harley, 'Maps,

which 'political power is most effectively reproduced, communicated, and experienced through maps'. The third perspective is based on the premise that map knowledge, like other types of scientific knowledge, is a social product. The quest for knowledge is not an objective or neutral activity, but is intimately related to the 'will to knowledge' of the truth-seeker.[6] Therefore, even where it presents itself as a 'neutral' activity undertaken in a spirit of 'pure' disinterested scientific inquiry (as most official national surveys purport to do), cartography in fact 'replicates the territorial imperatives of a particular political system'. In some cartographic projects the political agenda is declared explicitly, for example, where cartographers publicly associates themselves with a larger imperial project that aspires to bring the blessings of a 'superior civilisation' to benighted lands. Although delivered in a more restrained way, the message from the 1862 edition of the handbook used by OS surveyors is the same, explaining to those destined to work in the colonies that 'in a new country only the natural lines and features exist, the rest has all to be created'.[7] Cultivation, towns, roads, provinces and boundaries are specified as the features that will have to be 'created' in an 'empty' landscape; the map-maker will play a direct role in this task, by surveying the land to be occupied. But even where there is no explicit political purpose, map-making cannot escape being implicated in the processes by which power is deployed. Connections between power and knowledge are readily made; connections between political power and cartographic knowledge are inescapable.

This guide aims to introduce the use of maps in researching, writing and presenting local history, with the illustrative material drawn from the Irish cartographic record. While it is essentially a practical guide, including notes on the map series most immediately useful for local history, major repositories, catalogues and 'finding aids', as well as ways of utilising maps in local history and the process of map-making itself, these larger theoretical questions must first be explored. An understanding of the nature of maps, the questions we must bring to them (Appendix 1), and the ways in which this coded world can best be entered (chapter 2) are thus essential pre-requisites to their practical use. References are made to illustrations which appear later in the text.

WHY STUDY MAPS IN LOCAL HISTORY?

Few documents can command the genuine engagement that a detailed map of one's home area will invariably do. The pictorial form, the sense of standing 'outside looking in' while simultaneously being part of the place, and the immediate personal challenge to 'make sense' of this depiction, to locate one's self

knowledge and power' in Denis Cosgrove and Stephen Daniels (eds), *The iconography of landscape* (Cambridge, 1994), pp 277–312. 6 Michel Foucault, quoted in Harley, 'Maps, knowledge and power', p. 279. 7 Edward Frome, *Outline of the method of conducting a trigonometrical survey for the formation of geographical maps and plans* ... (3rd edn, London, 1862),

(and the routes one uses daily, the buildings passed by or frequented, the junctions at which one is halted), are all part of the map's allure. In the research, writing and presentation of local history the map can provide an unrivalled introduction to a geographical area and local community, immediately engaging attention and indeed curiosity, but also promoting the abstract thinking necessary to move into complex areas of questioning. And maps continue to be created, overwritten and revised, with EU farm funding schemes, applications for planning permission, and motorway bypass schemes being only some of the developments for which large-scale updated local maps are required. Practical samples of the ways in which maps can be used in local history are provided in chapter 4 of this guide, but firstly a few general points must be made.

The most obvious use of the map is for ascertaining location. To recall the *History of cartography* definition quoted earlier, the map's primary purpose is to facilitate a spatial understanding of things, concepts, conditions, processes and events in the human world. Taking, for example, an extract from the six-inch or townland survey of Baltimore County Cork (fig. 1), the map efficiently conveys a huge amount of information in a small compass, including coastline, relief, distances, field patterns, road and railway lines, public buildings, placenames and antiquities. Physical environment and human endeavour are depicted simultaneously, a reminder that all history is acted out in 'a certaine habitation' (John Smith, *A history of Virginia*, 1624). Some of the information provided in the map is not readily available elsewhere, including placenames, routeways and river directions. Most usefully the settlement of Baltimore is seen along with its hinterland; it is not just *site* (a sheltered inlet at the entrance to Baltimore harbour), but also *situation* that is introduced. Points at which roadways intersect, where buildings adjoin, and how far beyond the early core the new terraces of houses advance can all be traced from maps. Archaeological features are set in their larger contexts: the ruined church with its cemetery and nearby glebe house, the derelict castle keeping guard over the harbour. In many cases the line of early ecclesiastical enclosures is retained in modern road patterns and property boundaries; the map helps to make sense of disparate relict features from the early Christian period such as round towers and high crosses. The village or town is connected, people are moving in and out and bypassing it altogether. The cluster of houses that makes up Baltimore from 1842 onwards is part of larger administrative and political worlds: the parish, barony, region, country, empire, EU. And yet this place is apart, with its own internal dynamism, significant personalities, contentious issues and daily routines. Local history is primarily concerned with how communities operated, what interests they shared, the resources to which they had access;[8] the map can provide an introduction to where these people were gathered and the place that they called 'home'. Each map has to be 'peopled' by the reader, but it does provide a canvas upon which the jottings can begin. The significance of landscape features represented

p. 132. **8** Raymond Gillespie, 'An historian and the locality' in Raymond Gillespie and Myrtle Hill (eds), *Doing Irish local history: pursuit and practice* (Belfast, 1998), pp 7–23.

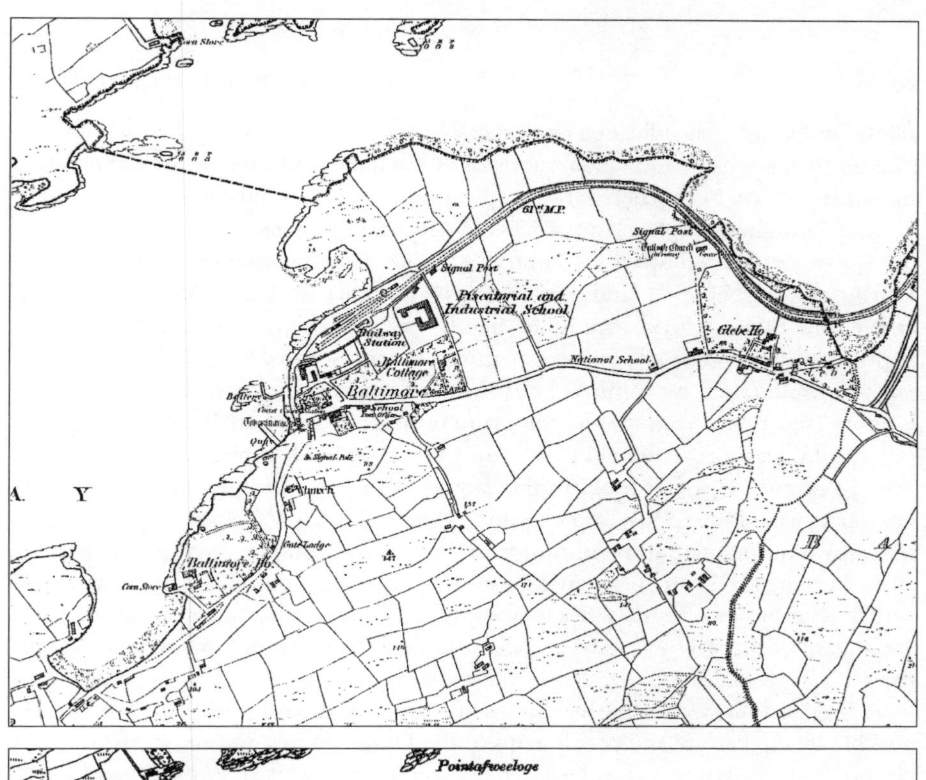

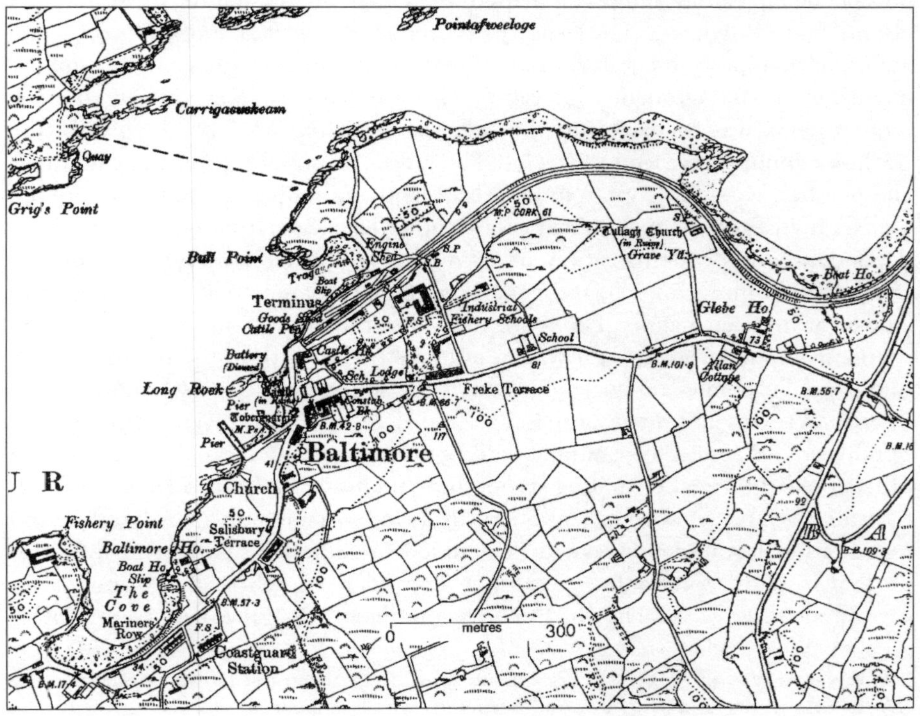

1 Baltimore, OS six-inch survey, County Cork, sheet 150, 1st edn 1842, and 2nd edn 1902.

symbolically here has still to be explored. What is omitted may prove to be of more interest than what the individual surveyor or centralised Ordnance Survey is determined to record. Nevertheless, the map can provide an invaluable starting point.

The 'bounding' of place, the process of determining what belongs and what lies outside a designated space, is central to both map-making and local studies. Distinguishing between land perceived as 'profitable' and 'unprofitable', pasture, arable and bog, in loyal Protestant hands or confiscated from rebellious 'Irish Papists', was the central purpose of the state-sponsored Down Survey (DS) of 1654–9 undertaken by William Petty. In the DS barony map of Ballyboy, King's County (fig. 2), for example the townland numbered 2 and 2b, Rathrobbin, parish of Killaghy, was formerly held by Dermot Dwiggin 'Irish papist', with a profitable acreage of 260 acres, but listed in the 'Books of survey and distribution' as having been transferred on 10 July 1668, to none other than 'Sr. Wm. Pettie' himself. Such transfers required careful recording of the townland and parish edges, with roads and rivers marked out in so far as they helped to fix the boundaries.[9] Lengthy verbal descriptions of boundaries to accompany both parish and barony index maps are a feature of the Down Survey, ensuring that there can be no confusion about the plots, townlands and parishes in question.

A logical follow-through to the confiscation and reallocation of land, made possible by the barony and parish maps of the Down Survey, was the division of the country into districts 'as His Majesty's Revenue is collected' (1689–1700).[10] The ambition to impose an equitable and universal property tax regime was behind the extension of the Ordnance Survey to all of Ireland in 1824; one of the first requirements was to mark out the boundaries of the *c*.60,000 townlands, the smallest administrative unit in Ireland. The tasks of computing area in acres, roods and perches, assigning a value to parcels of land and individual buildings, linking specific holdings to named individuals (occupier, lease-holder or lessor), were firstly dependent on the demarcation of townland boundaries (carried out by the Boundary Commissioners), and general public acceptance that these lines were arrived at in a fair manner. The work carried out by the OS in 1837 ascertaining municipal boundaries (fig. 46) was an essential preliminary to the bill to regulate municipal, corporate and borough towns in Ireland, proposing new boundaries for 67 named towns and dividing the cities and large towns into wards. The 'ancient boundaries' of Ardee, for example could not be ascertained on any map but the officers relied on oral testimony, noting how the town had expanded southwards over the river, beyond the ancient boundary, before deciding on a much tighter and more rational bounded area which included these 'new' suburbs and sufficient ground to allow for modest expansion. It must be noted that OS boundaries carried the force of law. They had to be reconsidered when the maps were revised,

9 There are a few differences in place-name spelling between the Down Survey and the 'Books of survey and distribution', some of which appear to be merely errors in transcription, as in 'doolekill' for 'Coolekill'. **10** *A map of Ireland divided into districts as His Majesty's Revenue is collected, 1689–1700* (NLI, MS 1437).

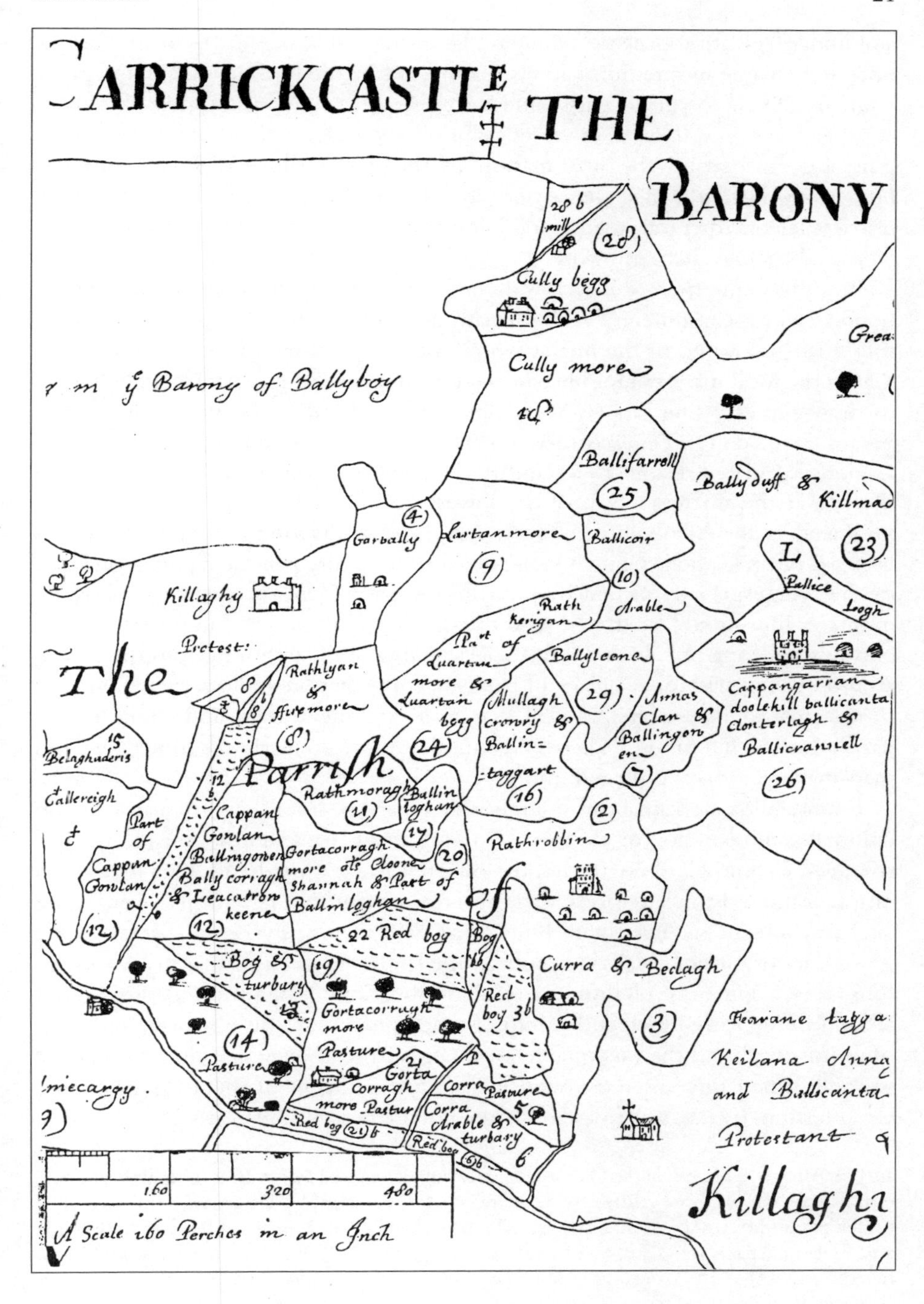

2 Down Survey barony map of Ballyboy, King's County, 1655

and under legislation of 1854, a different boundary could be shown on an OS map once the change in question had been approved by the lord lieutenant and privy council.[11] These townland, parish and barony boundaries were transferred to maps at other scales, and utilised, in various combinations, for a range of administrative purposes far beyond the imagination of the first Ordnance Survey officers. Dispensary districts, which were originally subdivisions of poor law unions, and the district electoral divisions used in local government and later national elections, can be traced back to this early work.

Boundary questions span very different types of map-making. Delimiting the boundary between different rock formations, circling where exactly coal deposits might be discovered or the line at which the city corporation's remit ended and that of the local manorial lord began, were matters of real import. Deciding where to draw the coastline was always a vexed issue. Should it be where vegetation begins (1:50,000 OS *Discovery* series) or as an average between high and low water levels as calculated relative to a standard datum (OS six-inch, fig. 1) or way out in the sea, at the furthest reach of the lowest possible tide, as on the early charts produced by the Admiralty? The formalising and re-drawing of parish boundaries by the Roman Catholic Church in the nineteenth century generated correspondence of the type especially valued by local historians.[12] Here the question of where exactly a line should be drawn (and the date of revisions) had important repercussions, affecting not alone the financial standing of the parish but also the official record of the parishioner's place of baptism, confirmation and marriage (as noted in the associated registers), as well as school catchment areas, and eligibility for parish-based relief funds.[13] These and countless more concerns are fundamental to map-making, as *space* is turned into *place*.

Bounding is an essential part of humanising space, be it for property management, colonial administration or elections. Coming to a place from the outside, there is the need to 'know', in terms that the outsider can comprehend. This may mean rationalising existing divisions, or even creating entirely new units, over-riding 'archaic', 'foreign', or 'inaccurate' boundaries with supposedly 'better' systems, as the newcomer or outsider seeks to construct the landscape in terms that make sense to him or her. Thus the plethora of subdenominations of land found throughout Ireland in the early nineteenth century, condemned for their 'obscurity and want of uniformity', and the inequitable local tax 'burthens' to which they gave rise,[14] were all tidied into discrete townlands with the advent of the 1824 Boundary Commission. Its task was to establish key boundaries through a lengthy process of

11 J.H. Andrews, *History in the Ordnance map, an introduction for Irish readers* (Dublin, 1974; 2nd edn, Kerry, Montgomeryshire, 1993), p. 12. **12** For example, see correspondence filed under 'parish box 1860', Cullen papers, which includes correspondence relating to St Paul's and St James' parishes, 7 June 1875 (Dublin Diocesan Archives, Drumcondra, Dublin 9, hereafter DDA). **13** For example, see Jacinta Prunty, *Dublin slums, 1800–1925: a study in urban geography* (Dublin, 1998), pp 279–316. **14** *Report of the select committee appointed to consider of the best mode of apportioning more equally the local burthens collected in Ireland, and to provide for a general survey and valuation of that part of the United Kingdom*, H.C. 1824 (445) viii, p. 79.

public consultation, to mark them on the ground (with wooden stakes, stones, ditches and earthen mounds where necessary), and to point out the information (quite literally, on foot), to the officer in charge of the Ordnance Survey party so that these boundaries might be included on the townland maps. The means by which space is bounded on the map are many and varied. Best known are the combinations of dots and dashes that make up the standard linear divisions of the traditional Ordnance Survey map, distinguishing townland, parish, barony, county, ward, urban district, parliamentary borough and other units. Varying the thickness, pattern or colour of line is part of the effort to bound and thus to know, and somehow control, the local place.

The naming and renaming of place, the recording (and ignoring) of placenames, spelling(s), derivation, and present understanding, are all central to the local history project. Toponomy (place-naming) is entwined with physical landscape, historical associations, religious devotion, and folklore; it spans worlds, cultures and centuries, linking a place to its past in an inimitable way. Urging local researchers to consider how a landscape was understood by those who lived there, Raymond Gillespie refers to the naming of place, and the folklore surrounding archaeological and topographical features, as evidence of 'the invisible world of saints and demons being related to the tangible, and religious belief being made concrete in particular times and places'.[15] The map holds the primary place in these explorations, as the painstaking scholarship of Breandán Mac Aodha, J.H. Andrews, Patrick O'Connor, Art O Maolfabhail and others demonstrates.[16] The Ulster Place-name Society has been to the forefront in deepening appreciation of placename heritage through its bulletin *Ainm*. Its millennium travelling exhibition on placenames, titled 'Celebrating Ulster's townlands', accompanied by a booklet of the same name (Belfast, 1999), is an attractive and authoritative introduction to placename study, the representation of placenames on maps, and further reading in this area. In the early 1990s, when heavy-handed 'simplification' of rural addresses was threatened by the post office in Northern Ireland, *Every stoney acre has a name: a celebration of the townland in Ulster*, edited by Tony Canavan (Belfast, 1991) was a most timely addition to the literature. An appreciation of the processes behind the recording of placenames on maps – most importantly the early nineteenth-century operations of the Ordnance Survey but also of their predecessors – is vital to an understanding of how and why placenames appear in the written form(s) we know today. The drama *Translations* by Brian Friel, which centres on the work of the Ordnance Survey in Donegal, has

15 Gillespie, 'An historian and the locality', p. 17. **16** Breandán Mac Aodha has published in practically every major local history journal, and also contributes regularly to *Ainm*; for example, see Breandán Mac Aodha, 'Ríocht na háitainmníochta in Éirinn' in G.L. Herries Davies (ed.), *Irish Geography: the Geographical Society of Ireland Golden Jubilee, 1934–84* (Dublin, 1984), pp 176–85; idem, *Sráidainmneacha na hÉireann* (Baile Átha Cliath, 1998); J.H. Andrews, 'The maps of Robert Lythe as a source for Irish placenames' in *Nomina*, xvi (1992–3), pp 7–22; Patrick J. O'Connor, *Atlas of Irish placenames* (Newcastle West, 2001); Art Ó Maolfabhail, 'An tSúirbheireacht Ordanais agus logainmneacha na hÉireann 1824–34' in *RIA Proc.*, lxxxix, C 3 (1989), pp 37–66.

generated welcome interest in placenames. Although of course intended as a work of fiction, its depiction of the operations of the OS has been exposed as historically unsound, a salutary warning to the local historian to beware of superficial cartographic analysis.[17] People name places, and the processes of naming and renaming, inclusion and exclusion on maps, experimentation with spellings and pronunciation, later consolidation and 'agreement' must all be considered. Andrews identifies five sources of toponymic change for exploration: substitution, translation, transcription, dictation and restoration.[18] At the intimate local level, tracing changes in the naming of city alleyways and courts has proved an important key to understanding some of the dynamics governing life in these small 'unseen' communities.[19] Dramatic changes of name can be recorded on maps, indeed they may be the only record of alterations that were inconsequential to all but the residents of these modest dwellings. In the south Dublin workhouse campus (site of St James's Hospital), the way in which names were struck through and new entries made by hand on the five-foot OS sheets produced for the use of the Valuation Office tells its own story: ~~Cut Throat Lane~~ Mount Brown Lane; ~~Murdering Lane~~ Cromwell's Quarter; ~~Dirty Lane~~ Bridgefoot Street Upper; ~~Pigtown~~ Ewington Lane.

Relict man-made features in the landscape are, for many enthusiasts, the starting point for local historical research. A ruined tower house, high cross, old church, portal tomb, a mile post or a pair of 'jostle stones', a gatehouse or the fragment of a town wall may continue to stimulate curiosity as it did among those antiquarians, 'collectors', archaeologists, surveyors and OS fieldworkers who played various roles in early map-making projects. Their combined (if sometimes unwitting) efforts have ensured that archaeological sites or sites of historic interest feature prominently in the cartographic record of Ireland. Round towers (RT), for example, were included in the railway commissioners' maps of 1838 (fig. 43), courtesy of George Petrie who was fascinated by them. The process of listing and locating archaeological sites on a map underpins current heritage protection legislation.[20] The purposeful inclusion of 'antiquities' and sites of historic interest, the classes selected, their symbolic representation and labelling, all provide stepping stones into the local past and its historiography. Close examination of a succession of maps of the same site is especially valuable to the archaeologist. In the example of the Four Courts on King's Inns Quay, Dublin (fig. 3) successive stages of the move from congested district with buildings of little distinction to monumental public building complex can be tracked. The rich detail of the large-scale town plan (fig. 3) has long been used by the local historian, archaeologist and geographer to explain the development of distinct urban units, the origins of placenames, the former line of the town wall and that of enclosed suburbs, the location of towers and gate-houses, the

17 J.H. Andrews, '"More suitable to the English tongue": the cartography of Celtic placenames' in *Ulster Local Studies*, xiv, no. 2 (1992), pp 18–21. **18** Ibid., pp 10–18. **19** Prunty, *Dublin slums*, pp 297–8. **20** David Sweetman, 'The man-made heritage: the legislative and institutional framework' in Neil Buttimer, Colin Rynne, Helen Guerin (eds), *The heritage of Ireland* (Cork, 2000), pp 527–33.

direction taken by water-courses now underground, and earlier field and property divisions.

As local history contributes immeasurably to the making (and re-making) of national, international and thematic history, the informed scrutiny of map-making at the local level can illuminate much larger pictures. The mapping of Ireland was truly an international affair in terms of origins, experience and the political loyalties of the first surveyors and cartographers, with certain maps first appearing as part of larger international atlases, produced by publishing houses in Amsterdam, London and elsewhere.[21] An exploration of the design, manufacture and distribution of surveying instruments used in Ireland provides one possible routeway into this larger world, based on the outstanding collection assembled in the National Science Museum, St Patrick's College, Maynooth, County Kildare.[22] Irish-made circumferentors (used to determine direction and measure horizontal angles), dating from 1688, and clinometers (or 'inclinometers', used to measure vertical angles) can be viewed in Maynooth, along with theodolites (which superseded the circumferentor), measuring both horizontal and vertical angles by the aid of a telescope and graduated circles. Very large theodolites made possible the triangulation of Ireland undertaken by the Ordnance Survey from 1824. While concerned with the recreation of operations 'in the field', such a study raises wider questions regarding the transmission of scientific ideas, the development of professions including surveying, civil engineering and cartography itself, and the patronage of scientific learning. The map itself bears testimony to changing technology, political priorities, artistic fashions, and the mobility of skilled personnel; by purposely focusing on the local level, cartography can provide new openings into the wider world of ideas and ambitions within which the local is embedded.

Where a series of maps of varying origins, types, scales, and dates of publication can be assembled for a particular area, the local historian is well equipped to undertake important work. The wider the chronological range the better. Early maps often stimulate questions that we might hesitate to ask of later, more familiar and 'authoritative' OS maps, while changes in the relative status of 'local' places will only become evident over a long time-span. The production of a new survey very often signals an important shift in the life of a local community, though this may be due to either national or immediately 'local' causes. Land-grabbing in the post-Cromwellian era (fig. 2) and the establishment of a local tax base country-wide (figs. 1, 3c) can be tied to specific map projects. By carefully comparing successive maps of an area (for example, figs. 3a–3d) the researcher can identify some matters for further exploration at the intimate local level: topographic growth, changes in field systems, archaeological inheritance, economic activity, institutional land use, changing administrative boundaries, drainage and reclamation works, the opening

21 J. H. Andrews, *Irish maps* (Dublin, 1978), pp 1–3; idem, *Plantation acres an historical study of the Irish land surveyor and his maps* (Omagh, 1985), pp 224–55. **22** Charles Mollan and John Upton, *The scientific apparatus of Nicholas Callan and other historic instruments* (Maynooth, 1994), pp 14, 260–85.

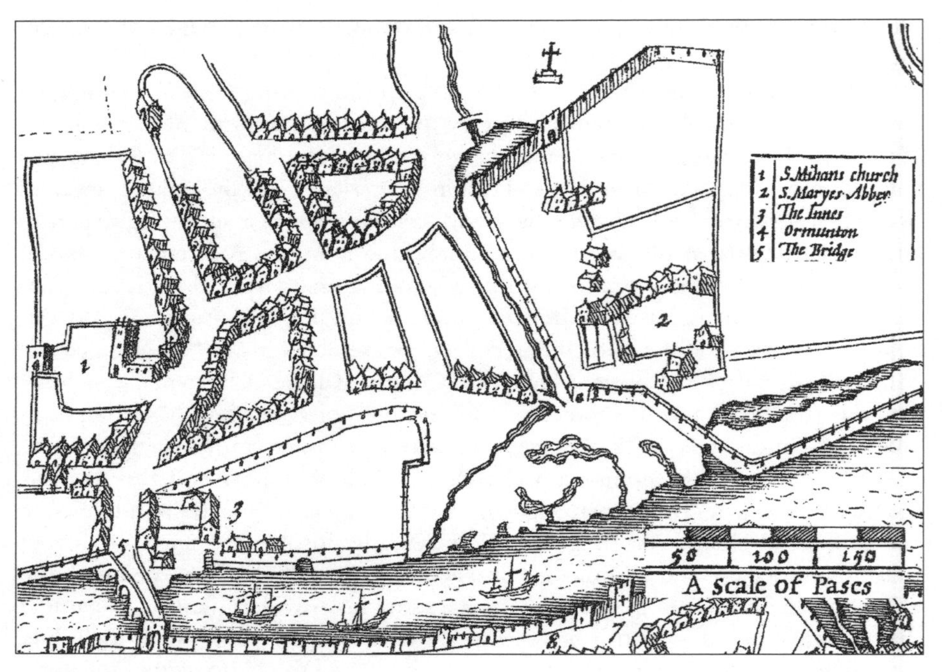

3 Four Courts, North Dublin city — a) John Speed, *Dubline*, 1610

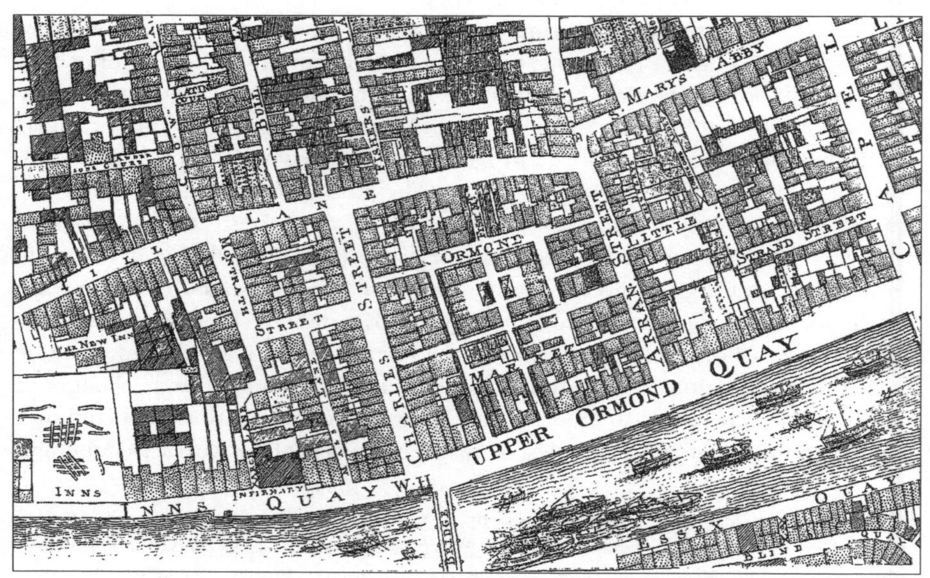

b) John Rocque, *Exact survey of the city and suburbs of Dublin*, 1756

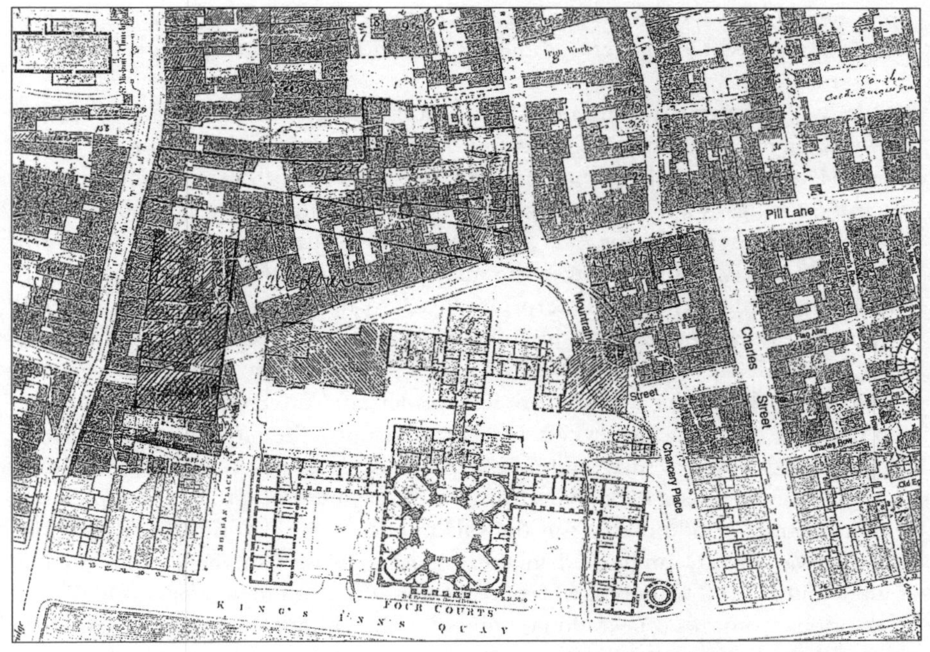

c) VO five foot, Dublin, sheet 8; *c.*1854

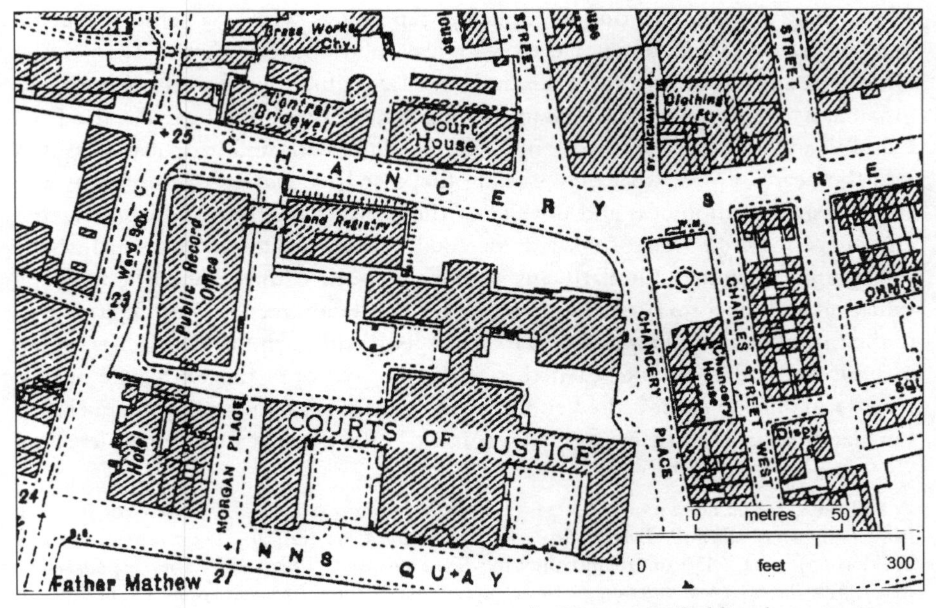

d) OS 1:2,500, Dublin, sheet 18.xi, 1939

up of new routeways and the closing off of others. The map can act as an 'anchor' source around which further complementary and contradictory information can be assembled.

Many of the maps that are most useful in studying local history are in essence part of larger document collections and must be considered in conjunction with them. Map-making in the plantation period resulted in both maps and 'terriers' (as discussed for figs. 5 and 6), naming the original occupier of each numbered plot, 'to whom disposed of by the Acts', its acreage and quality. In the case of the Down Survey, the official requirements of the survey, Petty's instructions to his staff (1655), and his own accounts of the enterprise (1656, 1659), also survive.[23] Maps of battle sites, fortifications and noteworthy military advances (such as the battle of Kinsale, 1602, or the defeat of rebels in Wexford, 1798), were accompanied by explanatory notes so that the reader would more fully understand the course of the campaign and the scale of the recent success. J.H. Andrews notes the movement from Petty's time onwards towards 'dual publishing projects', whereby maps and text were complementary but separate. The Ordnance Survey memoir project (Templemore, which includes Londonderry, was the sole publication, 1837) is the best example, but this was already prefigured in the scientific surveys of the Dublin Society (founded in 1731). The published reports of the Office of Public Works (OPW) and of the Bogs Commission, would be unintelligible without their maps. Estate maps are another obvious example, since the map record cannot be analysed without reference to the tenant lists, accounts and landlord correspondence. The formal reports of countless commissions of inquiry, submitted to parliament and published in the House of Commons series, were often accompanied by maps. And the power of mapping is such that this preliminary step brings with it a sense of having the problem in hand, amenable to a logical solution, as with the 1846 Select Committee on poor relief and medical charities, which mapped fever hospitals, infirmaries and dispensaries.[24] Map-making initiated by bodies such as the Bogs Commission and the Railway Commission brought educated, articulate outsiders into the heart of places that they would otherwise have had no reason to traverse, and the sheer remoteness and novelty of these places to the commissioners and engineers inspired them to write detailed commentaries to accompany the cartographic products. Similarly, any attempt to work with the published general valuation and the associated field books and cancellation record without attending to the maps ('marked and numbered to correspond with such field books'), is certain to limit the understanding reached (fig. 3c). The riches of the Ordnance Survey record can only be harnessed when the working materials upon which the published townland or six-inch maps (fig. 1) were based – the content field books, levelling

23 Andrews, *Plantation acres*, pp 63–72; Thomas A. Larcom (ed.), *The history of the survey of Ireland commonly called the Down Survey, by Sir William Petty* (Dublin, 1851; reprinted New York, 1967). **24** 'Map of Ireland shewing fever hospitals, infirmaries and dispensaries', *Report from the select committee on poor relief and medical charities (Ireland)*, appendix, H.C. 1846 (694) xi, part 2.

registers, plots, place-name lists, traces, 'fair plans', and accounts of trigonometrical stations – are consulted. The 'stand-alone' character that map collections assume is compounded by library and archival practices which insist on separating these records; consequently J.H. Andrews urges, what a librarian has divided, the researcher should seek to reunite.[25] That uniting process must go beyond the realm of the literal and utilitarian into the intellectual sphere, the researcher constantly looking for the associated documentation that will throw light on the context within which this map originated and was first used.

Some historical records, though not bound up with map production, are nevertheless most usefully considered in conjunction with maps: the census record, street directories, relief lists and municipal government surveys of the period 1830–80 are some of the more obvious examples.[26] Furthermore, there is an array of historical information, including civil survey, landlord, valuation, census, directory, charity and municipal sources, that can best be understood when mapped. The spatial distributions and relationships thus revealed can elevate the local study onto a new plane of inquiry. Mapping is a fundamental research tool in all the social sciences, but there is a tendency (even among some geographers) to leave mapping to the 'experts', the skills of map-making being regarded as too arcane to even contemplate embarking on that adventure. Local historians, far from merely 'reading' maps as one of their historical sources, should also engage in map-making themselves. For the presentation of research findings – making the map serve one's own purpose – there are few more accessible media.

The task of first locating and then trying to make sense of maps of one's local area can in itself act as a stimulus to serious local history study. The process of working out who might have found it in their interest to map a certain place (canal or railway company, landlord or church authority, harbour commissioners, enterprising visitor or local corporation), will generate new questions, and help disentangle the complex processes and overlapping chronologies behind the making of the local 'place'. And there are few better ways of entering into the spirit of place, past and present, than to 'walk the land', urban or rural, with the largest scale OS map (or series of maps) that can be procured. With map extracts in hand, the eye is keener in reading and questioning the field evidence. With this new 'angle' on the familiar place, the researcher can better investigate matters such as boundary divisions and underlying physical geography, and assess their historical significance.

The urge to 'let the evidence speak for itself', without analysis, without critical commentary, without considering what it omits to say, is a particular temptation for the local historian. The charm of the newly-found map can subvert the critical faculty, and the rigorous questioning that one would bring to a 'regular' historical document can be temporarily suspended. As with other primary material, mere reproduction is not history. 'Doing' local history requires asking questions, with

25 J.H. Andrews, 'Maps, prints and drawings' in William Nolan and Anngret Simms (eds), *Irish towns, a guide to sources* (Dublin, 1998), pp 27–40. **26** For example, Prunty, *Dublin slums*, pp 274–335.

maps as with other sources. Appendix 1 is a lengthy though still incomplete list of the questions which may be put to a map. It is offered to assist the newcomer – and remind the expert – of the complexity surrounding map origination, funding, creation, and technical production, not to speak of the background, skills, and personal agenda of the cartographer. Applying some of the questions in Appendix 1 to the 1756 map of Dublin city by John Rocque (1704–62), for example (fig. 3b)[27] throws light on the world within which it was created, and affects its assessment by historians. The international standing and previous experience of the cartographer is worthy of note; he had already completed maps of Paris, Rome and 'the cities of London and Westminster' (1747), and advertised his Dublin map as being 'on the same scale' as these prestigious projects. He actively sought and enjoyed aristocratic patronage in Ireland, most successfully that of the premier peer of the realm, James Fitzgerald, earl of Kildare, who employed him to survey his extensive estates in County Kildare, 1755–60. This was a boom time in Irish estate mapping, due to the consolidation of estates and economic development. Rocque had already produced an 'exact survey' of the 'beautiful and magnificent house, garden and park of the Rt. Hon. The Earl of Pembroke' with the 'ancient town' of Wilton, an ideal recommendation as the Fitzwilliam/Pembroke family also owned vast estates in Ireland, including valuable tracts from Dublin city southwards to Blackrock which they were in the process of developing. The importance Rocque attached to his association with the Fitzgeralds is reflected most strikingly in his map of the county of Dublin (1760) where he extends coverage into the adjoining county of Kildare to include the grand estate of the Fitzgeralds at Carton in Maynooth and that of the Conolly family (related to the Fitzgeralds by marriage) at nearby Castletown in Celbridge. The audience for which he produced the Dublin city map may be gauged from the dedication, a ruling elite with knowledge of continental fashions who by the 1750s were investing heavily in the development of gracious town houses, handsome squares and wide malls. Associating the Dublin map with those of Paris, London and Rome was no doubt a shrewd advertising move. In the final engraving, great attention is lavished on the private gardens of these high quality residences and on the communal (but still private) gardens where people assembled to be seen, as in the 'New gardens' of the Lying-in hospital. The Fitzwilliam/ Pembroke estate sector adjoining the Fitzgeralds' town house, Leinster House (now the Dáil) is, predictably, characterised by fine ornamental gardens. From Rocque's description of himself some years earlier as a '*dessinateur de jardin*', and advertisements in which he seeks 'topographical' commissions,[28] it may be inferred that he was keen to display this skill to the fullest, and no one, least of all wealthy clients, will complain if their property appears better cartographically than is the reality. Without an aerial survey there is no way Rocque and his assistants could have

27 Rocque's map has been reproduced as a facsimile over four sheets, and case bound as 'The A to Z of Georgian Dublin', by Harry Margary, Lympne Castle, Hythe, Kent CT21 4LQ, UK. **28** Andrews, *Plantation acres*, p. 164.

accurate knowledge of the layout of the gardens and reres of every single premises. Indeed this is clear from one section where Rocque inserts a plausible network of walls and buildings parallel with the city wall, when in fact (as later surveys reveal) the city wall cuts diagonally across. Born in France, Rocque betrays his continental origins in the classification he employs, distinguishing between churches, Presbyterian meeting houses (PMH), French churches (FC), Quakers' meeting houses (QMH), Dutch churches (DC), and 'Roman Chappels' (†). Rocque also notes several 'nunerys', institutions that in 1756 operated quietly, even clandestinely under penal legislation that was not revoked entirely until 1860.[29] How aware was he that such places did not officially exist? What else did he see that British Isles surveyors did not? What other features were present that no one mapped? Rocque's use of stipple to distinguish dwelling houses from public buildings (dark cross-hachuring) and from 'warehouses, stables, &c.' (lighter cross-hachuring) required high-level drafting skills as well as immense care in surveying. This is the work of a very professional practice, and in an entirely new league when compared with previous Dublin surveys. Rocque's pride in the scientific accuracy of his work is evident from the care he takes to show magnetic variation ('North west 20° in 1756'), the elaborate scale bar (in British feet, Irish perches, English perches, and '*echelle de toises*'), and the drawings of the instruments of his profession, most often playfully carried by cherubs. The quality of Rocque's surveying may be judged by overlaying sections of the Dublin map on the OS five-foot sheets and the later 1:2,500 sheets (figs. 3b–3c); the 'fit' is remarkably good on a block-by-block basis, considering the difference in resources available to the two projects. The Revd James Whitelaw, vicar of St Catherine's city-centre parish, Dublin, may be considered an authority on the Rocque map as he used it in his census of the city and survey of living conditions undertaken in 1798 (published 1805). Throughout five months of intense field work, the Rocque map was 'generally found minutely exact'.[30] The marginal note 'published according to Act of Parliament' explains why it was so readily available to Whitelaw and to others; the Irish Parliament, sitting at College Green, and composed of the very people whom Rocque flattered, was its collective sponsor. No doubt such sponsorship served him well when he continued with his other commercial ventures: surveys of Cork (1759), Kilkenny (1758), Newry and Armagh (1760) were to follow.[31]

In the case of John Rocque's 1756 survey of Dublin city, some familiarity with the rest of his portfolio, reference to contemporary advertisements, knowledge of the locality he was mapping and of the state of cartography during the period in

29 Séamus Enright, 'Women and Catholic life in Dublin, 1766–1852' in James Kelly and Dáire Keogh (eds), *History of the Catholic archdiocese of Dublin* (Dublin, 2000), pp 268–70; *Report from the select committee on conventual and monastic institutions*, 23 June 1871, p. v (DDA, Cullen, 328/4 file 1, laity, Jan-June 1871). **30** James Whitelaw, 'An essay on the population of Dublin being the result of an actual survey taken in 1798 …' (Dublin, 1805), facsimile reprint in Richard Wall (ed.), *Slum conditions in London and Dublin* (Farnborough, Hants, 1974), p. 52. **31** Rocque also produced 'A plan of the camp near Thurles in the county of Tipperary', 1755, dedicated to the Rt. Hon. the Earl of Rothe, and also showing the town

which he operated, enables the researcher to engage with the map more creatively and indeed confidently. However, there are times when cartographers are transparent about their purpose, sources, audience, and surveying methods. In the introduction to his Dublin Bay manuscript atlas of 1762, George Semple states that these reproductions of early maps, together with his own 'two different designs for compleating [the new pier], and the further improvements necessary to make it an exceeding safe and a commodious harbour', should be printed, 'not only to transmit the same to posterity, but to give present and general satisfaction to the publick, particularly the members of both houses of Parliament in time being as a convinceing proof that the indefatigable labours of your board [the Ballast Office] is already productive of publick utility' (*sic*).[32] While most map-makers are more reticent, preferring to keep the public in the dark as to their map-making techniques, they are no less influenced by the expectations of the patron or customer, and just as quick to use whatever cartographic resources come to hand, acknowledging – or suppressing – sources as they choose. The fact that it will not be possible to answer many of the questions listed in Appendix 1 with reference to an individual map is no reason to hesitate in posing them. The very process of questioning will stimulate the historical imagination, and assist the map-reader in thinking creatively about the source in hand. In that alone it will have served its purpose.

The requirement to consider critically all stages of the complex process of map production has been established (see App. 1). The cartographic historian R.A. Skelton maintains that this will enable the reader somehow to enter into the map-maker's work-space, to look over his shoulder and perceive why he drew an outline, or attached a place-name to a feature, and will throw light upon the pressures to which he was subject.[33] There is surely no more effective way of entering into the spirit of critical inquiry proposed at the opening of this guide than by pulling oneself up to the cartographer's table, and consciously becoming part of the map-making community.

of Thurles; for this and a selection of other Rocque town surveys see the atlas held by TCD (TCD, Maps 00 a 60). **32** Six manuscript charts of the harbour and bay of Dublin, atlas collection by George Semple, 1762 (Dublin Port Archives); Gerald Daly, 'George Semple's charts of Dublin Bay, 1762' in *RIA Proc.*, 93 C, no. 3 (1993), pp 81–105. **33** R.A. Skelton, *Maps, a historical survey of their study and collecting* (Chicago and London, 1975), pp 4–5.

CHAPTER 1

Map-making in Ireland: an historical overview

At least a general knowledge of the basic cartographic history of Ireland, and familiarity with the major classes of cartographic material, are essential prerequisites to the sensitive and effective use of maps in local history research. The discussion that follows is intended to give a flavour of the riches to be found, to highlight the energies and politics surrounding specific map-making projects, and to provide the researcher with direction to further reading at each point. Key texts are listed in full in the bibliography. And as with most categories of archival material, an appreciation of the context in which maps were created will greatly assist in the search of national and private repositories.

A concise introduction to map creation in Ireland, direction with respect to the main holdings, full-colour sample extracts, and a bibliography is provided by the cartographic historian J.H. Andrews, 'Maps and map-related sources' in William Nolan and Anngret Simms (eds), *Irish towns: a guide to sources* (Dublin, 1998), pp 27–40. Although obviously dealing with urban matters, much of what is presented in that essay is also applicable to the rural record. A succinct and well-illustrated introduction to the mapping of Ireland is provided by J.H. Andrews in *Irish maps* (Dublin, 1978). For a more comprehensive treatment, two of Andrews' most celebrated texts are essential reading: for pre-Ordnance Survey mapping, *Plantation acres: an historical study of the Irish land surveyor and his maps* (Omagh, 1985), and from the 1820s, *A paper landscape: the Ordnance Survey in nineteenth-century Ireland* (Oxford, 1975; 2nd edn, Dublin, 2002). *History in the Ordnance map: an introduction for Irish readers* (Dublin, 1974; 2nd edn, Kerry, Montgomeryshire, 1993) was produced expressly for the local historian or historical geographer and is indispensable. A good general introduction is an article titled 'Maps and the Irish local historian' in *Bulletin of the Group for the Study of Irish Historic Settlement*, 6 (1979), pp 22–31 by Andrews. His more recent study, *Shapes of Ireland: maps and their makers, 1564–1839* (Dublin, 1997), is also required reading for an understanding of map-making in Ireland up to the time of the Ordnance Survey. While *Shapes of Ireland* is concerned with all-Ireland maps, these can prove invaluable to the local historian, as noted below. The discussion which follows on selected map categories in Ireland relies heavily on these texts by Andrews. The Winter 1992 edition of *Ulster Local Studies* (xiv, no. 2), devoted to 'the Ordnance Survey and the local historian', is another essential reference work. It includes articles on placenames (J.H. Andrews), Thomas Drummond (A.J. Malley), and on the OS collections in PRONI (Trevor Parkhill). Also worth consulting at the outset is the special Irish issue of *Sheetlines* (vol. xxx, 1991), a journal devoted to the study of the Ordnance Survey. William Nolan's

Tracing the past: sources for local studies in the Republic of Ireland (Dublin, 1982) reviews map sources for different historical periods; as it is out of print library copies must be relied upon, but it continues to be a valuable handbook.

There are a number of UK guides to maps in local history which the Irish researcher could profitably consult, including J.B. Harley's *Maps for the local historian: a guide to the British sources* (London, 1972). David Smith's *Maps and plans for the local historian and collector* (London, 1988), although firmly centred on the British record, incorporates Irish examples (from both Northern Ireland and the Republic), and is well illustrated. *English maps, a history* by Catherine Delano-Smith and Roger J.P. Kain (London, 1999), is a comprehensive and generously-illustrated text. Irish researchers will learn much from its focus on the cultural context of map creation, map use and map users, and the original way in which it groups and regroups material. The cross-channel input into the mapping of Ireland compels the serious local historian to consider the larger British Isles and European contexts. The substantial bibliography in the Delano-Smith and Kain text provides stepping stones to further reading, and is in itself testimony to the many areas of study which are embraced under the heading of 'cartographic history'. The *Dictionary of land surveyors and local map-makers of Great Britain and Ireland* by Sarah Bendall (London, 1997) opens with a general historical overview of land surveying 1530–1850; the dictionary entries themselves make clear how many of those who practised in Ireland had wider British Isles experience. *The British Library companion to maps and map-making* by Rebecca Stetoff (London, 1995), purports to be a general guide to cartography; words, concepts and names (of places, cartographers, important maps, map types, organisations, expeditions, regions) are dealt with alphabetically, and the reader is referred to a substantial corpus of further reading. Targeted at a popular audience, its explanations are succinct and non-technical. The Map Room of the British Library regularly mounts exhibitions with published catalogues. The catalogue titled *What use is a map?* (London, 1989) provides an entertaining introduction to maps from the standpoint of the consumer, while *The lie of the land: the secret life of maps* (London, 2002) demonstrates how maps have been cleverly used to further particular political and military ambitions. Members of the public are welcome to visit the map displays mounted by the British Library in the exhibition areas at St Pancras; details of BL publications in the area of cartography and selections from its holdings may also be viewed online [http://www.bl.uk/]. Further routeways into the larger British Isles story of map-making may be followed by examining the cartographic holdings of the National Library of Wales, Aberystwyth [http://www.llgc.org.uk/dm/dm0067.htm] and the National Library of Scotland, George IV Bridge, Edinburgh. The NLS mounts exhibitions on site and online; its websites include 'Maps of Scotland, 1560–1928', 'Military maps of Scotland' and 'Ordnance Survey town plans 1847–95' [http://www.nls.uk/digitallibrary/map/]. *The nation survey'd* edited by Ian C. Cunningham (Edinburgh, 2001) was produced to accompany the Pont exhibition and website.[1] 'Charting the nation', a collaborative

1 77 manuscript maps of Scotland *c.*1583–96 attributed to Timothy Pont are held by the

digital imaging and cataloguing project aimed at widening access to maps of Scotland and their associated archives 1590–1740 was originally based in the Geography Department at Edinburgh University and is now managed by the university library [http://www.chartingthenation.lib.ed.ac.uk/index.html]; its partners include the National Library of Scotland and the National Archives of Scotland. This project is of interest to students of cartographic history throughout the British Isles, as skills learned in one map-making project were readily transferred to other geographical areas. Late eighteenth-century surveyors who worked in both Scotland and Ireland included Alexander Taylor, his brother George Taylor, and Andrew Skinner. Careful perusal of Bendall's *Dictionary of land surveyors and local map-makers of Great Britain and Ireland* discussed above will reveal many more cross-channel connections.

The most comprehensive and up-to-date bibliography and guide to secondary work on the history of cartography in Ireland has been compiled by Paul Ferguson, map librarian at Trinity College Dublin, and can be consulted in the Glucksman Map Library (limited opening hours and access, see http: //www.tcd.ie/ Library/ RR/ map lib.htm). This annotated guide is constantly updated and superbly cross-referenced with a detailed index, including county names. The map librarian always welcomes notification of other works on Irish map history that may not yet be included. Headings include bibliographies and guides, general works on the history of cartography, general works on Irish map history, and there are sections dedicated to the various classes of Irish maps using both chronological and stylistic divisions. This author is greatly indebted to Paul Ferguson for making the typescript available in the preparation of this book.

The best collection of facsimile maps at a local level is the *Irish Historic Towns Atlas* (see Appendix 2), which at the time of publication covers thirteen towns, with work well advanced on a number of other major urban centres, including Derry, Armagh, Enniskillen, Galway, Trim, Wicklow, Dublin part II and Belfast part II. It is part of a series of European national historic towns atlases, founded by the International Commission for the History of Towns in 1955 [http://www.ria.ie/projects/ihta/ihta.html]. Since then, fascicles for over 300 towns and cities in fifteen European countries have been published, along the lines specified by the Commission. The *IHTA* is an ideal starting point for the study of maps in local history, as it brings together a wide range of map types, along with models of how maps can be approached, in this case, in the detailed reconstruction of the topography of the town over time, and identification of the forces and individuals driving the change. As part of an international project with a unified concept and following strict guidelines, the fascicles in the *IHTA* can be set alongside those from any of the other fifteen participating European countries, allowing comparative study in a way that is most stimulating for Irish local history. The *IHTA* reproduces maps sourced from a wide variety of repositories throughout the British Isles and

NLS; these formed the basis for Blaeu's *Atlas novus* iv (Amsterdam, 1654). Project Pont was set up at the NLS to disseminate information on Pont's life, work and maps.

in some cases held in private hands. These include pre-OS military surveys and estate maps, some drawn from a 'bird's eye' perspective and others which are planimetric. Early outlines are overlaid on later OS maps in an effort to reconstruct line of defences, town walls and other landscape features. Following the guidelines of the International Commission for the History of Towns, the *IHTA* produces its own colour 'reconstruction' map, representing the town in question as it appeared *c.*1830s at the scale of 1:2,500, just before the industrialisation or other mid-nineteenth-century developments. The bibliography which concludes each fascicle includes note of the maps available for that town, and in the case of manuscript or unpublished maps, the repository in which each will be found.

With regard to maps of walled towns, Avril Thomas' two-volume gazetteer *The walled towns of Ireland* (Dublin, 1997) is an important resource. The evidence for town walls, along with their gates, towers and other defensive structures, is listed for each site under the following headings: field evidence, documentary evidence (maps and plans, written descriptions and illustrations), town records and property documents, and finally, murage grants and other charters. Fifty-six towns have been identified with certainty by Thomas as walled, another 35 towns are 'unproven' but have some evidence for walling, while there is a further group of 20 towns for which Thomas classes the evidence as 'doubtful'. These groupings are provided by Thomas as a basis for study rather than as definitive listings.[2] The gazetteer therefore deals with the large majority of towns and cities of significance in Ireland today, with the important exception of those which developed as industrial, resort, dormitory or 'new towns' in the nineteenth and twentieth centuries. The only maps selected for mention are those only which show or name defensive structures. The real value of Thomas' work lies in the association of the map evidence with a range of other sources, each shred of information incomplete and perhaps incomprehensible on its own but in combination allowing the reconstruction of a key feature in Irish urban history. While sketch maps showing the probable line of the town wall have been drawn for each entry, the gazetteer proper (vol. ii) is preceded by a comparative study of the walled towns of Ireland (vol. i), including a discussion of the cartographic and other evidence for this specialised study. A generous number of historic map extracts have been reproduced in this volume.

The National Library of Ireland has produced two sets of facsimile maps based on its own extensive holdings of both manuscript and printed maps: *Ireland from maps* (1980) and *Historic Dublin maps* (1988), with explanatory notes by Noel Kissane. The document collection, *Ireland from maps*, covers 16 examples of mapping from Ptolemy (*c.*150) to the Ordnance Survey and later, several of which are reproduced in full colour. It makes an excellent starting point for any map-based local study, and also provides an expert introduction to the cartographic history of Ireland. *A catalogue of the maps of the estates of the archbishops of Dublin, 1654–1850* (Dublin, 2000), edited by Raymond Refaussé and Mary Clark, reproduces (at a reduced scale) a selection of maps from this outstanding collection, providing an

2 Avril Thomas, *The walled towns of Ireland* (Dublin, 1997), table 1.1, p. 3.

insight into the variety and high quality to be found in pre-OS estate mapping in Ireland. Some of the Belfast maps held by the Linen Hall Library have been reproduced in the *Irish Historic Towns Atlas* fascicle, *Belfast, Part 1: to 1840* (Stephen Royle and Raymond Gillespie, Dublin 2003); the Linen Hall Library also produces its own series of facsimile maps of Belfast and other parts of Ulster. An earlier endeavour at bringing selections from our map heritage into the public domain was the monumental *Facsimiles of the national manuscripts of Ireland* ... prepared by John T. Gilbert (1874–84);[3] 13 maps, dating from the 1500s and 1600s, are reproduced in volume iv (parts 1 and 2), and have been widely used by historians since. Here, for example, the Goghe map of 1567 is reproduced.[4] To mark the Tenth International Conference on the History of Cartography held in Dublin in 1983, a set of facsimile maps, with a handbook, was compiled by J.H. Andrews and Arnold Horner, and the maps at least are still widely available in university and some public libraries. Facsimiles, at a reduced scale, of early Ordnance Survey maps of selected towns in Ireland have been produced by Phoenix Maps (Dublin), while there have also been several once-off facsimile projects, including the reprinting of Taylor and Skinner's *Maps of the roads of Ireland* (1778, 2nd edn, 1783), and Alexander Taylor's *Map of the county of Kildare*, 1783, edited by J.H. Andrews (Dublin, 1983). *Historical maps of Ireland*, by Michael Swift (London, 1999), provides an impressive sample of maps from the Public Records Office, Kew (now the National Archives). Though the reproductions are small, they are of high quality and in full colour, and despite a limited commentary (heavily reliant on the published TNA catalogue), will enthral any person not yet acquainted with historical maps. This publication also serves to remind the reader of the dominant role played by the English government in the mapping of Ireland; the forfeiting and redistribution of land, the creation of baronies and counties, the defence of towns and coastline, and ensuring safe anchorage for the ships sailing under her flag, were all concerns that led Britain to invest hugely in mapping Ireland.

At an all-Ireland level, a useful starting point for historians and geographers is the Royal Irish Academy's *Atlas of Ireland* (Dublin, 1979), regrettably out of print but it will be found in the reference section of most major libraries. This atlas includes distribution and network maps from prehistory through to the 1970s including redrawings of the road, canal and rail networks over time. The boundary maps include poor law unions, baronies, provinces, counties and liberties, dioceses, and their various shifts over time, and are especially useful to local historians, as are maps delimiting the congested districts and other specially designated areas. This is also the most accessible collection of maps showing the physical heritage, namely geology, drainage, relief, bogs and climate. In tandem with the RIA *Atlas of Ireland*, the researcher is directed to a collection of over 100 maps in volume ix of *A new*

3 John T. Gilbert (ed.), *Facsimiles of the national manuscripts of Ireland, selected and edited under direction of the Rt. Hon. Sir Edward Sullivan, Baronet, Master of the Rolls in Ireland* (4 vols, Oxford, Cambridge, Edinburgh, Dublin, 1874–84). **4** This early map is more recently reproduced in Andrews, *Shapes of Ireland*, p. 43.

history of Ireland (Oxford, 1984, reprinted 1998), edited by T.W. Moody, F.X. Martin and F.J. Byrne. This comprehensive collection of mostly small-scale maps includes several which will be found in the RIA *Atlas of Ireland* (where they are in colour, and at a larger scale). Most of the maps in the *New history of Ireland* collection were redrawn or otherwise revised for that publication; the notes (pp 97–120) direct the reader to the place where each map first appeared, the data upon which it is based, and, in many cases, key historical writings or parliamentary papers associated with the theme under scrutiny, such as parliamentary constituencies, plantation schemes, mail coach routes and railways. Also at the all-Ireland level, Andrew Bonar Law (Neptune Gallery, 41 South William Street, Dublin 2) has produced a carto-bibliography titled *The printed maps of Ireland up to 1612* (Dublin, 1983), followed by *The printed maps of Ireland, 1612–1850* (Dublin, 1997). These catalogues aim to assist map owners and purchasers to identify maps of Ireland wherever they are found – in atlases, in parliamentary papers or accompanying other published reports, illustrating commercial directories and traveller's guidebooks, as part of gazetteers and school geography texts, and in innumerable other places. This text allows the researcher to identify at least one repository where a copy of a particular map may be consulted (excluding maps in private hands); to ascertain where the map was first published and (in some cases) the authorities upon which it was based, the names of publishers and map-sellers, the dates of successive editions, and how to identify which edition is in hand. Where it is part of a larger record, such as a gazetteer or parliamentary inquiry, full details are given. The illustrations, though small, greatly increase its usefulness to scholars; the 'changing shape' of Ireland up to the OS is captured in these reproductions. Bonar Law's carto-bibliography opens up many topics of interest to students of the eighteenth and nineteenth centuries and acts in itself as an excellent introduction to the many purposes for which maps were created. In the study of defence, roads, canals, railways, parliamentary and municipal government, tourism and land reclamation, coastal navigation and fisheries, ecclesiastical administration, the postal system and population density, political agitation, education, trade statistics and the poor law, maps have been used, as noted in this index.

At county and diocesan levels, a small number of atlases have been produced over the years; the county library or diocesan headquarters should be contacted in the first instance. The Representative Church Body Library, Braemor Park, Dublin 14 (RCBL) holds an atlas of the diocese of Meath commissioned *c.*1870 which was based on the first edition OS six-inch sheets.[5] Patrick Duffy's *Landscapes of South Ulster, a parish atlas of the diocese of Clogher* (Belfast, 1993) is a more recent and innovative example of diocesan mapping which also relies on the ground-breaking work of the Ordnance Survey. Duffy's work is an exemplar of how to knit the rich oral and written heritage of the local place (in this case, the Roman Catholic (RC) parish), together with its geography, through the construction of parish maps. These

5 *Atlas of the diocese of Meath, c.*1870 (RCB Library, D7/6/9); I am indebted to Susan Hood for this information.

are designed and drawn by James Keenan, based on the 1:50,000 maps produced by the OSNI, and include townland boundaries and churches within each RC parish unit. This atlas is of the greatest utility to students of local history throughout the island, as an introduction to the rich heritage of place-specific material and the role of mapping in local studies. In addition, it introduces the researcher to more complex issues that must be explored in the effort to understand the origin and re-making of 'home places'.

The commercial map-making company *Léaráid* (159 Glenageary Park, Dún Laoghaire), headed by a former OS officer, Eoghan O'Regan, produced a number of maps, in English and in Irish, that are useful to local history work. At all-Ireland level, *Saoirse*, a map of the Anglo-Irish war 1919–21 (1986), depicts the sites of selected military engagements, and panels of statistics (casualties, numbers of IRA actions, and reprisals), along with a chronology of the main events in the 30 months of the conflict. There are two ecclesiastical maps: 'Map of ecclesiastical Ireland 1618–1988' (1989), based on a map produced by Matthew Kelly in 1859, with additional research, and a 'Historic map of the Church of Ireland, 1536–1990' (1990). For records of diocesan boundary changes, these Léaráid maps are especially useful. Historic town maps were produced for Dún Laoghaire (1986), Kilkenny (1986), and Tralee (1988).

Digital mapping is a novel means of making historic maps available to a wider audience; the NLS site titled 'Maps of Scotland, 1560–1928' is one of the best exemplars of this approach, allowing local historians to zoom in on individual places and compare how any single place was presented in a series of maps [http://www.nls.uk/digitallibrary/map/early/index.html]. As yet there is no comparable project in Ireland. Historic maps have been scanned for display in other contexts, as in the CD-ROM *Counties in time* produced by the NAI as a general introduction to its holdings. This is certain to be an area which will develop rapidly. One routeway to keeping abreast of this area is the map history gateway site at http://www.history.ac.uk/maps/ which includes an updated guide to national and international digital projects.

In any study of cartographic sources, it is impossible to categorise strictly. The heading 'military maps' is not at all adequate to describe maps of the plantation period. Late eighteenth-century county maps, road maps and maps produced to accompany visitors' guides and directories are all overlapping categories, while maps to accompany parliamentary inquiries and reports cover a massive range of cartographic styles and themes. 'Early modern maps' ranges far beyond the category of military fortifications to include portolan charts, plantation maps and road maps. The move towards scientific accuracy was not a linear progression, as map-makers often continued to use out-of-date or archaic materials even when more accurate surveys were available to them.[6] Ordnance Survey maps were used in commercial

6 Mercator's 1564 map of Ireland was widely used even after he had superseded it with a new map in 1595. The DS maps continued to be popularly regarded as incontrovertible, because their use was required by law in certain land transactions, even after they were overtaken by much more accurate surveys, including the OS. Andrews, *Shapes of Ireland*,

directories, for private estate management and land transfer, for the redevelopment of military property, to elucidate official reports and in countless other ways. The headings which follow are therefore pragmatic rather than authoritative cartographic divisions, and are intended to introduce the reader to the vast range available rather than to act as a definitive listing. It is also difficult to separate notes on the major repositories in which maps will be found (chapter 3) from an historical overview of map-making in Ireland, so that there is some overlap between the aims of this chapter and of chapter 3.

EARLY MAPS OF IRELAND

The fact that Ireland lacked a native cartographic tradition prior to the 1600s is widely acknowledged. Although ready to adapt the defensive architecture and military tactics of the outsider, Irishmen apparently felt no great need to emulate the cartographers of England or continental Europe.[7] While certainly holding a massive store of landscape knowledge, including territorial boundaries, placenames, routeways and historical associations, the early disinterest in mapping among the Irish was perhaps, as Andrews posits, quite simply because the island was small enough for people to keep an adequate knowledge of it in their heads.[8] In a heavily localized society, with little travel beyond one's immediate or adjoining lordships, there was no great urgency to 'know' the country as a whole. In addition, the fluidity of the lordship political structure, which was to the advantage of strong lords, could be better maintained without maps. But early maps of Ireland, produced by non-natives, are deserving of the local historian's attention. At the very least they establish the view of the island presented internationally, and the status afforded (or denied) particular places. More usefully, early all-Ireland maps can contain information on local places that is not readily available elsewhere, or indeed that may be unique to the cartographic record. J.H. Andrews' text *Shapes of Ireland*, opens with a discussion, and reproductions, of maps of Ireland previous to the 1560s. The NLI facsimile collection (*Ireland in maps*), used in conjunction with this book, will allow the reader to move from the AD 150 Ptolemy 'map' (as constructed from the latitude and longitude lists in his *Geographia*), through the map by Giraldus Cambrensis dating from *c.*1200 which is of the 'mappa mundi' type (NLI MS 700),[9] through to the portolan tradition of practical sea-charts from about 1300 in which field Italian and Catalan cartographers were the acknowledged experts, intrepidly plotting coastlines from compass bearings and estimated distances, with the safe passage of vessels the over-riding concern. Further facsimiles, several of which are also considered in *Shapes of Ireland*, are to be found in G.R. Crone's *Early maps of the British Isles, AD 1000 to AD 1579* (London, 1961); the early mapping of Ireland is

p. 53; idem, *Plantation acres*, pp 89–92. **7** Andrews, *Shapes of Ireland*, pp 22–3. **8** Ibid., p. 22. **9** Thomas O'Loughlin, 'An early thirteenth-century map in Dublin: a window into the world of Giraldus Cambrensis' in *Imago Mundi*, li (1999), pp 24–39.

4 *The kingdome of Irland 1610 by John Speed, included in his atlas Theatre of the empire of Great Britaine* (London, 1612)

best considered alongside that of Britain, sharing the seas to the north of the European continent.

The cartographer's knowledge of the island of Ireland in the Tudor period (1485–1603) was limited to the eastern and south-eastern regions; by the time of James I (1603–25) the general outline and much of the interior of the country was well-known, excepting only the north-west and west coast (fig. 4). The progression from rude semblance to a quite 'realistic' looking map is vividly presented in J.H. Andrews' *Shapes of Ireland*, chapters 1–4. Not alone was Ireland better mapped by the time of Speed's *Theatre of the empire of Great Britaine* (1612), but the complex process of map-making had itself played an important role in the subjugation and partial Anglicisation of Ireland under Elizabeth I (1558–1603). Mapping was part of the 'knowing' that was essential to military and political conquest: topographical country-wide surveys, maps and plans of forts, battles and sieges; and 'birds' eye-views' of walled towns, bastions of English power, proved indispensable to the extension of the English crown's authority.

Mapping the island of Ireland in the early modern period was driven by very practical considerations. English observers complained of the way in which the Irish exploited their superior knowledge of the countryside, 'a flyinge enemye hidinge him self in woodes and bogges from whence he will not drawe forth but into some streighte passage or perilous forde wheare he knowes the Armie must nedes passe'.[10] To combat this enemy, information was needed on the relative location of towns, forts, castles and other sites which might be defended or attacked in the course of the conquest. Identifying obstacles to movement, where the enemy might take refuge, and the best routes for safe passage was of paramount importance to map-makers drawn from or employed by the military. The surveying undertaken by Robert Lythe (1569–70) is typical of that approach; his work is widely known through a successor and copyist Baptista Boazio, 1599.[11] Treacherous terrain was highlighted, including dark bogs, impenetrable forests, deep water, snaking eskers and waterways (unconnected, or rising in lakes). The coastline was beset by frightening waves, hazardous currents, sand-bars and rocks.[12]

Figure 4 is taken from *The kingdome of Irland* 1610 by John Speed; this was first included in his atlas *Theatre of the empire of Great Britaine* (London, 1612) where it is followed by four provincial maps. As J.H. Andrews demonstrates, Speed's geographical knowledge came mainly from other people's research, not all of whom are named in the *Theatre* as his authorities. However, Speed does preface the atlas with an admission that 'it may be objected that I have put my Sickle into other men's corn, and have laid my Building upon other men's Foundations', explaining

10 Edmund Spenser, *A view of the state of Ireland as it was in the reign of Queen Elizabeth, written by way of dialogue between Eudoxus and Ireneus* (1633), reprinted in Andrew Hadfield and Willy Maley (eds), *Edmund Spenser, a view of the state of Ireland* (Oxford, 1997), p. 96. **11** Andrews, *Shapes of Ireland*, pp 61–75; idem, 'The Irish surveys of Robert Lythe' in *Imago Mundi*, xix (1965), pp 22–31. **12** J.H. Andrews, 'Paper landscapes, mapping Ireland's physical geography' in John Wilson Foster, *Nature in Ireland: a scientific and cultural history* (Dublin, 1997), pp 199–218.

that in a work of this scale and originality it would be impossible to do otherwise. Representations of the midlands and south of the island are derived from surveys by Robert Lythe (*c.*1567–9), those of central and eastern Ulster draw upon Francis Jobson's work (1590), and there were other surveyors, whose names are unknown or uncertain, on whom Speed relied for other parts of the country. However, the background to the Speed maps is further complicated as the material he relied upon came though the work of copyists such as Gerald Mercator (1564) and Baptist Boazio (1599), as well as directly from the maps of those with first-hand survey experience of at least parts of the island.[13] And Speed's own maps (all Ireland, four provincial maps, and inserts on the appropriate provincial maps for Dublin (fig. 3a), Cork, Limerick, Galway, and Enniskillen castle) were very quickly taken up by others to suit their own ends. These included the renowned Amsterdam map-makers Willem and Johan Blaeu who incorporated an all-Ireland map based on Speed (but without naming him as an authority), in their elaborate atlas *Theatrum Orbis Terrarum* (1635).[14] The Russell Library, Maynooth, holds a Spanish language Blaeu atlas titled *Nuevo Atlas de los Reynos de Esocia e Yrlanda* (Amsterdam, 1654), volume v in the *Theatrum* atlas, in which Speed's Ireland and provincial maps feature. This shows the status afforded the Speed map internationally, but also demonstrates how later copyists and publishers could bequeath their own set of errors on an unsuspecting overseas audience.[15] An extract from the original Speed map (fig. 4) is included in this guide to illustrate some lines of inquiry the local historian might follow when dealing with all-Ireland maps from the early modern period. Without an all-Ireland mathematical framework to 'fix' points, the country's outline is still uncertain, most especially along the Atlantic seaboard which was at the farthest remove from those most interested in mapping, namely the English colonial authorities. With land transport underdeveloped, bays, estuaries and navigable rivers loom large in the map-maker's world. Islands take on a much greater significance when viewed from a ship or from the shore, while trying to distinguish promontories from islands requires endless circumnavigation or very good local authorities. The temptation to 'snip off' headlands is difficult to resist. The realistic-looking symbols employed for mountains and forests can be taken too literally by the map-reader; they certainly give the impression of a land with too many hiding places for 'wild' Irishmen and women, portraits of whom are included along the margins of the map, alongside depictions of the 'civill' and 'gentle' Irish man and woman. Making cartographic sense of Ireland's drumlin belt (fig. 4) poses a particular challenge, considering that drumlins extend from land into the sea to make the countless islands of Clew Bay. Castles and forts are noted, even where

13 Andrews, *Shapes of Ireland*, pp 94–6, 113. **14** The *Theatrum* or *Novus atlas* was first published 1635 in two volumes but additional volumes were quickly added, culminating in the launch of a newly-organised 12-volume *Atlas major* in the 1660s. Andrew Bonar Law, *The printed maps of Ireland, 1612–1850* (Dublin, 1997), p. 35. **15** For direction on the many editions of Speed's map of Ireland see Andrew Bonar Law, *John Speed's maps of Ireland* (Dublin, 1979); idem, *The printed maps of Ireland to 1612* (Morristown, New Jersey, 1983); idem, *The printed maps of Ireland, 1612–1850*, pp 27–45.

they are not particularly substantial. Contractions can slow down identification, as in *C.*, *Ca.*, and *Cas.* all employed for castle, *Ile* and *I.* (island), *flu.* (river), *can.* (channel). *C.* is also used for cape. All 'proper' towns, however small, were mapped by the British-born or continental surveyor who had learned his trade in a more urbanised setting. Connacht pre-1700 provided very few examples of that class of settlement. The question must be asked: in a place with few significant towns, were small forts, villages and even lesser settlements upgraded to ensure a more even (and artistically pleasing) spread? If strict population thresholds were in place, which 'towns' would the cartographer have to delete? While the symbols used by Speed on the 1614 edition (fig. 4) are really too small to comment, on later copies of this map (as in Blaeu, 1662 edn), many settlements are shown as walled, a point that has been questioned by Avril Thomas in her study of the walled towns of Ireland. Tiny, highly-stylised pictures showing walls may not be based on field observation at all. An earthen fosse may be upgraded to an impressive wall by an enthusiastic engraver, or indeed a place which was certainly walled (as in Inistioge, County Kilkenny), can appear on sixteenth- and early seventeenth-century maps without this distinction.[16] Such errors are unlikely to be picked up if the author of the map is at some remove from the sphere of operations or reliant on maps of indifferent quality. A greater challenge to the map reader is trying to judge whether a small circle relates to a village, a district such as a townland without nucleated settlement, the homestead or merely the sphere of authority of a local lord. Alternatively, the circle could indicate a headland, a cape, or may be entirely misplaced, being simply the letter 'O' which has become separated from a proper name (such as the O'Donnells). Later copies of the Speed map (as in the Blaeu atlas), reduce practically all settlement symbols to these small circles, exacerbating the problem for the reader. In the international language that is the map, placenames are given in a mixture of Latin, English and Irish. But placenames must be 'heard' as well as seen; thus 'Fin Lough Garrogh' can be recognised as Lough Carra, and 'Mollogh neven' as Nephin mountain (here misplaced). The fluvial geomorphology of Ireland is admittedly complex. Most rivers have countless tributaries, while oddities include rivers which disappear underground in the limestone pavement areas and then reappear elsewhere, and lakes which dry up entirely for several weeks in the year (turloughs). The overwhelming desire to 'connect' sections of rivers and lakes into a coherent whole led to strange disjunctions, while the determination to allow only one river name to apply along its full length could also create problems. In Speed's *Irland* (fig. 4), the river Suck is extended from the *Synan flu.* to Lough Carra, taking over several other lesser watercourses (including the Dalgan) to make this believable. Mapping lakes is notoriously difficult; the work of the OS (Plots of rivers and lakes, NAI, OS 102), will convince the sceptical on this point. Most of the lakes in figure 4, such as Lough Carra, are far too large, and the number of islands greatly overstated. Early ecclesiastical sites (for example, as marked on the OS 1:50,000 *Discovery/Discoverer* maps and on the OS*i* 1:210,000 road atlas), are useful to help

16 Thomas, *Walled towns of Ireland*, pp 16–17.

'pin down' some placenames on pre-1700 maps. Thus from figure 4 Crogh McDarra (St Macdara's island), Balentober (Ballintober Abbey), Killaloy (Killala), Elpin (Elphin) and Corcome M. (Corcomroe) can all be identified with confidence. The problems of incorporating information from maps made by predecessors or competitors are innumerable, as scale, coverage, quality and reliability of the source maps were all variables. Where information is contradictory, which source is to be believed? And choosing what to leave out is especially difficult without first-hand knowledge of the place being mapped. The reliability of the map can vary even within itself, as updated information is used for one region and long out-dated and patently inaccurate data utilised elsewhere. This is apart from problems of transcription and misinterpretation, which can lead to names which make no apparent sense ('logh flu', 'Slewe' and 'Wreike' in fig. 4), while family names (Carew, Burk, O Coner Don, earl of Clanricket) were interspersed with placenames to fill up parts of the map that would otherwise be suspiciously empty. The cartographer was also free to insert notes that might interest his audience; 'Y^e^ Sanctuary Grany O Male', the great pirate queen of County Mayo' is a typical annotation (fig. 4). Pre-OS all-Ireland maps, such as the work of Speed (fig. 4) and its forbears, provide the historian and geographer with a wealth of information. However, unravelling what the map 'says' requires perseverance and the intimate knowledge that comes from long association with a local place. It is a particularly rewarding, though always incomplete study.

PLANTATION MAPS

Maps were an essential tool in the policy of confiscation and plantation, commenced in 1556 with the Leix-Offaly plantation, followed on a larger scale in Munster (1586), and most successfully of all in Ulster (1607–22). The richest cartographic record is to be found for Ulster; however, maps have survived relating to the other plantations also. Examples of the work of Francis Jobson, who was employed as a plantation surveyor in Munster in 1586 and later combined government contracts with freelance work in several parts of the country, are held by the NLI and are particularly impressive.[17] In working with maps created in such contexts, the researcher needs to consider whether they are records of 'real' scenes and events (allowing always for the multiplicity of perspectives that can be contained within even one map), or planning or speculative documents, depicting idealised fortifications or possible land subdivisions, which may or may not have been followed through in reality. The many questions to be asked of maps (see Appendix 1) created during the Tudor and Stuart eras bring the local historian directly into a critical period in the reshaping of Ireland culturally, politically and

17 Francis Jobson's map of Munster 1592 is reproduced in Noel Kissane (ed.), *Treasures from the National Library of Ireland* (Drogheda, 1994), p. 183; that of Ulster 1591 is simplified and redrawn in Andrews' *Shapes of Ireland*, p. 104.

geographically. Many of the maps are intimate local documents, but the researcher needs to consider the crucible within which these were produced, and the larger projects afoot, as documented in the wealth of polemical and state papers of the period. The city of Londonderry, for example, spectacularly sited on a schist outcrop between the river Foyle (east) and bogland (west), was planned as a show-piece of the entire plantation. Maps were commissioned by both the crown and the London companies (the investors) to evaluate the progress being made on the ground, or indeed the lack of progress from the perspective of those backers who were expecting a generous and rapid return on their investment.[18] Thomas Phillips, who commissioned a major mapping exercise completed in 1622, was himself a servitor-grantee or beneficiary of the Londonderry plantation and a member of the 1622 commission of inquiry.[19] On the top right hand corner of the original 1622 edition 'The plat of the cittie of Londonderrie as it stand built and fortyfyed', he notes the numbers of families 'dwelling in stone houses slated' (109), with a further twelve 'families of poure soldiers and poore labouring men dwelling w'thin the walles in cabbins'. The key (which is on a separate sheet) gives the names of streets and of buildings planned or proposed, with the purpose of the map made most clear in notes such as 'Ranges left where houses may be built in tyme to come' (N) and 'a place where the new Key were fitt to bee built' (L). The steeply-sloping nature of the site is not readily apparent to the map reader, whose attention is directed to the eight impressive bulwarks and the 'verie stronge wall excellently made and neatly wrought being all of good lyme and stone'.[20]

The intense map-making activity of the sixteenth and early seventeenth century in Ireland must be studied within an international framework, as introduced in J.H. Andrews' 'Colonial cartography in a European setting: the case of Tudor Ireland'.[21] Colm Lennon's *Sixteenth-century Ireland: the incomplete conquest* (Dublin, 1994), and Steven Ellis' *Ireland in the age of the Tudors* (London, 1998) are good introductions to the political struggles of the period. For an introduction to sources for the study of plantation towns, with map reproductions and suggestions for further reading, the reader is directed to Raymond Gillespie, 'Plantation records' in Nolan and Simms (eds), *Irish towns: a guide to sources*, pp 79–83. R.W. Dudley and Mary O'Dowd, 'Maps and drawings' in *Sources for early modern Irish history, 1534–1641* (Cambridge, 1985), pp 106–28 should also be consulted.

Margaret Gowen has published a union list of late sixteenth and early seventeenth century military plans held in the BL, TNA, PRONI, TCD, and NLI, arranged by collection, in the Military History Society's journal, the *Irish Sword*.[22] This journal should also be consulted for plans and discussions of individual fortification

18 Avril Thomas, 'Derry, a spectacular maiden' in H.B. Clarke (ed.), *Irish cities* (Cork, 1995), p. 74. **19** He is to be distinguished from the military surveyor, Thomas Phillips, who was active in the 1680s. **20** Nicholas Pynnar, 'The cittie and countie of London Derry' (TCD, MS 864, par. 181). **21** J.H. Andrews, 'Colonial cartography in a European setting: the case of Tudor Ireland' in *The history of cartography*, vol. iii, *Cartography in the age of renaissance and discovery* (Chicago and London, forthcoming). **22** Margaret Gowen, 'A bibliography of

projects, such as Spike Island, Smerwick and Charles Fort (Kinsale), while the *Royal Engineers' Journal* and the Defence forces publication *An Cosantóir* also feature occasional discussions of early maps and map-making. Vignettes of seventeenth-century maps regularly feature along the margins of eighteenth- and nineteenth-century town plans, as with Belfast city, where a map dated 1660 appears in the margins of a succession of maps published 1819–79.[23]

One of the most impressive collections of late sixteenth-and early seventeenth-century maps is that preserved by Sir George Carew, who held the office of master of the ordnance (1609–23) and was later lord president of Munster. His collection of over fifty maps, deposited in TCD (MS. 1209), is sometimes known as the Hardiman atlas, because it was first listed by this Galway historian in the *RIA Transactions*, xiv (1821–5), pp 55–77. A number of these maps were used in *Pacata Hibernia*, 1633 (reprinted 1810, 1896), an account of the wars in Munster while under the government of Carew.[24] Some plantation maps were copied at the time of their production, most importantly those of the county and city of Londonderry in the Ulster plantation. The survey undertaken by Thomas Phillips (1609–29) to expose the incompetence of the London companies, and submitted to the king, was reprinted, with maps, in 1928.[25] Other facsimiles of Ulster plantation maps (mostly by Barthelet) will be found in G.A. Hayes-McCoy, *Ulster and other Irish maps, c.1600* (Dublin, 1964). The PRONI education facsimile volume *Plantations in Ulster, c.1600–41* (1975, 1989) is a valuable source for maps, associated documents and explanatory notes, featuring a facsimile of the 1622 map of Londonderry by Thomas Raven for Phillips is found in this collection. *Maps of the escheated counties of Ireland* published by the Ordnance Survey (Southampton) in 1861 (and sometimes known as the 'Irish historical atlas') includes all the 'escheated counties maps' made for the distribution of lands in the Ulster plantation known to have survived.[26] This collection opens with *A generalle description of Ulster* which in itself is an excellent testament to the map-maker's craft at this tubulent period. Dotted with forts, castles and heavily defended towns, the maps include several annotations: 'Here Shane o neal was slaine'; 'The high hills of Benbolbin, where yearly limbereth a Falcon esteemed the hardiest in Ireland'; 'Three Spanish shipps here cast awaie in An°. Dom. 1588'. This set of facsimiles consists of provincial, county and barony maps, with townland names recorded on the latter; extracts from this collection have been deservedly reproduced in many studies. The set concludes with an action-packed depiction of the siege of 'Enneskillin Castell', with local details such as the use of 'cotts' or currachs as well as aspects of military matters

contemporary plans of late sixteenth and seventeenth century artillery fortifications in Ireland in *Irish Sword*, xiv (1980–81), pp 230–6. **23** *A union list of Belfast maps to 1900* (Belfast, 1998). **24** [Thomas Stafford] (ed.), *Pacata Hibernica: Ireland appeased and reduced* (1633); reprinted with a new introduction and notes by Standish O'Grady (2 vols, London, 1896). **25** David A. Chart (ed.), *Londonderry and the London Companies, 1609–29; being a survey and other documents submitted to King Charles I, by Sir Thomas Phillips* (Belfast, 1928). **26** Ordnance Survey, *Maps of the escheated counties of Ireland*, facsimile reproduction, introduction by Hans Hamilton (Southampton, 1861).

including camp layout, firepower, 'the house of munitions', the scaling of the walls, 'the bote ankered to breach' and the very public display of the heads of enemies, on spikes, being recorded. Volume iv parts 1 and 2 of the national manuscripts facsimile collection edited by John T. Gilbert (1874–84) includes 18 map reproductions from the plantation period.[27] Selected titles in the *Irish Historic Towns Atlas* series also include facsimile prints of this period, most notably the fascicles of Bandon, Carrickfergus and Athlone.

Siege plans were another product of the early modern or plantation period, and were exquisitely drawn to accompany battle accounts. Richard Barthelet (many writers modernise to Bartlett) was one of several surveyors who recorded the strongpoints and fortifications that featured in the campaign of the English commander Mountjoy as he brought the earl of Tyrone, Hugh O'Neill, to final submission (1603). Barthelet's surveys included drawings of houses, raths, woods, crannógs and farmland; reproductions can be found in Hayes-McCoy, *Ulster and other Irish maps*. The Barthelet map of Armagh *c.*1602 is also reproduced in *Treasures from the National Library of Ireland* (Dublin, 1995), edited by Noel Kissane. In addition, the 'Early Printed Books' Department of TCD Library holds an excellent collection of late sixteenth-century atlases and maps. M. Swift's *Historical maps of Ireland* (London, 1999) includes 57 colour reproductions of maps dating from 1558 to 1691, held in the National Archives, Kew, and provides an introduction to the cartographic priorities and styles of the period.

Typical of all-Ireland, regional and local maps during the early modern period is the attention given to towns as walled entities. Raven's map of Londonderry most obviously celebrates its walled status, and the Petty maps (discussed below) include elaborate drawings of walls in several (but by no means all) instances. The Speed map (fig. 4) depicts as walled only a select number of towns ('Knockfergus', Dublin, Limerick, Cork, Kinsale, Waterford, Kilkenny, and Carlow). *Plans of the principal towns, forts and harbours in Ireland for Mr Tindal's continuation of Mr Rapin's history* (1744–7) reproduces on a single sheet a mixture of seventeenth- and eighteenth-century plans; town wall, defences, and strategic siting are the chief considerations in these simplified reproductions. The importance of walled towns, imaginatively and politically, is reflected not alone in the cartographic record but in other polemical documents of the Plantation period. Edmund Spenser's *A view of the present state of Ireland* (1596) sketches out a plan for the creation of fortified towns in Ulster which would stand as centres of culture and civility, protecting settlements and agriculture around them.[28] The town is advertised as a place which will keep watch over the surrounding countryside, stimulate agriculture and trade, and act as the means through which the lawless but rich interior will in time become subject to English customs and law. Its sturdy walls symbolised the wealth of the town and the degree of protection that citizens and foreigners alike might enjoy therein.[29] The market function is especially cherished: 'there is nothing doth

27 Gilbert (ed.), *Facsimiles of the national manuscripts of Ireland*. **28** Spenser, *A view of the state of Ireland*, pp 194–5. **29** Thomas, *Walled towns of Ireland*, i, p. 11.

sooner cause Civility in any Countrey than many Market Towns, by reason that People repairing often thither for their Needs, will daily see and learn civil Manners of the better Sort'.[30] However, the concept of 'town', as mapped by Lythe, his contemporaries and successors, is the subject of continuing debate. Despite the use of rather convincing symbols (as in fig. 4), some of the places mapped in the late 1500s and 1600s are merely small, undefended settlement clusters, without church or castle to recommend them as 'towns', and without the charter status (with market privileges) that commentators such as Spenser regarded as defining a 'town'.[31] There is also the difficulty, as noted before, of deciding if pictorial symbols showing the town as walled are based on field observation or other first-hand knowledge, or whether they should be treated as merely stylised (and very tiny) symbols. The close attention of local historians to these sites may go some way towards discovering the reasons behind their inclusion.

DOWN SURVEY (WILLIAM PETTY), 1654–9

The Down Survey (DS) maps are the earliest large-scale maps to which many local historians will have access and are thus of especial interest. These are the maps that made possible the most comprehensive land transfers of the plantation period. The Down Survey was preceded by the Strafford Survey of Connacht, Clare and North Tipperary (1636), the Gross Survey (1653–4) and the Civil Survey (1654–6) which covered 27 counties in whole or part.[32] The Strafford Survey was undertaken as a terrier (a register or roll of landholding), accompanied by parish, county, and index maps. Regrettably only one unsigned, undated index map of the survey area has survived. A fire in the surveyor general's office in 1711 destroyed the originals, and no copies of the entire series as a whole appear to have been made.[33] Occasionally copies of odd townlands appear among estate papers, and a purported Strafford Survey extract for County Roscommon was reproduced by R.C. Simington in the Irish Manuscript Commission's edition of the book of survey and distribution for Roscommon (discussed below). The Civil Survey, which commenced in 1654, was a government inquiry into the property valued over 10*s*. held by Roman Catholics with a view to confiscation. Under the Act of Satisfaction (1653), which effectively brought the wars of mid-seventeenth century Ireland to a close, these confiscated estates were to be given as payment to soldiers and the adventurers who financed the war. Special 'courts of survey' under the civil (not military) authorities were set up, and by means of inquisition rather than by mapped measurements, local jurors drew up a record of the barony boundaries, proprietors' names in 1640, the extent and value of each holding, the type of tenure, and the number and type of buildings

30 Spenser, *A view of the state of Ireland* pp 250–1. **31** Ibid.; Andrews, *Shapes of Ireland*, pp 67–9. **32** W. O'Sullivan (ed.), *The Strafford inquisition of County Mayo* (Dublin, 1958); the Sligo survey will be found in BL, Haley 2048. **33** This map is reproduced in Andrews, *Shapes of Ireland*, pp 120–1.

upon it.[34] Each barony was divided into parishes and then into 'towneshipps or villages', while the record for urban areas includes notes on the dimensions and use of individual premises. Of the 27 counties surveyed, the records of only fourteen now survive. These have been edited by R.C. Simington and published in 10 volumes by the Irish Manuscripts Commission 1931–61; they are widely available in county and university libraries, as well as in all the major state repositories.[35] However, though a geographical study, the Civil Survey was not map based.

The extensive land confiscations after Oliver Cromwell's military campaigns led to an urgent need for a large-scale cartographic survey, which would measure land units and tie verbal descriptions to individual plots and their proprietors. The scale was unprecedented: almost half the country, embracing an area later estimated at nearly 8,400,000 acres, being all the land held by Roman Catholic proprietors at the outbreak of the 1641 rebellion.[36] Dr (afterwards Sir) William Petty, physician-in-chief to the Cromwellian army, sought out this challenge at great personal profit, and accomplished the first major country-wide mapping in which measurements were made rather than estimated. It was a truly herculean undertaking, not to be surpassed until the time of the Ordnance Survey. The distinction between parish maps and barony maps is explained below, but researchers should be aware at the outset that a set of barony maps titled *Hibernia Regnum* (published in 1908) was intended to record all barony boundaries regardless of whether the land was to be confiscated (fig. 2). These are well known, as all 220 maps were reproduced for public sale by the Ordnance Survey in Southampton in 1908 (fig. 2); this two-volume loose-leaf atlas is in black and white, but some sets (as in Cambridge University Library) were coloured by the OS. There are also barony maps (fig. 5) which were intended as indexes to the parish maps (fig. 6).

The originals of the completed DS parish maps have not survived. A few contemporary copies of the barony maps ended up in the Quit Rent Office to be eventually deposited in the National Archives of Ireland. The British Library set of barony maps is fully described in *The British Library catalogue of additions to the manuscripts: the Petty papers* (London, 2000). The most complete set of DS barony maps is to be found in the Bibliothèque Nationale in Paris. It is largely from this set that copies have been made, including those made by the military engineer Charles Vallancey on behalf of the government in 1786.[37] Another set of copies (of parish and barony maps), also made in the 1780s, became known as the Reeves collection, and is now in the National Library of Ireland (fifteen volumes, covering Munster and Leinster) and in the Public Record Office of Northern Ireland (two volumes, covering Ulster counties).[38] Microfilm of the NLI (Reeves) collection

34 J.G. Simms, 'The Civil Survey 1654–6' in *Irish Historical Studies*, ix, no. 35 (1955) pp 253–63. **35** The counties for which the Civil Survey has been published are: Tipperary (vols i and ii), Donegal, Derry and Tyrone (vol. iii), Limerick (vol. iv), Meath (vol. v), Waterford, Muskerry barony (County Cork) and Kilkenny city (vol. vi), Dublin (vol. vii), Kildare (vol. viii), Wexford (ix); vol. x is titled 'miscellanea' and includes part of Kerry and Louth. **36** Andrews, *Shapes of Ireland*, p. 122. **37** Andrews, *Plantation acres*, pp 87–9. **38** Charles McNeill, 'Copies of the

(barony maps, parish maps and terriers) is available in many college and county libraries. Figures 5 and 6 are extracts from that set, taken from microfilm. The original field books do not survive, but the Down Survey (including the key to proprietors' names), and the career of Petty himself were both the subject of early studies. Thomas Larcom, director of the Ordnance Survey office at the Phoenix Park, Dublin, edited Petty's own history of the Down Survey (1851), while a three-volume work in French by Yann Goblet (1930) is still the major reference for serious students of Petty.[39] Biographies of the polymath include Edward Fitzmaurice, *The life of Sir William Petty* (London, 1895) and Eric Strauss, *Sir William Petty, portrait of a genius* (London, 1954).

The methods employed by William Petty on the Down Survey were not extraordinary. Andrews notes that the chain and circumferentor method of surveying was already well established in Ireland. It continued beyond Petty's period, albeit with better instruments, and with some further refinements, most notably bringing the traverse stations closer together and paying more attention to offsets.[40] But Petty succeeded in covering an unprecedented amount of ground to a very high standard due to his organisational skills and the specialisation which he employed among his staff. Those recruited had to be able 'to endure travaile, ill-lodginge and dyett, as also heates and colds, beinge also men of activitie, that could leap hedge and ditch, and could also ruffle with the several rude persons in the country, from whom they might expect so often crossed and opposed'. Several soldiers who were sufficiently literate for the purpose were engaged:

> Such therefore (if they be but headful and steddy minded, though not of the nimblest wits) were taught… how to make use of their instruments, in order to take the bearinge of any line, and alsoe how to handle the chaines, especially in the case of risinge or falling grounds; as alsoe how to make severall markes with a spade, whereby to distinguish the various breakings and abutments which they were to take notice of; and to choose the most convenient stations or place for observations …[41]

The Down Survey was therefore organised on efficient military lines, in a hostile and 'foreign' environment (several surveyors lost their heads before completion), requiring minimal literacy among at least some of those employed, and above all a willingness to follow instructions exactly in the simple, repetitive, but vital work of measuring distance and direction. Petty certainly regarded the whole exercise as a significant contribution towards the ongoing 'civilising' (that is, Anglicisation) of the country, a theme that is elaborated on repeatedly in his political writings.[42]

Down survey in private keeping' in *Analecta Hibernica*, viii (1938), pp 419–27. **39** A. Larcom (ed.), *A history of the … Down Survey*; Yann M. Goblet, *La transformation de la géographie politique de l'Irlande au XVIIe siècle dans les cartes et essais anthropogéographiques de Sir William Petty* (3 vols, Paris, 1930). **40** Andrews, *Shapes of Ireland*, pp 121–2, 135. **41** See Larcom (ed.), *A history of … the Down Survey, by Sir William Petty;* on Larcom see: Thomas E. Jordan, *Imaginative empiricist: Thomas Aiskew Larcom, 1801–79, and Victorian Ireland* (Lewiston, NY, 2002). **42** William Petty, *The anatomy of Ireland*, 2nd edn (London, 1719), pp 25–6.

5 Down Survey barony map of Gowran, County Kilkenny, 1655

In little over one year, the soldiers had finished their surveying and the mathematical calculations turned into maps drawn by parishes (fig. 6), some at about three inches to one mile, others at about six inches to one mile (40 or 80 Irish perches to an inch, where the perch contained 21 feet), with index or barony maps (fig. 5) on a smaller scale (at 80, 160, 213 or 320 perches to an inch). Where a student seeks to take precise measurements based on these maps, it is obviously essential to be certain whether one is dealing with a parish or barony map, and the scale employed originally. Orientation varies with each map; the north point (magnetic north) is the first item to locate. The barony index map (fig. 5) must be studied in conjunction with the written terrier (fig. 5a). The barony of Gowran, County Kilkenny, for example, is

The Baroney of Gowran in the Countie of Kilkenney. —

Is Bounded on the East with the River Barrow that divides this Baroney from the Countey of Caterlagh on the Northeast with the said County of Caterlagh on the North with the Baroney of ffassadinenge upon the west with the Liberties of Kilkenny & the River Neor that divides this Baroney from the Baronies of Sheiloghter & Knocktofer and on the South with the Baroney of Ibercan: —
Theres Contained in this Baroney the Parrishes of Graige — Inisteoge Jerpoynt Ullard Collumkill Powerstowne Dungarvan Kilfane Tulloghelim Killerney Dunbill Thomastowne upper= =grange Gowran Blanchfeildstowne Rathcoole Churchclaragh Blackrath Killderry Teighcoffin Killmacahill Shankill & Killma= =dum & part of St Martins which is Joyned to Blackrath; — The middle of this Baroney and Especially that part next to Kilkenney and the River Neor is plaine and ffertile & for the most part arable The South part is more hilly & Mountain= =ous and not soe fruitfull, the East part towards the River Barrow is more woody both Timber wood and Shrubbs inter= mixt with verie good Arable The Northwest part of it is mountainous and plaine, the Northeast is well stored with Timber Wood:
This Baroney is Scituate betwixt the Rivers Barrow and Neor both of them Navigable with Boates from Waterford upon Neor to Kilkenny & upon Barrow to Caterlogh both of them plentifull in ffish and Severall sorts of ffowle in the Winter. —

An Index of Parrishes names and Quantitey of Acres Contained in Each of them.

1.	Graige	7066–0–0	5343–0–0	1723–0–0
2.	Ullard	1733–0–0	1733–0–0	
3.	Powerstowne	3034–0–0	3034–0–0	
4.	Jerpoynt	2584–0–0	2584–0–0	
5.	Innisteoge	3836–0–0	3836–0–0	
6.	Thomastowne	1840–0–0	1840–0–0	
7.	Cullumkill	2101:0:0	2012:0:0	89–0–0
8.	Killfane	2213–0–0	2173–0–0	40–0–0
9.	Dungarvan	2764 0.0	2764 0:0	
10.	Tulloghelin	2759–0–0	2759–0–0	
11	Killerney	1070–0–0	1070–0–0	
12.	Dunbill	1200–0–0	1200–0–0	
13.	Blanchfeildstowne & Smithstowne	1278–0–0	1278–0–0	
14.	Blackrath and St Martyns Parr	1529–0–0	1529–0–0	
15.	Teighscoffyn	2439–0–0	2439–0–0	

An Index of Proprietors names and pages where to finde them

A.	
Redmond Archdeacon	3:9
Henry Archer	15:16:5:21:3
John Archdeacon	9
B	
Sr Edmd Butler	1:2:3:4:10
Peirce Butler	1
Edmond Blanchfield	1:3
Sr Edmd Blanchfield	3:8:13 14:2:16
Sr Richard Butler	4
Rich: Butler	5
Burgers of the towne	6
James Blanchfeild	15:16
Peirce Burren	21
C.	
James Cooley	7
Peirce Coddy	7
John Cantwell	8:18
Nicholas Cantwell	18
John Cantwell Junior	18
John Comberford	21
D.	
Nicholas Dobbin	5
Thomas Dennbi	7:10:14
Robert Dobbin	6
Edmond Dobbin	7
Peirce Dobbin	9
F	
Oliver Fennell	6
Robert ffforstall	8:9
Patrick ffforstall	9
G	
Gerald Grace	4
Gerald Grace	4:11
Lord Mt Garrett	5
Mr Fitzgerald	5
Mary Fitzgerald	10

5a Down Survey terrier, barony of Gowran, County Kilkenny, 1655

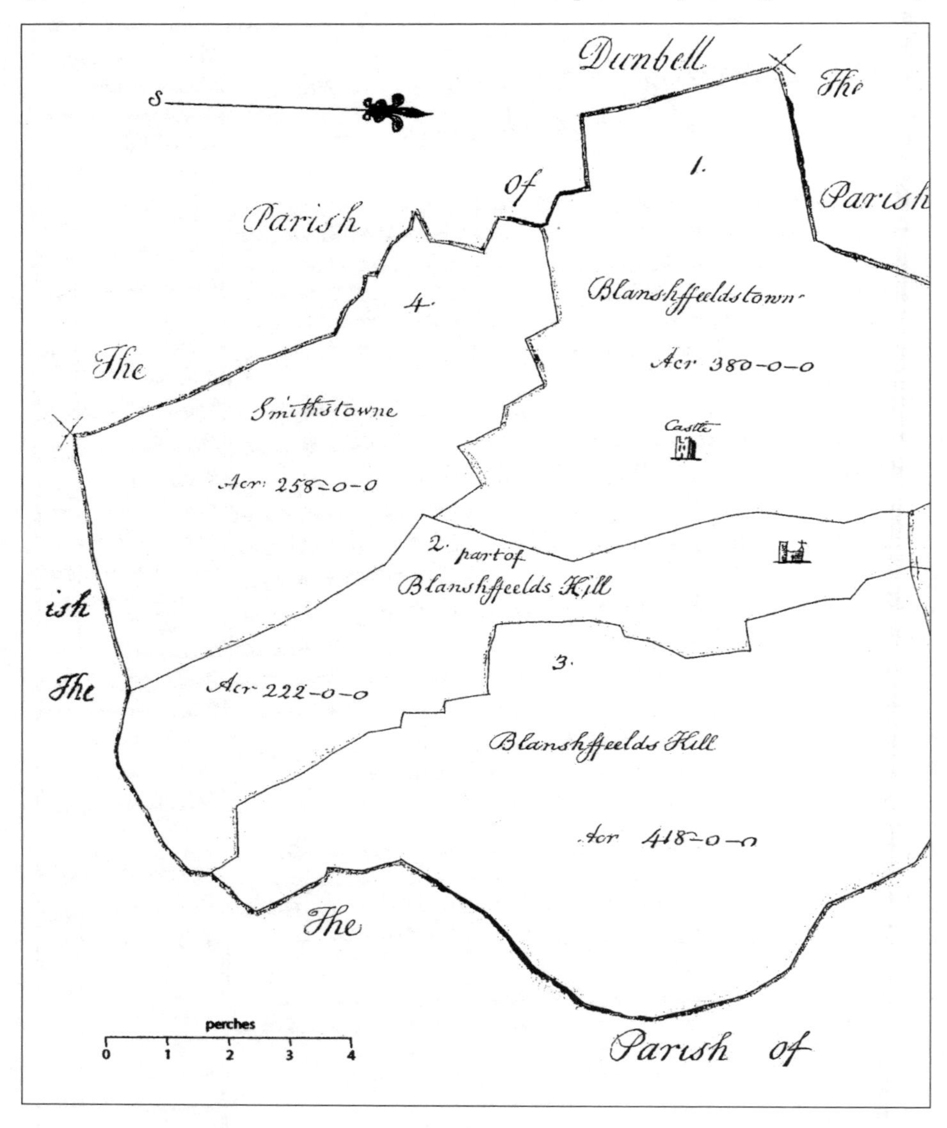

6 Down Survey parish map of Blanchfields, barony of Gowran, county Kilkenny, 1655

bounded by the river Barrow on the north-east and on the south by the river Nore, and contains 22 named parishes (Smithstown is mapped with Blanchfieldstown, figure 6, and 'part of St. Martyn's' is mapped with Blackrath). The number of the parish is drawn on the barony map, making it clear that the latter is really an index to the former. There are other small variations between the maps and terriers produced by different surveyors, with some districts enjoying the attention of more

Blanchfeilds & Smithstowne Parrish

Is Bounded on the East with the Parrish of Goran on the North with the Parrishes of Church Claragh & Teigecoffin on the West and South with the Parrishes of Dunbill & Tullohelim: —

It doth Containe these ensueinge Towne Landse, viz.: Blanchfeildstown Blanchfeildskill in Goran Blanchsfeildskill in Blanchfeildstowne and Smithstowne Parrish — & Smithstowne.

The Soyle is Generally good Arable and the rest good Pasture.

There is in Blanchfeildstowne a large Castle in repaire and on that part of Blanchvillskill which belongeth to Blanchfeildstowne Parrish the ruinous walls of a Church and nothinge else observable except Cabbins in this Parrish.

Note though Smithstowne bee put in this Parrish it hath beene reputed to bee a Parrish of itselfe.

Numbers in Plot	Proprietors Names	Denominations	Number of acres by Admeasurement	Landes
1.	Sir Edmd Blanchfeild	Blanchfeildstowne	380-0-0	Ar: R p
2.	The Same	Blanchfeildskill	222-0-0	Same
3.	The Same	The other part of Blanchfeildskill	418-0-0	Same

Retyped by J. Prunty

6a Down Survey parish map of Blanchfields, barony of Gowran, County Kilkenny, 1655

accomplished or painstaking map-makers than others. In Limerick, for example, the DS barony map is enhanced by a large-scale inset of the walled city and 'the Irish towne'. The information on most DS maps is sparse, but is supplemented by the written record (although spelling may vary). In all cases the core information is standard. In the case of Gowran (figs. 5 and 5a) for example, there is a brief description of the land quality: the part watered by the river Nore 'is plaine and fertile and for the most part arable, the south is 'more hilly and mountainous and not soe fruitfull', while there is also a special eye for areas 'well stored with Timber wood', resources such as fish and wild fowl, and possibilities for navigation. The barony map (fig. 5) shows parish, barony and county boundaries, names of these units (and of some townlands also), principal rivers and bridges, and stone or otherwise substantial buildings, such as 'one castle and one very lardge house built English wise both in good repaire' in the townland of Ballyfoyle, Kilmodum parish. Profile symbols are used for forest, mountain and bog on some barony maps, in others the verbal description must suffice. Larger-scale parish maps (see Blanchfieldstown, fig. 6), show very little that is not already shown in the barony maps. The townlands which make up the parish are bounded, each is named and its acreage noted and, as before, the few major buildings to command the attention of the surveyor are drawn in profile. But the parish terrier makes up for the sparseness of the graphic content. At the opening of each terrier (fig. 6a) the parish bounds are described, followed by a summary of the contents of each townland, noting 'cabbins', 'thatcht houses', woods (an important resource), streams, mills, markets, and (occasionally) local antiquities. The Blanchfield case is unremarkable: 'the soyle is generally good Arable and the rest good Pasture. The only significant buildings here are 'a large castle in repaire' and 'the ruinous walls of a church', with 'nothing else observable except cabbins in this parish'. The first column of the terrier ('number in plott') links the named proprietor (in this case 'Sir. Edmd. Blanchfeild') with the numbered units on the map. Throughout the Down Survey, it is land liable to confiscation (or 'soldiers' land') which is the focus of attention. Church land (including glebe land), the lands of Trinity College, crown lands and the property of Protestant occupiers are merely noted, not surveyed. The first column is headed 'proprietors' names'; in some DS terriers the term 'Irish Papist' (abbreviated to IP) is added while in other cases this is simply presumed. This is followed by 'denominations of lands' (townland name), 'number of acres by admeasurement', 'landes profitable', and 'landes unprofitable' (such as 'lough' or 'bog'). In cases where Protestant proprietorship is noted and the land 'admeasured', the amount is squeezed into the column featuring the townland name, lest it be included in the final row, 'the totall of forfeited land'. Under 'lands profitable', arable, meadow, pasture and wood are all further noted; abbreviations being used (*ar: me: pa:* or *past. w^{d}*). To supplement the information in the DS maps, the reader should also consult the 'Abstracts of various surveys, 1636–1703', for the purpose of establishing an official record of landed proprietors and their respective estates, the date of the 'decrees, certificates and pattents' under which they held them, and clarifying who exactly was liable for quit rent (payable to the government annually on lands granted under the terms of the new Acts of Settlement (1662) and of Explanation (1665)). The entries here are keyed to the Down Survey map, as well as to the number of the 'book or roll', and the 'page or skin' of the Quit Rent Office records wherein the land titles have been entered; the original manuscript is held by the RIA. The Down Survey was an improvement on any previous large-scale cartographic undertaking in Ireland in terms of accuracy, excepting perhaps the Strafford Survey. Based on a direct, measured survey, the Down Survey covered confiscated land throughout the island, excluding most of the area already covered by the Strafford Survey. However, the resultant maps are limited (figs. 5 and 6): they are concerned with the townland boundaries, one named proprietor, and the classification of forfeited land as cultivable, bog, mountain or wood. The method of measurement ensured that each barony outline was fairly accurate as a stand-alone exercise, but joining separate units together to make a larger picture reveals the limitations of the survey as a scientific exercise. Only with triangulation (see under Ordnance Survey) could country-wide accuracy be achieved.[43]

Predictably, the problem which was to confound Petty to the end was the failure to survey the land which was not forfeited ('innocent Protestant & church lands'), leaving him with numerous empty spaces that in a less scientific age would be glossed over with 'floating' text or otherwise imaginatively filled. This deficiency

43 The later use of the Down Survey by Petty in his famous atlas, *Hiberniae delineatio* (1685), is recounted by Andrews in *Shapes of Ireland*, pp 118–52.

could only be made good through another massive surveying operation, or by local proprietors volunteering estate survey material that would allow him to fill in the blanks. Comprehensive state-sponsored surveying did not happen again until the Ordnance Survey launched its townland survey in the 1820s, and while Petty may have gained access to some private maps, there were certainly far too many blanks to cover through this route. The DS maps are not topographical maps, as there is no attempt to make a complete or 'internal' record of roads, buildings, drainage or relief. The sparseness is disappointing to the local historian; to expect more is unreasonable.

REGIONAL/COUNTY MAPS

The earliest regional or county maps, mostly in manuscript form, date from the late sixteenth and early seventeenth century; a small sample may be identified through consultation of the Hayes-McCoy catalogue, *Ulster and other Irish maps, c.1600*. While Hayes-McCoy catalogues the 27 'Bowlby' maps held in the NLI (MS. 2656), he includes incidental mention of many others. A typescript of John Andrews' preliminary catalogue numbering over 500 pre-1630 regional maps of Ireland may be consulted in the Glucksman Map Library, TCD. Andrews' paper on Richard Bingham's map of Connacht 1591, accompanied by a full colour facsimile reprint, introduces the local researcher to late sixteenth century cartography at the provincial scale and makes available an important survey which will repay closer scrutiny.[44] The most peaceful part of the island, the east and south-east, or what might be termed the Pale in an extended sense, has remarkably few original maps of this period, mainly those of Robert Lythe (1569–70). In the provincial maps of John Speed, published as part of his *Theatre of the empire of Great Britaine* (1612) discussed above, early modern and largely political mapping of Ireland reaches its highest point, with contemporaries regarding his maps as definitive.[45] William Petty's county atlas *Hiberniae delineatio* (1685), based on the Down Survey, replaced Speed's map as the definitive all-Ireland map. An extract from the province of Munster (fig. 7) shows how the atlas maps differ from those of the Down Survey. Making reference to the questions posed in Appendix 1, the researcher might ask how does this compare with previous output at a local scale, specifically the Down Survey maps? What other sources were to hand? Who was responsible for its final production? According to J.H. Andrews, the *Hiberniae delineatio*, despite its imperfections, is a landmark in the cartographic history of Ireland, and 'ranks as a primary topographic and toponymic document, whose message must be patiently disentangled from the misrepresentations of draftsmen and engravers' who were working in Amsterdam but with insufficient editorial supervision.[46] In figure 7

44 J.H. Andrews, 'Sir Richard Bingham and the mapping of western Ireland', in *RIA Proc.*, ciii, 3 (2003), pp 61–95. **45** Andrews, 'Maps, prints and drawings', p. 28. **46** J.H. Andrews, 'Introduction to *Hiberniae Delineatio* and Geographical Description', introductory notes to the facsimile, William Petty, *Hiberniae Delineatio* (Shannon, 1968), p. 13.

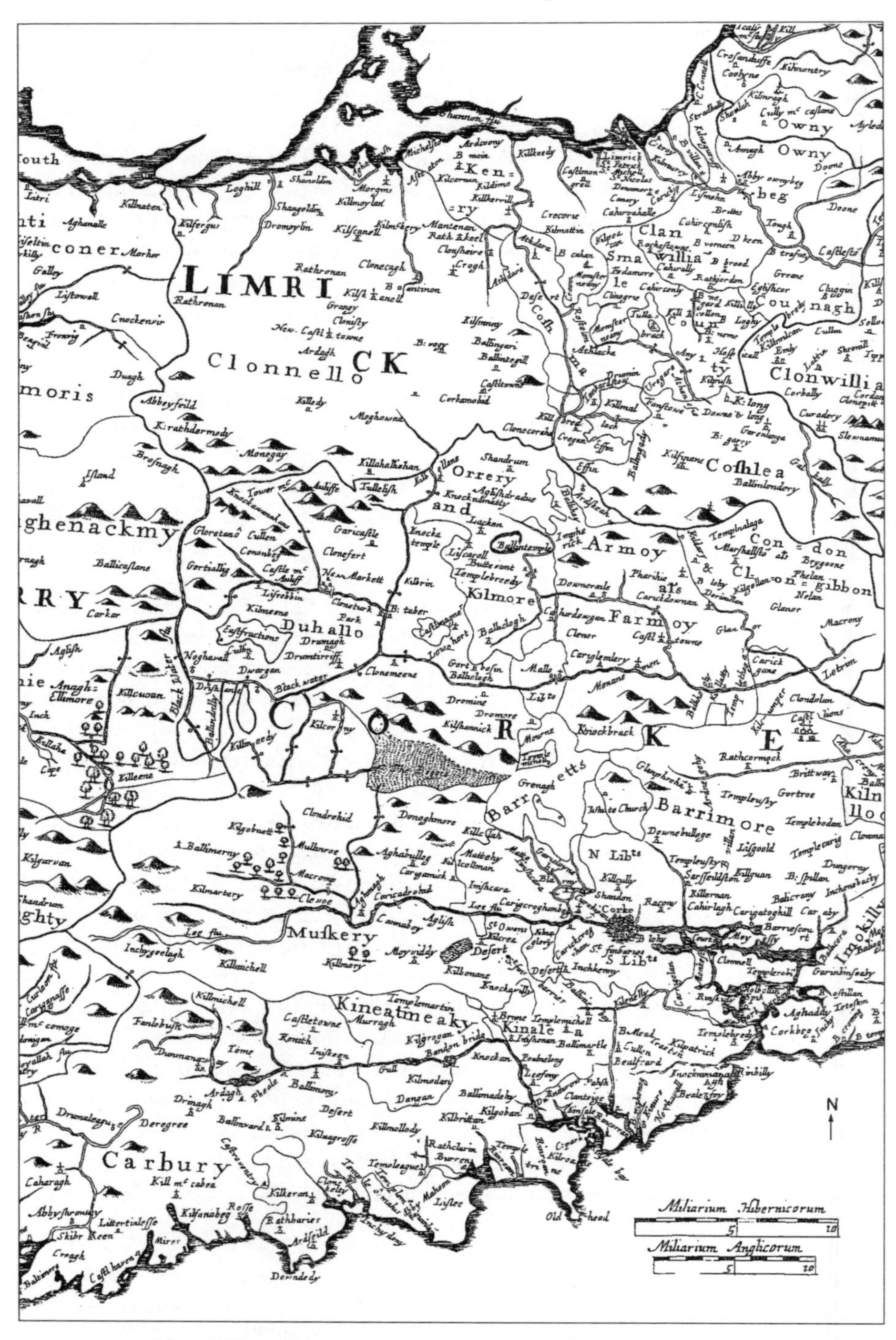

7 From William Petty, Munster, in *Hiberniae delineatio* (London, 1685)

some of these errors can be spotted, such as the boglands shading for the *Boggra* (Boggeragh mountains), resulting in an implausible landscape where rivers rise in bogland and then flow uphill. There is also a lot of uncertainty about the fluvial geomorphology, with the cartographer, as before (fig. 4) loath to leave tributaries unconnected. Territorial hierarchy is clear from the change in typeface for townland, barony and county, but some major placenames are crowded out by lesser names, as in Kinsale, and there are many placenames left 'floating'. Fewer names appear in the atlas than were available to Petty from the Down Survey; while the smaller scale makes this inevitable, the choice of name can be somewhat arbitrary, a point which study focused on the local area will make clear. In addition, small-scale maps often include information not recorded on larger scale maps, making a strong case for the local historian to study the fullest possible range of maps available for any given area, at all scales. A full discussion of this atlas is found in Andrews, *Shapes of Ireland*, pp 118–52.

While there are county maps for the pre-1700 period, it should be noted that the county itself was not an important unit of mapping until the late eighteenth and nineteenth centuries, when county maps become important for civil administration and road planning. John Andrews in *Plantation acres* itemises county maps individually.[47] Original surveys were readily reworked, often to great commercial advantage, as in Bernard Scalé's *Hibernian atlas*, 1776.[48] Those of his predecessors who produced stand-alone county maps (as listed by Andrews), contributed more to the advancement of county mapping than he, but Scalé's work is better known perhaps due to clever marketing and the handsome appearance of the final product. The preface advertised the work as an 'actual survey of this ancient and of late cultivated fertile kingdom', while the title page depicts the fruits of careful husbandry (honey, wheat and vegetables), and of home industry (from the processing of wool and flax to cloth manufacture). The kingdom of Ireland is seen literally as a cornucopia of good things bursting with trading potential, its rich landscape awaiting ravishment. Careful note is made of administrative divisions, of parliamentary representation, and official post roads, for the information of his majesty's officers and loyal subjects. The scale and orientation of each county map varies, as does the length of the accompanying written commentary. Within Ireland a fundamental regional division is noted: 'The Northern and Eastern Counties are best cultivated and inclosed and the most populous, flourishing and industrious.' The county of Wexford is among the more detailed maps, over-represented in the Irish parliament (each seat noted by an asterisk), with a dense network of settlements, roads, navigable rivers, military barracks, and heavily-defined baronies and counties. The outsider will be impressed by the wealth of placenames and strong military presence; the local scholar will quickly pick up the difficulties, as some places refer to landlord houses (Loftus Hall), others to places that are truly insignificant in terms of nucleated settlement (Tarah-hill, Prospect). 'Ballymun' presumably means Ballymurn, Old Ross is entered

47 Andrews, *Plantation acres*, pp 350–1. **48** See also *County atlases of the British Isles published after 1703: a bibliography compiled by David Hodson*, iii, *Atlases published 1764 to 1789 and their subsequent editions* (London, 1997).

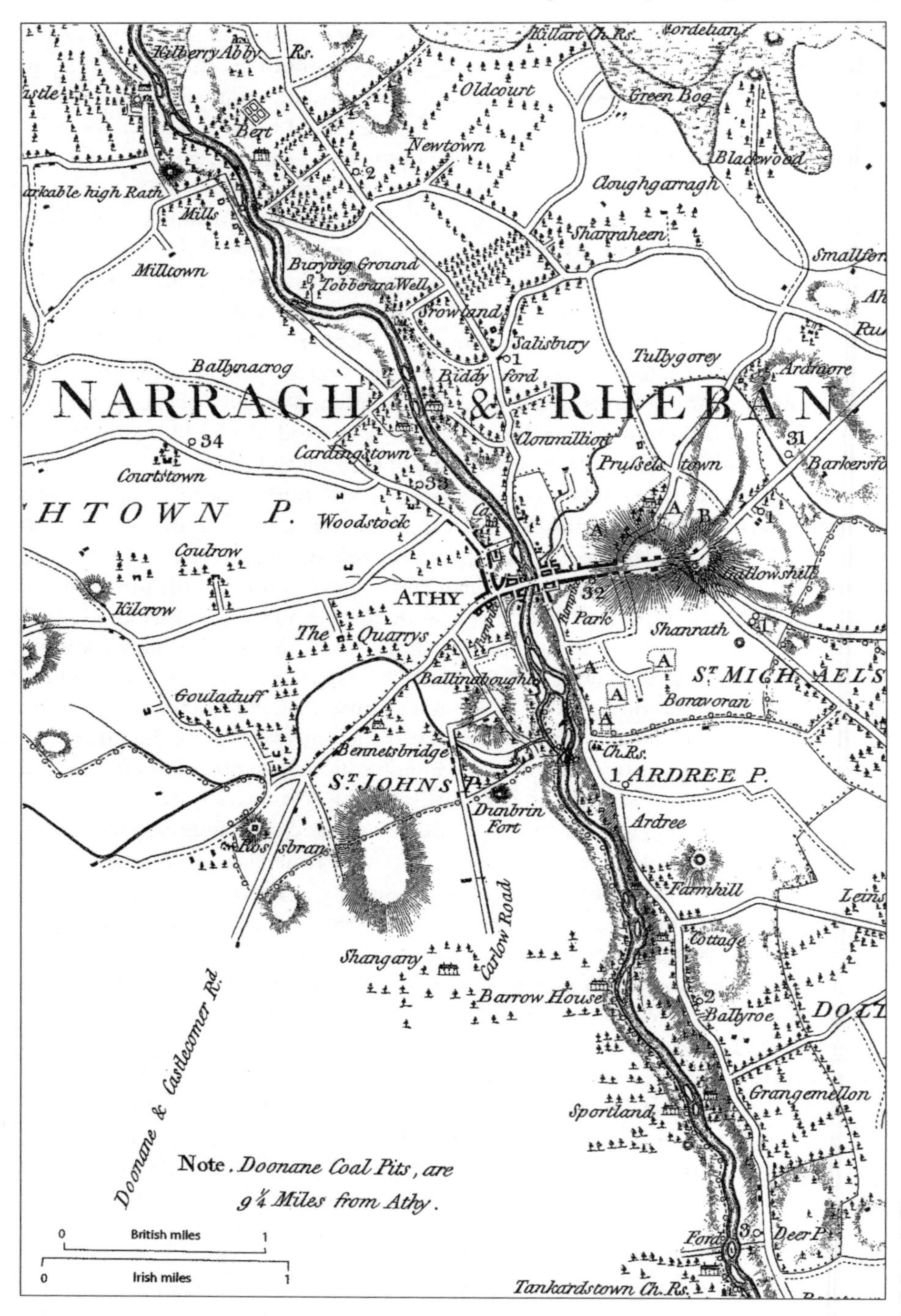

8 Athy, from Alexander Taylor, *A map of the county of Kildare* (Dublin, 1783)

twice, and Wexford town appears as a heavily-fortified star-shaped bastion, an overstatement, to say the least, of its defences at any stage of its history. In contrast, there is little cartographic detail for the county of Leitrim, its poor standing in military or political terms underlined by the brevity of the commentary: 'the air is cold and sharp but wholesome, and though it is a very mountainous bleak county, it is noted for grazing vast herds of cattle'.

Occasionally, experience gained as a military surveyor could be profitable in the private sector, as exemplified in the career of Lieutenant Alexander Taylor, who chose to map the wealthy county of Kildare at a period when it was enjoying substantial landlord investment in roads, canals, demesnes and town enhancement (fig. 8). His 1783 county map of Kildare (facsimile reprint by the RIA, 1983) was financed by gentlemen subscribers from the county, who paid in advance for what was advertised as an 'actual survey' including 'Noblemen's and Gentlemen's Seats, the Towns, Village, Churches, Ruins and the principal townlands' (house clusters), as well as roads, relief, drainage, woods and bogs, all 'exactly laid down, so that it shall be a just and true Representation of the Country'.[49] The extract from around the town of Athy (fig. 8) shows how the road network, gentlemen's houses and wooded demesnes dominated Taylor's cartographic style, while the promise that the map be 'elegantly engraved' and thus ideal for display was certainly fulfilled. The critical map-reader might delve deeper here (see Appendix 1) and question what issues were uppermost in county or local politics *c.*1780 at the time this map was commissioned, and which might have influenced its style or content. The concern with showing landscape 'improvements' was certainly topical. That year the English-born agriculturalist Arthur Young published his *Tour of Ireland* in which he celebrates the steady advance of 'civilisation' (by which he understands the culture and customs of his homeland) through the island. Investment in forestry for 'picturesque' reasons, intensive cultivation, 'scientific' animal husbandry, rebuilding or new construction of grand houses, scenic settings and vistas were all matters which drew approving comments. In County Kildare, Castletown is praised for its 'fine plantations disposed to the best advantage', its shrubbery 'prettily laid out', and the 'noble apartments' and 'beautiful gallery' of the house. At nearby Carton the parkland 'spreads on every side in fine sheets of lawn, kept in the highest order by 1100 sheep, scattered over with rich plantations and bounded by a large margin of wood through which is a riding'.[50] The increase in urbanisation over the preceding twenty years was celebrated by Young as 'a strong mark of rising prosperity'. 'Towns are markets which enrich and cultivate the country' he argued, promoting cultivation in the 'remotest corners' of the countryside, and by their own growth (in 'manufactures, commerce and luxury') towns stimulated 'riches and employment' and increased population within its very sphere of influence.[51] County maps of the

49 Advertisement in *Dublin Evening Post*, 4 Mar. 1780, quoted in J.H. Andrews, 'Alexander Taylor and his map of County Kildare', introductory notes to the facsimile, Alexander Taylor, *A map of the county of Kildare, 1783* (Dublin, 1983). **50** Arthur Young, *A tour of Ireland, with general observations on the state of that kingdom made in the years 1776, 1777 and 1778, and brought down to the end of 1779* (2 vols, Dublin, 1780), i, pp 22–3. **51** Ibid., ii, p. 195.

late 1700s (as fig. 8) are very much part of the desire to show the 'improvements' recently effected or in hand, including the landscaping of demesnes, the creation or remodelling of 'estate' towns or villages by their landlords, and the laying out of new canals. The landlord-controlled grand jury, as the principal promoter of road-making (discussed below), played a major role in county mapping during these very hopeful decades. Taylor's Kildare map of 1783 was followed by a commission for County Down in 1785 (though not completed). The demand for surveying to a high standard of accuracy was fuelled by the work of masters such as Taylor who popularised detailed maps at the county scale among the landed classes.

There was a boom in county surveys during the period from the 1790s down to 1824. Researchers should be mindful that date of publication may be many years later than the survey itself. Alphabetically, Andrews has identified sixteen counties for which major new surveys were made in the period. In Ulster these were Antrim (Lendrick and Williamson, 1808); Londonderry (Sampson, 1814); and Tyrone (McCrea and Knox, 1815). Larkin was particularly productive, creating maps for Westmeath (1808), Meath (1817), Waterford (1818), Sligo, Galway and Leitrim (1819). Bald produced a map of Mayo (1830, but surveyed 1809–16). William Edgeworth produced a highly-regarded county map for Longford (1814) and then collaborated with Richard Griffith to produce a similar map for Roscommon (1825). Other major county map projects were William Duncan's map of Dublin, completed for the grand jury (1821); Arthur Neville's map of Wicklow (1798); Valentine Gill's map of Wexford (1811); and the grand jury map of Cork by Neville Bath (1811). Copies of some grand jury maps will be found in the records of the Quit Rent Office (NAI, file VII, 2B–46 nos. 55–9), while the county councils, which inherited the powers and papers of the grand juries, also hold contemporary copies. Individual maps are widely dispersed, featuring among private estate records, in the NLI, in TCD and elsewhere.[52] Correspondence between the cartographer, William McCrea and the military authorities – Generals Knox and Lake, and Sir George Hill – from December 1796 to the summer of 1797, reveals something of the pressures under which these county map-makers operated. Trying to protect the copyright in their newly-surveyed maps, using data from existing maps with and without permission, and trying to get paid for what was a labour-intensive and time-consuming occupation were typical strains. Under pressure from the military authorities, McCrea had already allowed Captain Taylor of the Royal Engineers to make two copies of some of his county maps of Ulster, for which he now feared he would never be paid. In addition, McCrea explained that he made it his business to take possession of the manuscript maps of those counties he had surveyed directly after the county assizes (when requests for road works at the county's expense were presented). Recognising the urgent need for a regional map, he offered, on payment of £800, to make 'one general map of the counties of Armagh, Tyrone, Monaghan, Donegal and Derry, to which I will add Antrim, Down and

52 For example, see map of the Queen's County from an actual survey made for the grand jury in 1805, TCD, S:00.9.23.N+.3.

Louth improved from materials now in my possession, together with as much of Fermanagh and Cavan as I can with accuracy, the whole to be laid down by a scale of one inch to an Irish mile, which shall be finished with the utmost expedition'.[53] The reworking of maps produced by different surveyors and for varying purposes was clearly a standard part of the map-maker's trade prior to the Ordnance Survey.

The ambitious county statistical surveys, commissioned in 1800 (and completed 1833) by the Dublin Society for Improving Husbandry, Manufactures and other Useful Arts and Sciences (from 1820 the Royal Dublin Society), were accompanied by county maps. In some cases these were based on existing county maps, as in the case of the volume on County Armagh (1804) by Charles Coote, who acknowledges the Rocque map of Armagh as his source, or T.J. Rawson's survey of Kildare (1807) which reprints, in a much reduced format and with the canal network updated, the county map produced by Alexander Taylor in 1783 (fig. 8). In other cases, the statistical surveys include otherwise unknown maps, while most (as noted later) also feature geological surveys, or at least, geological references. On the 1808 map of County Clare produced by Hely Dutton, the karst landscape of the Burren ('boundary of Lime Stone') is clearly identified.

The end of separate county map-making, commissioned by Grand Juries 'or otherwise', was marked by the 1824 Spring Rice report setting up the Ordnance Survey, known more formally as the *Report from the Select Committee appointed to consider the best mode of apportioning more equally the local burthens collected in Ireland, and to provide for a general survey and valuation of that part of the United Kingdom*. Its assessment of the contemptible state of Irish county mapping is concluded with the note, 'it is evident that the ordnance survey will supply all that can be required for county purposes'.[54] For road making and for the collection of rates the maps of the Ordnance Survey were to dominate from the 1830s onwards.

ROAD MAPS

While roads or fragments of roads appear on pre-eighteenth maps of different types, the year 1710 may be taken as the beginning of modern road planning and associated specialist map-making, with the passing of the first of several grand jury road acts (9 Anne c.9). Road planning and road use produced a wealth of cartographic materials; the subsection 'Irish road maps' in the NLI maps catalogue bears witness to this activity. County libraries, archives and road planning departments of the local authorities are also major repositories for this class of map. The present local authorities inherited road-making and road maintenance duties directly from predecessor bodies, including the grand jury, and special commissioners (as in Dublin the Wide Streets Commissioners and the Paving Board). In many instances,

53 William McCrea to Sir George Hill, 3 July 1797 (NAI, Rebellion papers, 620/31/202); see also same to same, 2 May 1797 (NAI, Rebellion papers, 620/30/9); I am indebted to Lorcan Egan for bringing this source to my attention. **54** *Report from the select committee on the survey and valuation of Ireland*, H.C. 1824 (445) viii; p. 79; reprinted in Andrews, *A paper landscape*, appendix A, pp 301–8.

along with their maps, they now also hold the minute books, presentment lists and correspondence that are their essential accompaniments. The Royal Irish Academy, Dawson Street, Dublin 2, holds a fine collection of travel maps, including examples by most of the 'masters' in the field. Central government, in the form of the post office, relief works, and the Board of Works (OPW) also played a role, especially in the nineteenth century, and their published reports (see under reports and parliamentary papers) also feature road maps (fig. 9). Finally there are commercial road maps (fig. 10), user-friendly, marketed with style, borrowing freely (with and without acknowledgement) from other maps, and very often a commercial off-shoot from the surveying undertaken for public bodies. Two separate lines of inquiry therefore can be followed: maps created for 'official' road-making and post-office purposes, discussed firstly, and then maps made for the commercial or popular market.

Road making

Tracing the evolution and remodelling of the local and county road network, relative to topography, geology and settlement, is an instructive exercise for the local historian. The routes taken – straight steep line, disregarding gradient; circuitous lines, avoiding long-forgotten obstacles; sharp and apparently irrational bends; roads 'going nowhere'; parallel routes where the 'former' road continues unstopped alongside the brave new line; sweeping detours, facilitating yet another 'big house' – all deserve closer attention. An early article by J.H. Andrews (1964) explores these intriguing questions;[55] however, study at the local and county level is needed to tease this out further. And as road-making resulted in a useful collection of maps prior to the OS, it can be a fruitful avenue of research under several headings.

In 1739 the landlord-controlled Grand Juries were awarded powers of compulsory purchase, excepting built-up areas, private gardens, avenues and orchards (13 Geo.2. c.10), and the new roads made possible by this legislation were supposed to connect market towns in as direct a line as possible. In 1759, after further improvements to the legislation in 1759 (33 Geo.2 c.8) which included abolishing unpaid statute labour, road making accelerated.[56] The effectiveness of this local authority in getting roads made during the 1770s is praised by Arthur Young who approved of the way in which taxes raised locally were spent locally, and all was controlled at local level. As was to be expected, 'every person is desirous of making the roads leading to his own house', with the good roads 'all found leading from houses like rays from a center'; however, Young argued that serving private interest did in fact contribute to the public good, as within a few years, 'those rays, pointing from so many centers, met, and then the communication was complete'.[57]

Although the powers of Grand Juries were eroded from the 1830s onwards, they were not finally abolished until the creation of county councils in 1898. Their planning projects generated demand for small-scale county-wide topographical

55 J.H. Andrews, 'Road planning in Ireland before the railway age' in *Irish Geography*, v, no. 1 (1964), pp 17–41. **56** For a discussion of the presentment system under which public roads were funded see Andrews, 'Road planning in Ireland before the railway age', pp 19–22. **57** Young, *A tour of Ireland*, ii, part II, pp 57–8.

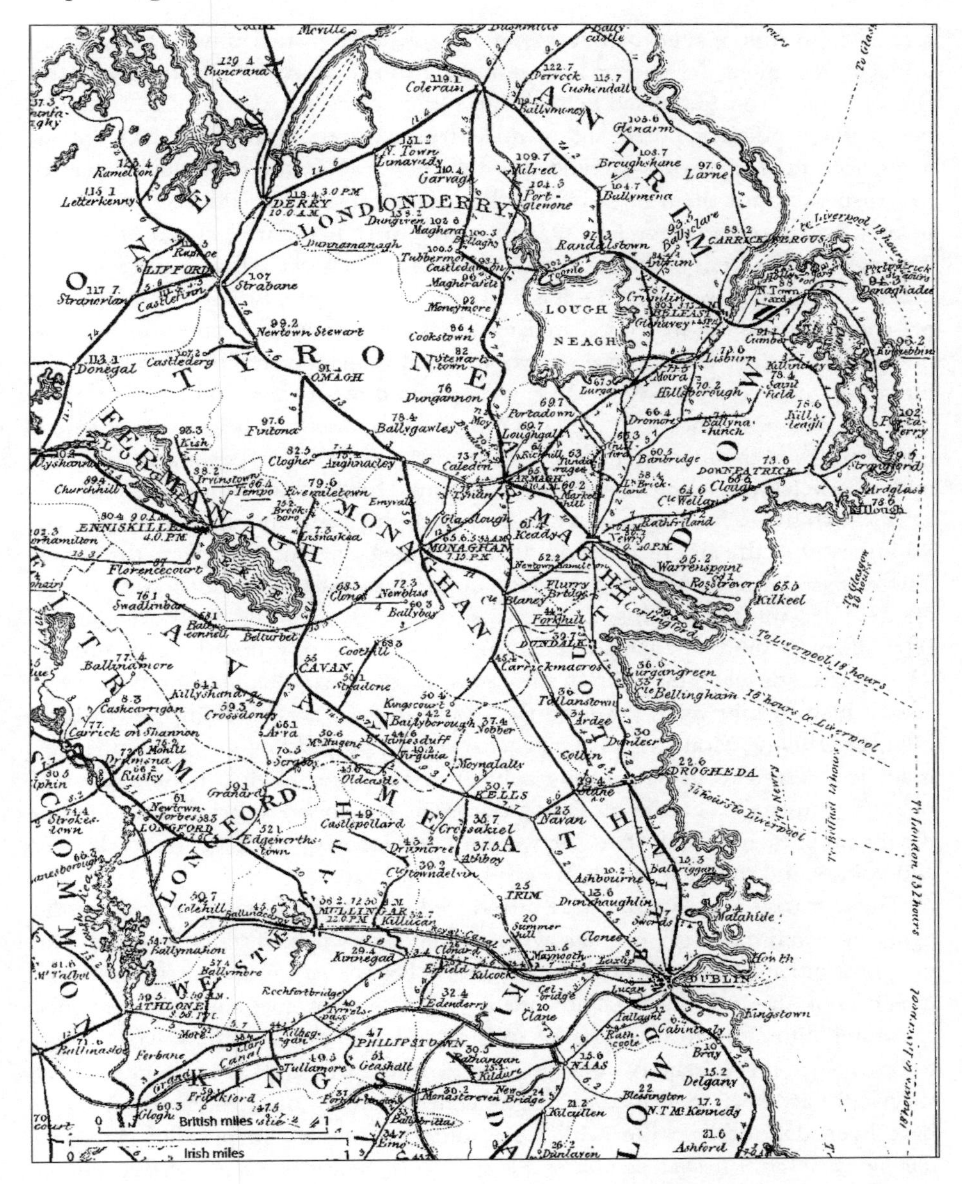

9 Section of 'A new map of Ireland, shewing the Post Towns and Mail conveyances throughout together with the distances in Irish miles and furlongs of each town from the next and the total distance from Dublin by the Post Road,' 1832 (*Report of select committee on state of roads under turnpike trusts in Ireland*, H.C. 1831–2 (645) xvii, p. 716)

maps, as can be noted in the elaborate requirement for maps and sections to accompany road applications under the grand jury Act of 1818. Large-size county maps (see above) were commissioned by grand juries; some were hung on the wall of the courthouse for public consultation. When an early grand jury map was revised to include recent road developments, the local historian has a particularly valuable document.[58] Other road-making and thus map-making bodies were the Bogs Commission (discussed below), the Office of Public Works (also discussed below under official reports), and turnpike trusts. The roads made by the turnpike trusts, from 1729 onwards, were funded by tolls. However, by the 1760s public opinion had turned against them in favour of the locally-funded but toll-free grand jury roads.[59] A map, published 26 July 1832 (one inch to 15 Irish miles) as part of a major inquiry into the matter of turnpike roads, distinguishes between different classes of mail coach road: turnpikes on which toll is paid (black) or not paid (yellow); grand jury or presentment roads (for mail coach, red; for car, horse or foot, blue); and a further class of turnpike (green) where horse or two-wheeled vehicles were not subject to tolls but an upgrade to a four-wheeled carriage would incur tolls.[60] This map closely resembles the post office map (dated 9 August 1832) discussed below (fig. 9). There is also a 'Plan of the metropolis roads north and south of the river Anna Liffey' (one inch to one Irish mile) produced for the County Dublin Toll Committee in March 1832 which shows how the seven major roads to the north and the one major road leading west of the city (via Maynooth and Kinnegad to Athlone) were turnpikes. There was only one south-bound turnpike (via Tallaght and Blessington to Graney), leaving the south and south-east sectors (Rathmines, Rathgar, and the Fitzwilliam/Pembroke estate) at an advantage in terms of property development. The matter was taken up again in the 1850s; an 1854 map ('reduced from the OS') at one inch to four miles covering the eastern region from Dublin stretching to Dundalk, Athlone, Monasterevin, Graiguenamanagh, and Arklow shows the turnpike roads in relation to general topographical detail, other roads (uncoloured), the railways (constructed and 'in progress'), canals and county boundaries.[61] In addition, the names of each turnpike trust are recorded, its extent, the mileage in each barony and county, and the total length of each turnpike road.

The first Irish mail coaches were introduced in 1789, accelerating interest in the mapping – and making – of roads which could take wheeled traffic at speed. The new post office undertook mail-coach surveys between 1805 and 1816, under the direction of Alexander Taylor. These made suggestions for lines and improvements, specifying minimum road widths (of 46 feet), and ideal gradients (less than 1 in 35) to better hasten His Majesty's business. The resulting MS maps (202 of which have survived) have been described as 'magnificent specimens of cartographic art', and have been deposited in the NLI.[62] The official summary map, titled *Post office communication with Ireland, post towns and mail conveyances throughout, together with the distances*, was published in August 1832 to mark the recent 'consolidation' of the post

58 W.H. Crawford and R.H. Foy (eds), *Townlands in Ulster* (Belfast, 1998), pp 30–3. **59** For a negative assessment of the turnpike roads see Young, *A tour in Ireland*, ii, part II, p. 56. **60** Andrews, *A paper landscape*, pp 25–6; *Report from the select committee on turnpike roads in Ireland*, H.C. 1831–32 (645) xvii.397–707, pp 16–17. **61** *Reports of the commissioners appointed to inquire into the Dublin turnpikes with notes of evidence and appendix*, H.C. 1854–5 (01) xix, pp 697–827. **62** Andrews, *Plantation acres*, p. 200; Irish road maps (NLI, 15.A.3–15).

offices of England and Ireland.[63] Not surprisingly, this map highlights the interconnections between cross-channel routes and the internal Irish postal network.

By referring to an earlier list of roads which were partly realigned on the basis of the post office surveys (1829), some progress can be made towards identifying what infrastructural improvements in an area can be credited to this body. However, the government could not force the counties to construct the recommended roads. There were many difficulties in road making under the grand jury system,[64] and only a few were in fact completed. From the coloured 1832 published post office road map, the local historian can work out where his or her place of study might be placed in the post office hierarchy, and the quality of the connecting routeways. The direct mail lines from Dublin are marked in black, the line of the 'riding posts' in blue, that of the mail carts in red and footposts, the most lowly connection, are marked in yellow. Post from Dublin to Nobber, County Meath, for example (fig. 9), followed the main post road westwards to Maynooth, was taken by riding post to Trim, from where a footpost connected with Athboy, then a riding post took the mail again onto Kells, where another footpost brought the letter to its destination in Nobber. At least seven mail routes converged in Armagh: by footpost mail could be brought on to Loughgall or Caledon, by mail cart to Lurgan, by riding post to Monaghan or Castleblayney, while the main route linking Coleraine to Newry and on to Dublin passed through Armagh. The total distance (in Irish miles and furlongs) of each town from Dublin is marked in regular font over the town, thus Nobber is 37 miles four furlongs from Dublin city and Armagh is 65 miles, both 'by the post road'. Distance between towns is marked along the line of road (Dunshaughlin to Navan 9 miles 2 furlongs). Postage rates depended on the mileage from Dublin as calculated 'by the post road'; within seven miles the cost was 2*d.* A letter from Dublin city to Nobber cost 6*d.*; to Armagh or to Belfast 9*d.*; to Dingle 1*s.* Times of arrival (over the name) and of dispatch (under town name) are given in 'sloping figures' for selected towns, for instance, the mail arrived in Newry at 10a.m., and was dispatched at 9.40p.m. Appendices list the post towns in Ireland and the routes 'by which letters from London are forwarded', along with returns of the rate of travelling of each horse post, mail car and mail coach respectively (place and time of departure, time of arrival, distance in miles and furlongs, time occupied).[65] Irish miles are entered on the maps and the scale bar shows both Irish and statute miles. However, English or statute miles only are used in the accompanying lists, an anomaly that map-readers need to be alert to in all pre-OS mapping.

Commercial road maps

The boom in the publication of commercial road maps from the mid-eighteenth century is tied in to increased prosperity, manifest in the remodelling of large country houses and entertaining on a lavish scale. There was now a pressing need

63 *Report from the select committee on the post communication with Ireland, with the minutes of evidence, appendix and plan*, H.C. 1831–32 (716) xvii. **64** Ibid., p. 28. **65** Ibid., appendices 18–20, 44.

to travel between one's town residence and country house in rhythm with the social season. And there were also the beginnings of modern tourism as spas, coastal villages and 'scenic spots' attracted the upper classes. Arthur Young, undertaking his agricultural researches in the late 1770s, was a typical map-user of the period, planning his itinerary from week to week:

> the effect of [the new legislation, so ably implemented by the grand juries], in all parts of the kingdom is so great, that I found it perfectly practicable to travel upon wheels by a map; I will go here. I will go there; I could trace a route upon paper as wild as fancy could dictate, and every where I found beautiful roads without break or hindrance, to enable me to realize my design.[66]

Responding to the market for road maps, George Taylor and Andrew Skinner published their *Maps of the roads of Ireland*, surveyed 1777, printed 1778, with a second (quite rare) edition appearing in 1783. It is this second edition that was reproduced as a facsimile (Shannon, 1969), with an introduction by J.H. Andrews. There are 288 pages of maps, arranged in double or single columns in a 'strip map' format (imitating the very successful strip maps of England published by John Ogilby in 1675), and covering the entire island. Andrews notes that there is general linear accuracy, but joining strips together – as the map frame would invite – reveals directional errors of several degrees. This is unimportant in cases when the maps are used for their intended purpose, stage by stage, but it would have made an accurate all-Ireland map impossible. The legend shows how 'noblemen and gentlemen's seats' are to be represented, but there are lots of examples of the map illustration going beyond the prescribed drawing, as with Carton (seat of the duke of Leinster) and Castletown (home of Speaker Conolly). Close scrutiny by local scholars of 'their' maps will reveal other felicities. While the scale has been calculated at about 1:126,720 (half inch to one statute or British mile), the scale bar on the page headed 'Explanation' is given in Irish miles only, and the distances from point to point on the maps themselves are given only in Irish miles and furlongs, with the note 'eleven Irish miles are equal to fourteen British'. The authors trumpet their careful calculation of distance, and, as later examination has proven, very rightly so, on the whole. The table at the top of each page lists 'distances from the last stage' (market house or cross roads), and 'distances from Dublin' in (Irish) miles and furlongs, although the column headings are not always included. A comprehensive alphabetical index measures the total distance of each place from Dublin Castle, and gives the page numbers that the traveller should consult in planning his or her trip. Similarities in style between the 'Roads' strip maps surveyed in 1777 (fig. 10) and the Alexander Taylor map of County Kildare, 1783 (fig. 8) are easily explained: Alexander and George Taylor were brothers, and while George was the original creator (along with Andrew Skinner) of the 'Roads' atlas published 1778, it was his

66 Young, *A tour in Ireland*, ii, part II, p. 57.

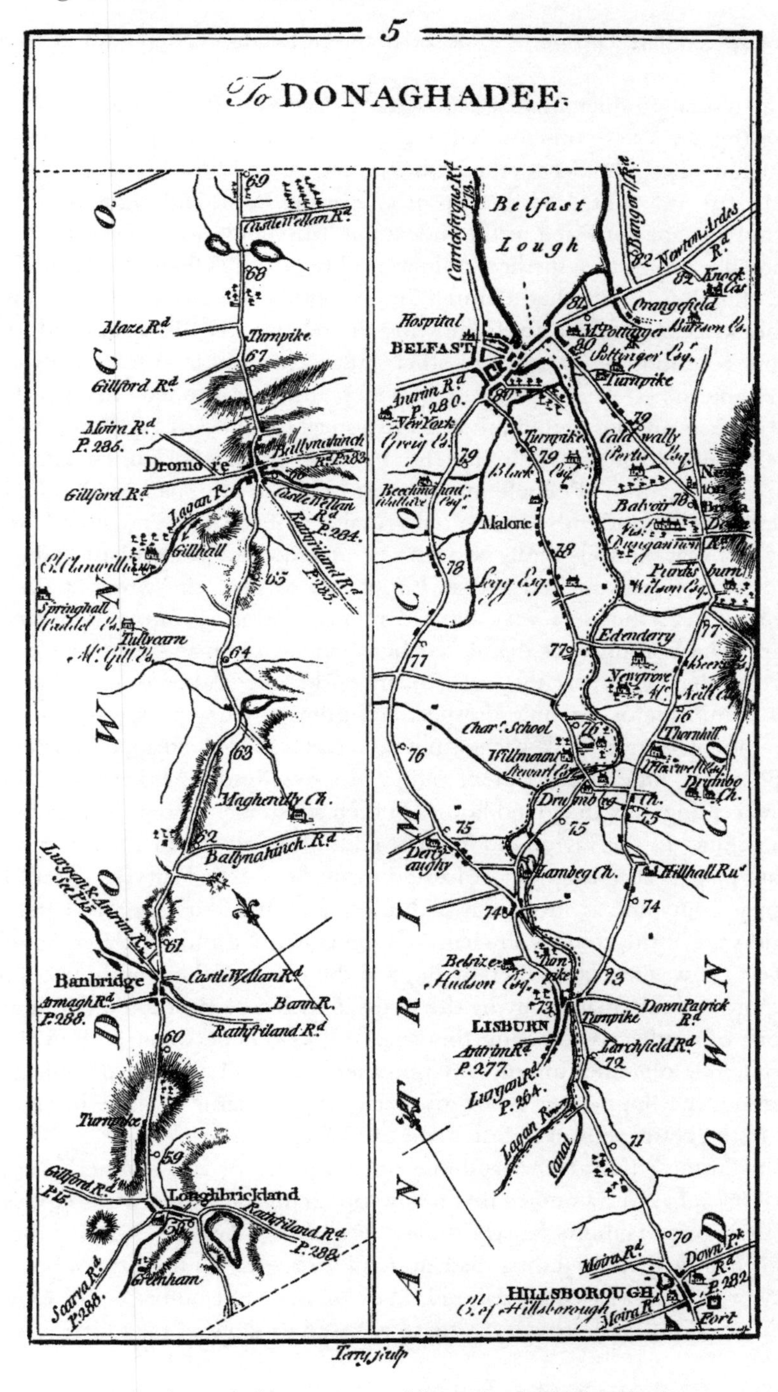

10 Taylor and Skinner, *Maps of the roads of Ireland*, 2nd edn (Dublin, 1783), map 5, To Donaghadee

brother, Alexander, who undertook the revisions for the second edition, published 1783.[67]

As a cultural product, *Maps of the roads of Ireland* (fig. 10), like the county map of Kildare (fig. 8), is self-consciously from the 'gentleman's viewpoint'. The name of the esteemed householder is entered beside each grand house. The legend lists the titles as earl, viscount, lord, bishop, rev., esquire, hon., and right hon. Women householders are not listed in the index, but 'Miss' and 'Mrs' appear on the maps occasionally, prompting further inquiry. Buildings other than 'gentlemen's seats', follies, churches (Established Church), mills, castles, and antiquities are entered as solid black rectangles, effectively showing the relative size of villages and towns *en route*, but confounding together all other settlement features. A miscellany of other features appears – racecourses, inns, barracks, and state-funded Charter schools – but all of these are also within the ambit of a small landed class. No Roman Catholic presence is in evidence, and the stretches between one landlord house and the next are apparently empty of people, and in most cases not even named. Forest planting – undertaken by landlords – is the only rural activity that is represented. Allowing for the fact that the scale limits what can be included, and much Irish industry was still cottage or home-based and so impossible to map, the work of Taylor and Skinner is nevertheless a very limited representation. County boundaries and parliamentary boroughs are delineated, but very few townland names are recorded, and where they do appear, they refer to landlord houses, castles or churches. English names dominate, for example Prospect and Edmondsbury, as do suffixes such as -town, -mount, -grove, -park. What is not even seen cannot be mapped, and these cartographers saw little to delay them as they progressed from market town to market town, excepting only the grand houses of their subscribers and potential customers.

An adjunct to the Taylor and Skinner editions of 1778 and 1783 is 'A new and accurate map of the Kingdom of Ireland made from actual surveys', an all-Ireland road map, contained as a fold-out within each volume. The contrast between the east and west of the country in terms of infrastructure is most striking. All of Mayo and west Galway are quite literally beyond the reach of these authors, a situation that improved only marginally by the 1830s. This can be contrasted with the dense network of roads criss-crossing the eastern corridor between Dublin, Armagh, Belfast and Coleraine, linking up innumerable small towns and villages, and reinforcing the opportunities for prosperity the residents of these places already enjoy in the better soils of Leinster and east Ulster.

In his introduction to the facsimile edition of Taylor and Skinner's *Maps of the roads of Ireland*, Andrews notes that many topographical or traveller's guides relied upon these maps. Wilson's *Post chaise companion, through Ireland* (1784), is a miscellany of information, presented on a road-by-road basis, gleaned from the subscribers to the Taylor and Skinner enterprise and other willing contributors.[68] *The Gentleman's and Citizen's Almanack* from 1781 to 1809 based its table of road distances on the

67 Andrews, 'Alexander Taylor and his map of County Kildare', p. iv. **68** Andrews, p. xiv, 'Introduction to maps of the roads of Ireland', George Taylor and Alexander Skinner.

calculations made by Taylor and Skinner.[69] Others constructed their commentary around the itineraries of Taylor and Skinner, and borrowed freely from their maps. George Tyner was especially enterprising, producing a guide to complement the map published in 1793 by Captain Alexander Taylor which was based (in its turn) on the Vallancey military survey upon which Taylor had been employed. This was advertised as *The travellers' guide through Ireland, being an accurate and complete companion to Captain Alexander Taylor's map of Ireland ... ornamented with a map of the roads* (Dublin, 1794).[70] Later imitators included William Curry, *Road maps for tourists in Ireland* (Dublin, 1844). It was only the advent of the Ordnance Survey's unrivalled quality and coverage that ended the dominance, directly and indirectly, of the Taylor and Skinner road maps, and the all-Ireland maps produced by Taylor in his own right.

Among the road maps appended to traveller's guides, topographical accounts, almanacks, and directories it is difficult to determine authorship and date. A dated map may be inserted into a directory volume which boasts of carrying the most up-to-date information, while only painstaking examination will reveal what might be the precursor or source(s) of the map where authorship is not acknowledged. Samuel Lewis' *Topographical dictionary of Ireland*, one of the most widely-known sources for local history, is accompanied by a handsome *Atlas comprising the counties of Ireland* (London, 1837). This enables the reader to locate the dictionary entries within counties and baronies, to see where each places fits within the road and river or canal network, and to calculate distances. Only the maps for the southern portion were original productions (though not based on original surveys); those for the northern part were derived from pre-existing county maps. R. Creighton is named as the cartographer.[71] Despite the dense road and river network represented on these coloured maps, and the scale employed (two inches to one mile), they feature comparatively few placenames, and represent relief rather ineffectively

The Ordnance Survey was, understandably, unwilling to allow the fruits of its surveying to be exploited by commercial users. Sales of its own products were needed to recoup some of its massive outlay, and it could not be expected to provide competitors with maps. One exception is the collaboration between the OS and the government printer Alexander Thom. The OS produced a Dublin city and district six-inch to one mile map to accompany the 1871 and later editions of *Thom's Almanac and official street directory*. This was a combination of sheets 18 and 22, extending southwards to include the highly respectable new township development and the homes of an important section of the directory-buying public.[72] The transfers (from engraved copper plates to a cheaper planographic printing medium) were held by the OS, who filled Thom's usual print order of 5,000 copies per year at the bargain price of £32 10*s*. The maps could only be sold when bound into

69 Ibid. **70** This is in the RIA; scale of 35 Irish miles to one inch. **71** Andrews, *Shapes of Ireland*, p. 282; see also copy in CUL, Hib.3.837.1. **72** J.H. Andrews, 'Medium and message in early six-inch Irish Ordnance maps: the case of Dublin city' in *Irish Geography*, vi, no. 5 (1973) pp 589–92.

the directory. However, the maps themselves were not dated, and there was nothing to stop the binder from using a sheet earlier than the one supplied for a particular year.[73]

MAPS OF THE BOGS COMMISSIONERS

A highpoint in the cartographic history of Ireland in the pre-Ordnance Survey era is the creation of the maps, plans, sections and wide-ranging written reports of the Bogs Commissioners, appointed in 1809 'to inquire into the nature and extent of the several bogs in Ireland, and the practicability of draining and cultivating them'. Only bogs occupying more than 500 acres were surveyed, though within these expanses several smaller distinct bogs were identified. The high prices for food (during the Napoleonic wars) had focused attention on expanding tillage. Ten civil engineers were employed, including Richard Griffith and John Longfield.[74] By far the most heavily involved of these men was Richard Griffith, who surveyed over 206,000 acres (perhaps 10% of the total bogland of Ireland) in minute detail, and a further 267,000 of mountain bogland. In Mayo, for example, Griffith was sent to undertake 'a general rather than an accurate survey of this mountainous and extensive district'. while there were similar overviews of Wicklow and 'Cunnamara', all published in the final (1814) report. The detailed instructions to these mostly very young engineers is given as an appendix to the Commissioners' first report (1810) and included the order to prepare a map of the district assigned to him. Each engineer was to record the extent and nature of the bog, whether its surface is 'black bog or shaking quagmire', wherever 'limestone, marle or other manures might be found' nearby, the waterbodies which 'appear to occasion the wetness of any of the bogs', the course of any streams, roads or canals 'by which the bog is already intersected', and distinguishing 'such lines of new roads as appear most proper for the carriage of manure, for carrying out the future produce of the reclaimed bogs, and for communication with the roads in the vicinity'.[75] In practice, they took care to map features that might have a bearing on the water table such as eel weirs, locks and sandhills. The maps were to be on a scale of four inches to an Irish mile, but reduced for presentation to parliament to a scale of two inches to one Irish mile (fig. 11); the maximum fold-out size possible for copper-printing was 20in. broad by 26in. long. Where the district could not be mapped as a unit the engineer was to further subdivide it until it could be mapped at a

73 Ibid. **74** The other civil engineers employed by the Bogs Commission were Thomas Colbourne, William Bald, Richard Lovell Edgeworth, Thomas Townsend, Richard Brassington, James Alexander Jones, David Aher and Alexander Nimmo. **75** *The first report of the commissioners appointed to enquire into the nature and extent of the several bogs in Ireland and the practicability of draining and cultivating them*, H.C. 1810 (365) x, appendix no 1, pp 10–12; the other reports are as follows: *Second report*, 1811 (96) VI, pp 206; *Third report*, 1819 (100) pp 1–166; *Fourth report*, 1813–14 (130) VI first part, pp 166; 1813–14 (131) VI second part, pp 215; 1819 (101) pp 215.

manageable size. It is this repeated subdivision that led to the cumbersome titles, as in 'A map of bogs in the Eastern Division of district no. 7, situate in the counties of Longford and Westmeath' (Richard Lovell Edgeworth, 1810).[76] All maps were to be orientated to magnetic north (not true north). In addition to surveying and map-making, the engineers also had to cost their proposed improvements. This major cartographic undertaking required Griffith and his colleagues to produce their own topographical surveys from first principles, wherever previous county maps were inadequate. To that end, each engineer notes the usefulness or otherwise of the maps which he had available for that particular district, and the extent of the new surveying which he had to undertake. The roads of the Wicklow boglands, for example, were taken from the county map by Neville (1798).[77]

Drainage works, the reclamation of marginal land, the laying out of roads and rail lines have obscured earlier environmental realities; the Bogs Commissioners' maps provide access to this world, at a time when population pressure on marginal land was immense. The survey commenced in the extensive Bog of Allen in County Kildare which, as well as being located near Dublin, was also traversed by the Grand and Royal canals. Because of the 'very considerable obstacles to improvement' which were well-known, it was regarded as an ideal place in which to demonstrate the practicability of draining the boglands in general. While the original ambition to connect all the surveys by a trigonometrical network of observations had to be shelved (that scale of co-ordination would have to await the organisational genius of the Ordnance Survey), the individual surveys were regarded as the most accurate of the time. Reports were completed in 1810, 1810–11, and 1813–14, and were published at various intervals up to 1819 in the House of Commons series of parliamentary papers.[78] Researchers should note the date of each individual survey, not merely the date of publication, as they were grouped on a geographical basis rather than strictly according to date of completion. Each report includes maps, sections (showing the bog cut through), and extensive commentary, ranging well beyond the strictly geological.

Figure 11, from the Bogs Commission map of the river Suck district by Richard Griffith (1813), illustrates some of the features most of interest to local historians. The estates of French Park (later known as French Lawn) and Fort William (outside Ballintobber) can be identified along the eastern margin. The bog areas are outlined by a light dotted line (and tinted orange-brown on the original), black lines are proposed minor drains, the arrows point out the fall of the bogs and also the fall of the rivers. As these were primarily intended for drainage and other engineering works, the map-maker needed to ascertain exact elevation; 'the upper line of figures denote the elevation above high water mark in Dublin Bay, and the lower the depth of the bog at that place'. Hachuring or shading is employed to show relief. The arrow head shows magnetic north. There is no official gazetteer or other placename authority which the surveyors can rely upon; Griffith explains that in this case

76 *Second report of the Bogs Commissioners*, appendix no. 8. **77** *Fourth report of the Bogs Commissioners*, p. 164. **78** See footnote 74 for full listing. The 1814 report was reprinted in 1819.

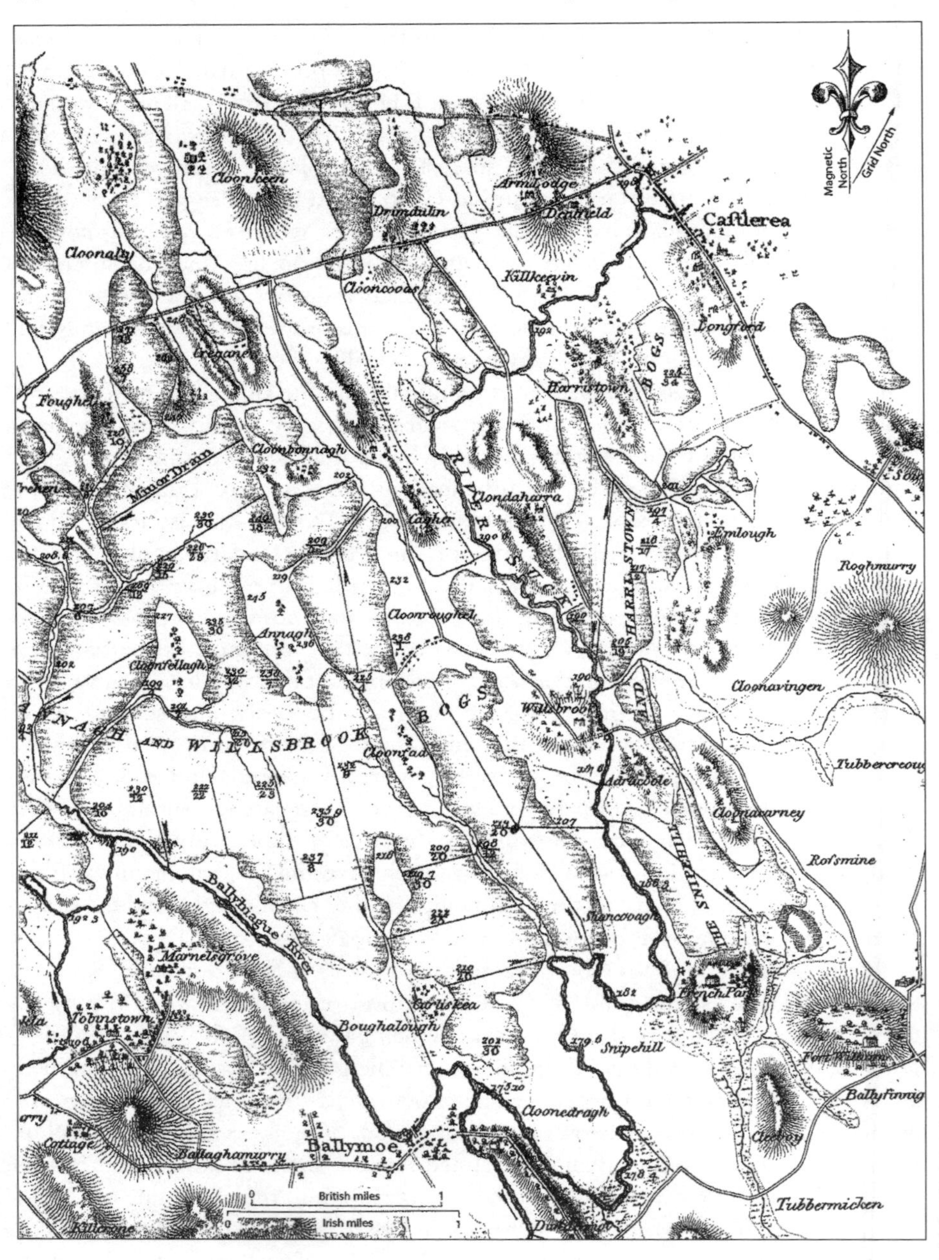

11 Map of part of the bogs belonging to the district of the River Suck situated in the counties of Galway and Roscommon, levelled and surveyed by Richard Griffith Jnr. Civil Engineer & c., 1813, at a scale of 2 inches to one Irish mile (*Fourth report of the commissioners on the practicability of draining and cultivating the bogs of Ireland*, H.C. 1819 (101) plate xiv)

(fig. 11) he has 'selected any prominent name and given it to an extensive tract of bog without intending it to be understood that the whole of the bog so named belongs to the proprietor of the land from whence the name has been taken'.[79] Having named and numbered every bog according to his own judgment, Griffith discusses each in turn, giving estimates for the expense of sinking and widening the streams, and making the minor, cross and surface drains, with the depth of sinking and proposed breadth of drains. The extent of the Snipehill and Harristown bogs (no. 3), illustrated in figure 11, was calculated at 711 Irish acres (1,152 statute acres), with an average depth of 15 feet, and the drainage work costed at £1,103. The comments throw some light on the local scene: 'that nearest the town of Castlerea is used as turbary, and a considerable part has been already been cut away; this might be reclaimed at a trifling expense but is suffered to remain useless and neglected'. Local proprietors (sharing even very small bogs) are named, those who have already invested in drainage improvements are commended, and the future potential is assessed.[80]

There are stylistic differences between the maps produced by different engineers on behalf of the Bogs Commission. Some put houses in profile, others in plan; figure 11 uses both, ensuring that grand houses are conspicuous while the small black squares cover town dwellings, roadside cottages and isolated *clachan* settlement on marginal land. Catholic chapels feature on several of these maps, as do cabin clusters in the most remote areas. The bog maps may be the only record, cartographic or otherwise, of the existence of these homesteads, before successive outbreaks of disease and crop failure, culminating in the disaster of the 1840s, ended a lifestyle already precariously balanced between survival and extinction. Roads – existing, planned and proposed – feature prominently on all of these maps. Footpaths or minor roads may also be shown, as in figure 11 where light dashed lines link up some of the more remote areas. The strength of hachuring, stippling, width and boldness of lines all vary between surveyors, while individuality is expressed in minor ways such as elaborate north points, forest symbols, and the ways in which bogs are bounded and named.

Direct comparison of the Bogs Commissioners' maps with the later one-inch Geological Survey of Ireland (GSI) maps (discussed below) can help explain settlement patterns, and identify the reasons why particular routeways were, or were not, taken. Considering the much more modest resources available to the earlier surveyors, and the lack of a comprehensive trigonometrical framework, the Bogs Commissioners' maps prove to be technically outstanding, with individual clachans and villages (as in Curliskea and Ballymoe), roads, rivers and big houses all located with an accuracy that rivals the OS.

For a later commentary on the Bogs Commission project, and a reproduction of the 1810 map of selected districts, the 1923 Comisiúin na gCanalach agus na mBóthar Uisce Intire[81] is worth consulting. The repository holding most of the

79 *Fourth report of the Bogs Commissioners*, 1819, p. 153. **80** Ibid., pp 152–3, 155. **81** *Comisiún na gCanalach agus na mBóthar Uisce Intire* (IPP 1923, iii, 607 (250 II.23)).

original maps of the survey and the engineering reports is the NLI.[82] Some of Griffith's extensive correspondence relating to the bogs survey (among other projects) are in the Larcom papers, NLI. Facsimiles of the maps produced by Alexander Nimmo of Iveragh (1811) and of the Kenmare river (1812), County Kerry, for the Bogs Commissioners have been published by Glen Maps (13 St. Catherine's Park, Glenageary, County Dublin), with extracts from Nimmo's reports and commentaries by Arnold Horner. Horner's major study of the Bogs commissioners will be published in 2005.[83] Though cartographically a flagship project, the bogs survey was unfortunate in its timing: its excellent maps were soon overtaken by the Ordnance Survey sheets, and it was completed just as tillage prices collapsed with the ending of the Napoleonic wars (1815). The removal of the pressure to extend tillage by reclaiming marginal land undermined the whole purpose of the Bogs Commission.

SEA CHARTS AND HARBOUR MAPS

The tradition of sea charts comes from a very different perspective to that of the landlubber; exact position (latitude and longitude), nearest landfall and its structure, and safe passage thereto over obstacles hidden to the naked eye override all other considerations. It is also an area of cartographic history where political divisions matter least. Charts of the Irish Sea, the North Channel and St George's Channel can all be rich sources for the local historian, along with charts used in plotting journeys across the North Atlantic. From about 1300 onwards, *portolano* of the Mediterranean and northern Europe (maps charting the way from port to port) were produced by the great commercial sea-faring countries, especially Italy, Portugal, Spain and the Netherlands. For the local historian, early charts provide new perspectives on place. Islands and inlets loom large on the horizon (quite literally), while the careful sequencing of placenames, despite the crude shape of the coastline, allows many places to be identified with confidence. As with all maps, close familiarity with the local place is invaluable in their interpretation, especially with placenames which have become garbled in transmission (oral and written, and recorded by non-natives). Samples of early portalon charts will be found in the NLI publication *Ireland from maps*, and in T.J. Westropp, 'Early Italian maps of Ireland from 1300 to 1600, with notes on foreign settlers and trade', in *Proceedings of the Royal Irish Academy*, xxxi C (1913).

A.H.W. Robinson's *Marine cartography in Britain, a history of the seachart to 1855* (Leicester, 1962) provides a useful introduction to the making of maps for navigation. A British rather than British Isles study, this text is nevertheless valuable to those working on charts of the Irish coast as it covers private and official surveys which

82 The maps were deposited in the RDS from whence they went to the NLI. They are part of the map collection, NLI MSS 308–20. **83** See also John Feehan and G. O'Donovan, *The bogs of Ireland: an introduction to the natural, cultural and industrial heritage of Irish peatlands* (Dublin, 1996).

extended into Irish waters also, such as those of Greenvile Collins and Murdoch Mackenzie senior, as explained below. The demand for increasingly accurate sea-charts came from merchant, military and political interests. The translation into English of the famous handbook for pilots produced by the Dutchman Lucas Jansz Waghenaer (1533–1606), under the heading *The mariner's mirrour* (London, 1588), with 45 coloured charts, was the first of a long series of popular publications of the most practical kind.[84] Insets of Galway harbour and Limerick feature in that first edition. In 1681 Captain Greenvile Collins was appointed by the Admiralty to undertake a complete survey of the coasts of Great Britain; in practice this survey included much of the Irish coastline also. In 1693 *Great Britain's coasting pilot* was published; there were 14 subsequent editions up to 1793.[85] Belfast Lough, Carrickfergus, Dublin bay, Cork harbour and Kinsale, for example, were all subjected to his careful attention. The 1792 edition carried 'a new and correct chart of the coast of Ireland'. A typical feature of such 'pilot books' was the inclusion of inset maps of harbours. *Le petit atlas maritime* by Jacques Nicolas Bellin (Paris, 1764), included (in vol. iv), *Plan de Galloway et ses environs; Plan de la ville de Dublin; Port et ville de Kingsail en Irlande; Carte de l'Irlande.* John William Norie's *The compleat British and Irish coastal pilot* (London, 1835), included as insets a plan of Dublin bay, 'Lough Carlingford' and Cork.

The surveys undertaken by Murdoch Mackenzie (senior) are among the most valuable map resources available to historians of coastal communities. His *Maritim survey of Ireland and the west of Great Britain* was published in 1776 (the Irish coast was surveyed 1758–1768); these charts and their inset maps of individual harbours were to feature in pilot books and chart collections over the next decades. His *Treatise on maritim surveying* (London, 1774) published when he had retired from active surveying, provides insights into current practice and thinking in this specialised field, such as the merits of different methods of triangulation, and how one might take soundings for the purpose of making sea charts. Figure 12 is an extract from the 1821 edition of the sheet titled 'A new Mercator's chart of the Coast of Ireland from Kerry Head to Kilmury' (surveyed 1764). This coastline survey (in figure 12 showing part of the Waterford coast) is supplemented by large-scale insets of individual harbours and safe havens, in this case including Kinsale (fig. 12), Castlemaine, Tralee, Dingle, Ventry, Valentia, Dunmanus Bay, Cobh and Cork. Profile views are also supplied of significant mountain ranges as they will appear to the mariner bearing in on the coast. From 'the Skilliks' or Dursey, Hungry Hill is a notable landmark, while the shape of distant 'Caperqueen' mountain' and 'Knockmeldown' will assist the pilot still three leagues out from the shoreline making for Dungarvan harbour (fig. 12).

Sea charts 'wherein may playnly be seen the courses, heights, distances, depths, soundings, flouds and ebs' (as explained by Waghenaer in the introduction to the *Mariner's mirrour*), are written in a specialised but unambiguous language of their

84 The best collection of pilots' books is to be found in the Caird Library, National Maritime Museum, Greenwich. **85** 1693 edition at TCD, S.PP.a.29.

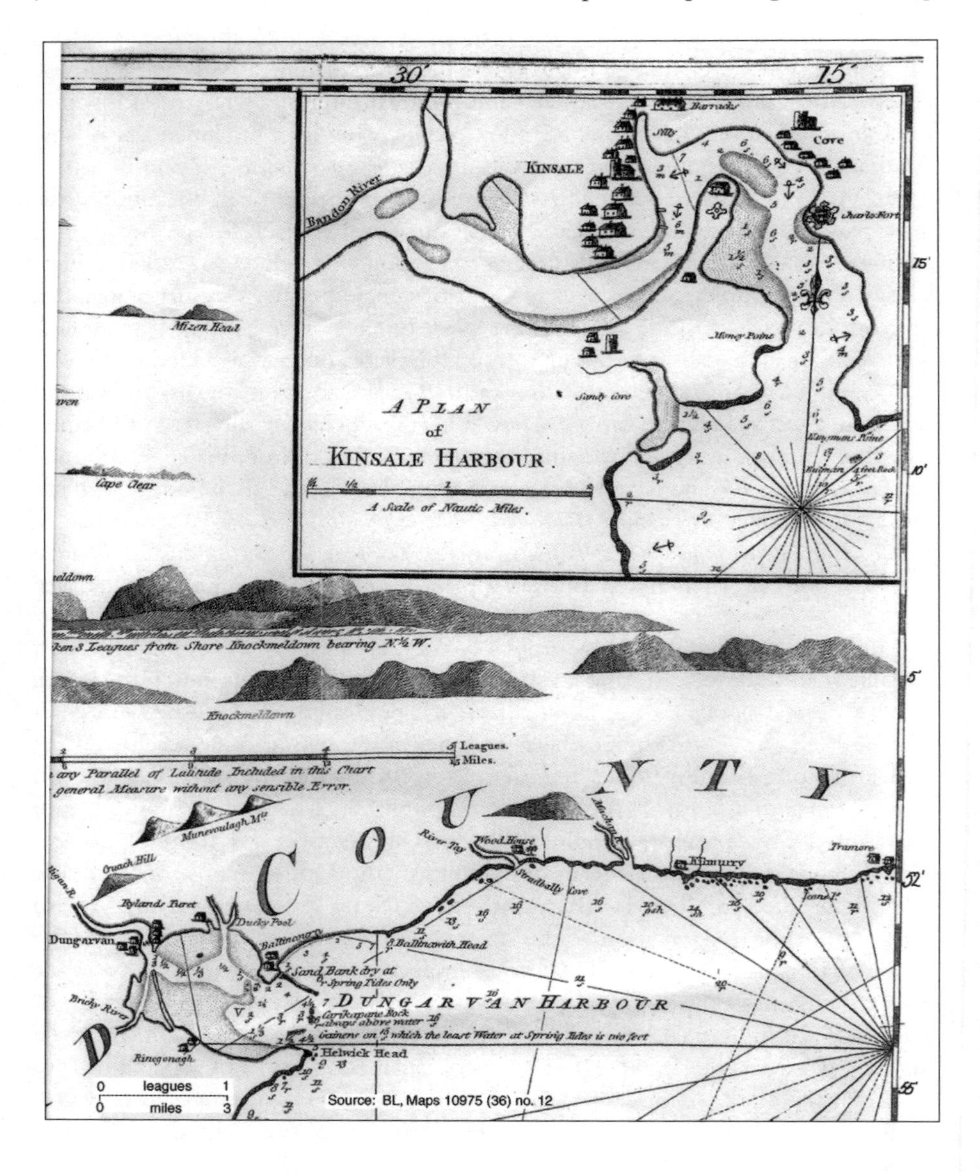

12 Murdoch Mackenzie, *A new Mercator's chart of the coast of Ireland from Kerry Head to Kilmury*, new edn (London, 1821)

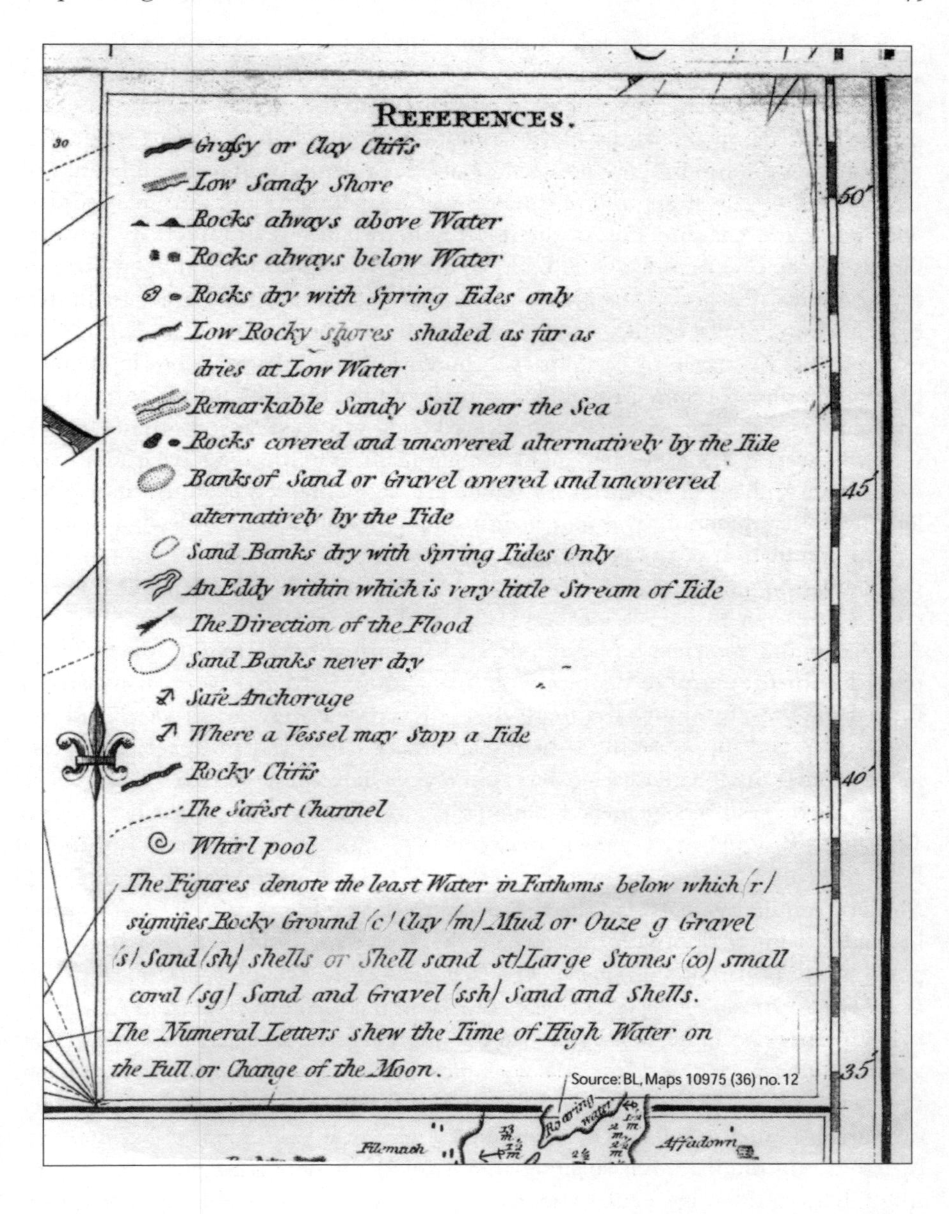

Source: BL, Maps 10975 (36) no. 12

12a Reference table, Murdoch Mackenzie, *A new Mercator's chart of the coast of Ireland from Kerry Head to Kilmury*, new edn (London, 1821)

own. Historians and geographers utilising these maps need to give particular attention to the code in which they are written. In Mackenzie's survey, the priorities of the marine surveyor were translated into the international language of the sea chart: depth and nature of the sea bottom, timing of high water ('on the full or change of the moon'), the position of, and least depth over, rocks and banks of sand or gravel; the strength and direction of tidal streams and currents (eddies, whirlpools and 'the direction of the flood'), the brightness and intervals between signals (in the case of a revolving light) of the nearest lighthouse, and conspicuous topographical features that may assist the mariner in recognising an unfamiliar coast, including hill summits, notable houses, church spires and towers. Kinsale, for example (fig. 12), has its forts and parish churches, while a mariner entering Dublin bay could utilise the spire of Coolock church and the obelisk on Dalkey hill to set an exact course. Even where a sophisticated code was used (as in figures 12–12a), surveyors were quick to further annotate their maps, as in the 1788–9 harbour maps of Captain William O'Brien Drury which are supplemented by authoritative but lengthy sailing directions warning against sunk rocks, shoals and 'gravelly points'.[86]

The foundation of the Hydrographic Department of the British navy in 1795 was a milestone in the mapping of the Irish coast. Its remit was 'to take charge and custody of such plans as then were, or should thereafter be, deposited in the Admiralty and to select and compile such information as might appear to be requisite for the purpose of improving navigation'.[87] Marine surveys became a formal state responsibility, the first systematic surveys were undertaken, and the production and supply of high-quality admiralty charts begun in earnest. These incorporated or updated earlier navy surveys, where they existed, such as those already carried out by Greenvile Collins (1681–7), Murdoch Mackenzie (1770s) and William O'Brien Drury (1788–9). Between 1829 and 1855, under the command of Francis Beaufort, the survey of the coastline of Ireland was practically completed. Modern Admiralty charts are the successors to these early surveys. They too must be read in conjunction with their legend sheet, which is published separately (and periodically updated) as *Explanation of signs and abbreviations adopted in the charts issued by the Hydrographic office, Admiralty* (London, 1866). Although current Admiralty maps are more accurate, standardised and comprehensive, and are (in most cases) at a much larger scale, the direct link with previous surveys is maintained by noting the names of earlier surveyors and the date of their work, while stylistically they owe much to their forbears. The changing nature of the coastline is still symbolised by letters – m(mud), s(sand), sh(shingle), st (stones); the depths are given in fathoms (though ascertained by regular and even-spaced soundings); and heights are given in feet. Lighthouses are still afforded the prominence they merit, but are described in shorthand.[88] For a comprehensive introduction to the requirements of a

86 Captain William O'Brien Drury, *1788–89 Surveys of the Harbours of Rutland and Road of Arran, Blacksod, and Broadhaven, Valentia, Bear Haven, and Corke, wanting Waterford, with soundings, observations and sailing directions*, 5 charts (Dublin, 1789). **87** *Admiralty manual of hydrographic surveying*, i, 3rd edn (London, 1965), p. 1. **88** For example, the beacon on Valencia is described on the chart as (U)occ.ev.3sec.82ft14M, which translates as unwatched,

hydrographic survey and the techniques and mathematics involved, the two volumes of the *Admiralty manual of hydrographic surveying* (3rd edn, London, 1965) is the standard text. For those seeking to use Admiralty maps in local history research, the instructions which this agency gave to its officers deserve close attention. With respect to placenames, for example, the surveyors were required to record the names of 'towns, villages, rivers, streams, points, shoals, rocks and islands, and of any prominent and important features of the coast', and were warned that 'correct nomenclature and spelling are matters of great importance on the published chart, and no effort should be spared to trace acknowledged and native names'.[89]

The navigation maps produced by the Royal Navy's hydrographic section take pre-eminence in the mapping of the coasts of the British Isles from the late eighteenth century onwards. However, harbour authorities in their own port development work had pressing reasons to collect existing maps and commission their own.[90] The Ballast Office in Dublin, for example, commissioned a number of important surveys following its creation in 1708. On behalf of the 'chairman and gentlemen' of this office, George Semple in 1762, 'industeriously endeavourd to collect and carefully laid down some of the most authentick surveys of this harbour for some hundred years past'; his atlas (held by Dublin Port) incorporates redrawings of a number of important maps of the harbour and bay, most notably that of Gabriel Stokes in 1725. The practice of reprinting early harbour maps alongside contemporary plans was repeated in the *Dublin Main Drainage Scheme souvenir handbook* 1906 (NLI Ir 6282 d 5), which included the 1756 survey of the bay and harbour undertaken by George Gibson.[91] Local anxiety to secure government packet commissions (where mail is carried along with passengers) was a particular spur to harbour mapping. Such maps were dissected, defended, promoted and glossed over by both Cork and Galway, in the competition to secure the lucrative North Atlantic Commission; some of these maps will be found in the published reports of the steam packet companies (held in the NLI), and their presentations to parliament.[92]

The Office of Public Works (Board of Works), established in 1831, held responsibility for the completion and development of the royal ports of Kingstown (Dún Laoghaire) and Dunmore (from 1831), Howth (1836) and Donaghdee and Ardglass (1838).[93] The cartographic record in these cases (held in the National Archives of

occulting, every 3 seconds, focal plane height of 82 feet above mean high water spring tide, visible from 14 nautical miles. **89** *Admiralty manual of hydrographic surveying*, p. 4. **90** George Evans, *Report on the present condition and relative merits of the harbours of Larne, Loch Ryan, Port Patrick, Donaghadee and Belfast, surveyed by order of the Lords Commissioners of the admiralty by Capt George Evans, 1846, previously reported in 1839*; NLI, P. 878. **91** *Dublin Main Drainage Scheme souvenir handbook* 1906, NLI, Ir 6282 d 5. **92** *Mr Le Fanu's report on the port of Cork as a packet station* (Dublin, 1851); *Report of the Commissioners appointed to inquire as to the proposal for an Irish packet station* ... H.C. 1851 (C.1391) xxv; *Transatlantic steam packet company nautical and statistical report* (Dublin, 1851); see also *Memorial to Admiralty, March 1854, from the Inhabitants of Drogheda, praying for a government Survey of East Coast of Ireland, for Harbour of Refuge*, H.C. [Cd. accounts and papers], 1854–55 (322) xlvii, p. 477. **93** Rena Lohan, *Guide*

Ireland) is part of much larger document collections which detail the day-to-day obstacles and annoyances that characterise the management of any major public project. From 1847 the OPW was responsible for the disbursement of state funds in loans and grants for the construction of piers, harbours and other works 'useful for the encouragement and promotion of the sea fisheries'. This infrastructural development required formal memorials (appeals), local hydrographic surveys and engineering plans, financial estimates, reports of engineers and fishery inspectors; by the time any one project was accomplished, a massive file of correspondence, along with labour lists and public notices, had been created.[94] The OPW therefore was directly or indirectly responsible for the large-scale mapping of significant numbers of harbours, predominantly along the west coast and including many small places that would not otherwise have merited such special cartographic attention; their 1852 report, for example, includes 43 harbour maps, at a scale of six inches to one mile.[95]

MILITARY AND FORTIFICATION MAPS

One of the greatest stimuli to map-making has long been the need of military leaders, not alone in preparation for assaults on specific targets (castles, towns, harbours) but also as a means of recording the progress of a particular campaign, plotting the movements and counter-movements of the participants (at land or at sea), working out the game plan and subsequent revisions, and apportioning credit for what is most usually a successful outcome – at least from the partisan map-maker's point of view. A good introduction to the sheer range of maps produced by the Board of Ordnance is provided for Scotland at the site 'Military maps of Scotland' [http://www.nls.uk/digitallibrary/map/military/index.html]. Ireland has a rich heritage of military maps from the later sixteenth and seventeenth centuries, ranging from plans and elevations of forts, barracks and towns to military sketches and topographical surveys. Those surveys that relate specifically to land confiscation and redistribution have been introduced above under early modern/plantation mapping. Eighteenth-century military science placed great emphasis on geometry and geography, and their point of contact – mapping. The evident 'correctness' of the map, the ease with which it could substitute for the territory, and the manner in which the map defined and created 'theatres of operation', where competing armies were represented as coloured squares, further allowed the armies to be dehumanised and their movements regarded as pure geometry.[96]

Strictly speaking, all maps may be used for military purposes, including topographical surveys, road maps, regional maps, coastal surveys and large-scale town plans. Most early maps of Ireland, such as surveys by Robert Lythe (1568–71) and

to the archives of the Office of Public Works (Dublin, 1994), p. 3. **94** Ibid., pp 125–7, 130–1. **95** Public Works (Ireland), *Twentieth annual report*, 1852; H.C. 1852–3 (1569) xli.407. **96** M.H. Edney, 'British military education, mapmaking, and military 'map-mindedness' in the later Enlightenment' in *The Cartographic Journal*, xxxi (1994), pp 18–19.

Francis Jobson (1586–90),[97] afforded overt attention to matters of state security, with an emphasis on defensive urban centres, forts and safe passes; Speed's map (fig. 4) has been introduced as an example of that style of cartography. He also produced a cartographic account of the major historic battles fought in Ireland, from the Anglo-Norman conquest of Wexford in 1169 through to the Blackwater battle of 1598. The work titled *The invasions of England and Ireland with all their civill warrs since the conquest*, was published as part of his *Prospect of the most famous parts of the world* (London, 1627). The whereabouts of original and facsimile reprints of early military maps has been discussed above in the section on early modern mapping. A good selection of maps produced or annotated for military purposes will be found in Michael Swift's *Historical maps of Ireland* (London, 1999) which is based on original material in the National Archives, Kew (PRO). An overview of key military surveys, with map extracts, is included in Paul Kerrigan's *Castles and fortifications of Ireland, 1485–1945* (Cork, 1995), while Andrews' *Shapes of Ireland* places the work of individual military surveyors, including Robert Lythe and Charles Vallancey, into the larger framework of the mapping of Ireland. The role of 'director general of fortifications in Ireland' is explored in biographical studies of holders of that office during the period 1612 to 1644 by W.P. Pakenham-Walsh.[98]

The map-making associated with military campaigns and political upheavals of sixteenth- and seventeenth-century Ireland was part of a Europe-wide obsession with planning ever more complex fortifications to withstand siege warfare, skills which the advent of gunpowder quickly rendered redundant. In the case of Londonderry city, for example, its rational, geometric layout closely resembles that of the French frontier settlement of Vitry-le-François completed about 1560, and the idealised military settlement described in *Theorie and praktike of modern warres* (London, 1598).[99] The 'patterns' for fortifications and new towns, which were one of the great selling points of military handbooks of the sixteenth and seventeenth centuries, inspired both engineers and military map-makers operating in Ireland. Star-shaped fortifications, complete with bastions, ravelins, redoubts and platforms, were very satisfying to draw, and armchair planners of the new frontier settlements were impressed by the rationality and durability that these widely-circulated plans conveyed. The plans of Irish towns and forts in *Mr Tindal's continuation of Mr Rapin's history of England* (1744–7), though small scale and greatly simplified, still manage to show all the features characteristic of this genre of map-making: walls and battlements, towers and gatehouses, ditches and moats, magazines for ammunition, platforms and bulwarks wide enough to take heavy ordnance. Escape and relief routes (by sea and overland) are noted, and the 'well of fresh water' within the citadel, without which no garrison could withstand a siege for any length. When

97 J.H. Andrews, 'Robert Lythe's petitions, 1571' in *Analecta Hibernica*, xxiv (1967), pp 232–41. **98** W.P. Pakenham Walsh, 'Capt. Sir Josias Bodely, director general of fortifications in Ireland, 1612–17' in *Royal Engineers' Journal*, viii (1908), pp 253–64; idem, 'Capts. Sir Thomas Rotheram, Knt. and Nicholas Pinnar, directors general of fortifications in Ireland, 1617–44' in *Royal Engineers' Journal*, x (1909), pp 125–34. **99** Brian Lacy, *Siege city, the story of Derry and Londonderry* (Belfast, 1990), pp 89–90.

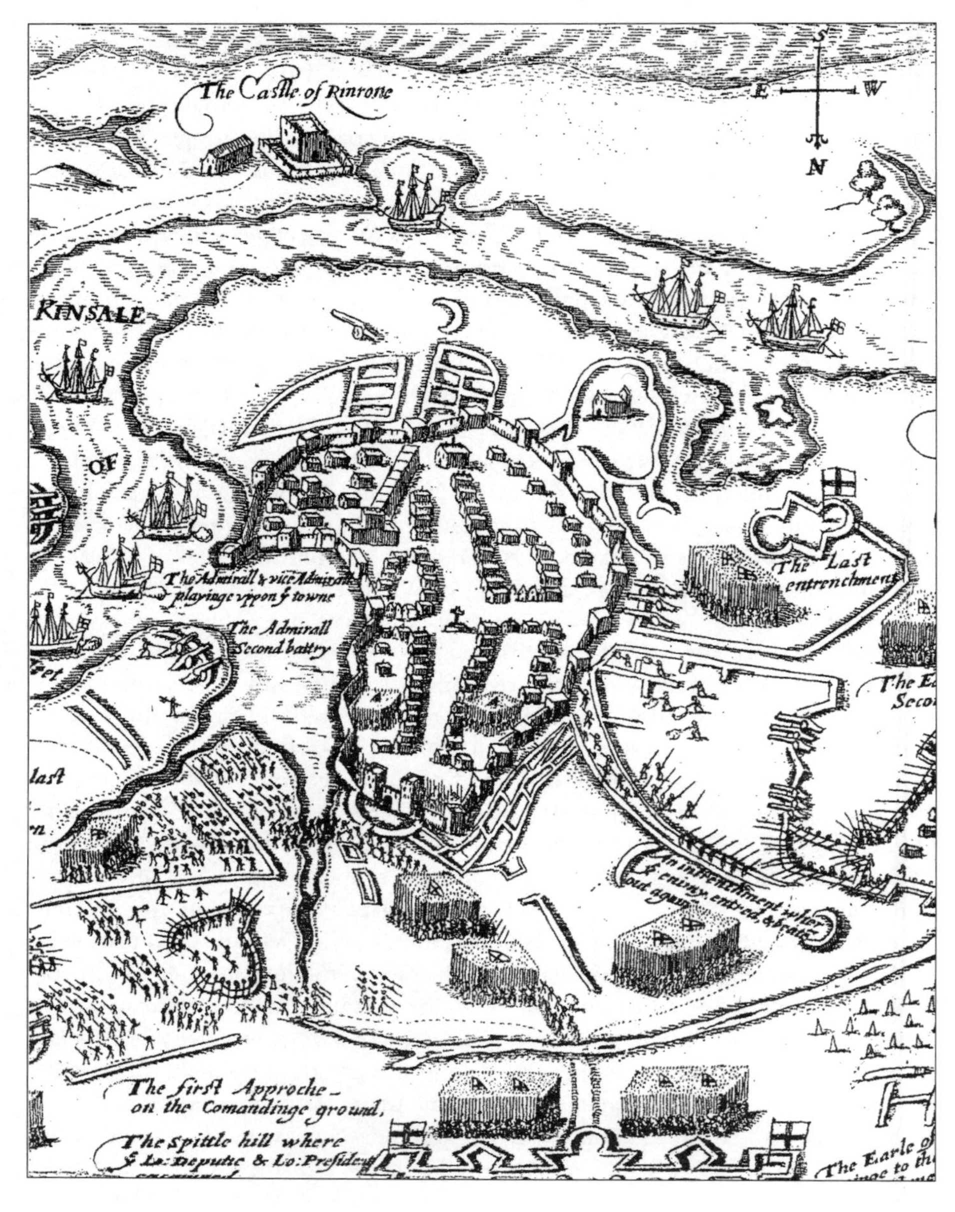

13 Siege of Kinsale, 1601, in William Stafford (ed.), *Pacata Hibernica: Ireland appeased and reduced* (1633)

such plans were produced to elucidate published reports or books, they had a wide appeal, and can be considered as part of popular political literature or 'propaganda'. One of the best examples is Thomas Stafford's *Pacata Hibernia: Ireland appeased and reduced* (1633, reprinted 1810 and 1896); this is 'illustrated with seventeene several mappes, for the better understanding of the storie', namely an account of the Munster wars under the government of Sir George Carew. The maps come from a variety of authorities, and include scenes of 'Corke', 'Haulbowlin', 'Beare County', Youghal, Limerick castle, Glin castle, Dunboy castle and Kinsale (fig. 13). Some of these maps bear little relationship to the text, but others such as those of Kinsale and Dunboy are an integral part of the discussion. By entering imaginatively into the milieu in which these maps were created (Appendix 1) the local historian will more fully appreciate what the first audience demanded. Armour, weaponry, heraldry, ships, and camp layout, battle formation, siege machines, displays of horsemanship and battle tactics were of absorbing interest, while army topographers, adept at producing plans of fortifications, could also turn their talents to scale drawings of individual buildings. Relative height, slope and direction matter; this is typically shown by well-judged shading. The siege and battle of Kinsale, when 4,000 Spanish forces, at the invitation of their Irish co-religionists, captured the fortified town of Kinsale but were themselves besieged by crown troops from 17 October 1601 to 9 January 1602, is presented in a typically elaborate and lively manner (fig. 13). The encampments, entrenchments, and location of each piece of ordnance 'to beat into the town', the 'stakes pitched to gale the enemies' horse', the names and exploits of the commanding officers on the English side, and the sequence of attacks, sorties and counter-attacks that marked the progress of what was a disastrous military campaign for the combined Irish and Spanish forces are all noted (fig. 13). The town of Kinsale, with its impressive wall, market cross and massed defenders, now hopelessly surrounded, is but one item in an action-packed canvas.

Despite mass confiscation of land, organised plantations, and the practice of rewarding loyal supporters and soldiers with estates that followed upon the military successes of the seventeenth century, the crown's authority was not at all secure. George Storey's *Impartial history of the wars in Ireland* (London 1691) again provides a combined cartographic and textual account, this time of the Williamite wars. The work of several of the most accomplished cartographers of the time is featured here, including surveys by Samuel Hobson (Londonderry and Limerick) and by Francis Neville (Londonderry). Despite the title, and the author's protestations that he has 'done our enemies all the justice that in every point that the merit of their cause would bear', this turns out to be a typically partisan account. Fears of further alliances between disaffected Irish Catholics and their co-religionists in Spain (1690s), and later between Irish 'rebels' and Republican France (1780s and 1790s) led to renewed mapping of vulnerable sections of coast, of fortifications (present or recommended), and of important centres of power, as in Carrickfergus, Dublin and Londonderry. A military survey was undertaken in 1685 by Captain Thomas Phillips, resulting in 46 water colour 'maps and plans for the refortification of

Ireland, with prospects of Irish towns'.[1] Two sets are held by the NLI and samples are reproduced in *Treasures of the National Library of Ireland*. Phillips provides contemporary commentary on the state of readiness of coastal towns and forts, as well as of Duncannon and Athlone. Phillips' maps are limited to forts and towns, so that excepting the cities of Galway and Limerick, the western coast is ignored, dismissed as 'so wild and so barbarious that there is scarce living for friend or foe, neither will an enemy attempt landing so far from his business'. In the case of Kinsale, recently a theatre of war, he advised that 'his majestie be at an extraordinary charge for the security of this place more than any other', which really could not be justified, as 'there are several other harbours of larger extent and deeper water which are quite neglected'.[2] Providing a further perspective on several of the same defensive sites are the Goubet maps *c.*1690–95, also held by the NLI, consisting of 28 sketches of fortified towns, forts and harbours of which 19 relate to Ireland.[3]

The Board of Ordnance, reconstituted in 1683 as a military department, was concerned with questions of fortification and defence, and by the 1760s commissioned surveys of areas of military importance. The military survey of Colonel Charles Vallancey (1776–82) was the most significant.[4] Under the orders of George III, Vallancey's survey

> embraced that part of Ireland lying south of the Bays of Dublin and Galway, comprehending the harbours of Dublin, Wexford, Waterford, Dungarvan, Youghal, Corke, Kinsale, Bantry, Kenmare and Galway, the coast adjoining the said harbours, the roads leading from each of these ports to Dublin, and the cross roads leading from each to the other.[5]

From the general plan of the work, it is clear that five divisions or parts were intended, stretching from Lourgh Corrib in the west to Drogheda in the east, and as far north as Granard and Kells, but only three (mostly the 'maritime parts') appear to have been completed. Despite being incomplete, it is still one of the most valuable pre-OS cartographic records of southern Ireland. Some of the plans are at the very large scale of four inches to an Irish mile (given as '3 2/16 inches to 1 English mile' or approximately 1:20,275), with close-up studies of selected harbours and forts, as in Cove and Duncannon (100 feet to one inch) and Kinsale (40 feet to one inch). The scale of two inches to an Irish mile (1:40,550) is favoured in later maps. The manuscript maps and their explanations, along with a handwritten essay also by Vallancey, titled 'On military surveys, accompanied with military itineraries' (1779), are held in the British Library; microfilm copies of the maps may be consulted in the BL and in the NLI. The 1779 essay provides some insights into the theoretical underpinning of this work.[6]

1 Thomas Phillips' maps (NLI, MS 3137; MS 2557). **2** NLI, MS 3137. **3** Goubet, MS plans, *c.*1690–5 (NLI, MS 2742); two plans relate to England, and seven to the Netherlands. **4** Monica Nevin, 'General Charles Vallancey, 1725–1812' in *Journal of the Royal Society of Antiquaries of Ireland*, cxxiii (1993) pp 19–58. **5** Charles Vallancey, 'On military surveys, accompanied with military itineraries', 1779, BL, Maps K.Top.51.31.2; a contemporary manuscript copy is in the Library of Congress. **6** Ibid., see also Douglas W. Marshall, 'Instructions for a military survey in 1779' in *Cartographica*, xviii, no. 1 (1981), pp 1–12.

This high-quality topographical survey was produced in response to the threat of a French invasion which the English authorities expected would focus on the southern coast. Vallancey reckoned that an invasion would consist of a very large force which would be 'as precipitate in landing as possible'. The French would probably divide their forces into two bodies: one directed to Cork because of 'the general magazine of provisions with which that city abounds', and the other branch to either Waterford or Limerick, 'the rivers of which will lead in a very considerable distance from the coast, towards the metropolis, with great facility and expedition'. The maps show settlement, in bright red, both as solitary premises and as clusters; some of the town plans, as of Kilkenny and Cashel, are usefully detailed. Significant landlord houses are surrounded by tree symbols and named according to occupier and house name, even where this is repetitive, as in Curraghmore, Lord Tyrone; Mount Congreve, Mr Congreve; Bessborough, Lord Bessborough. The transition from lowland to upland is emphasised by careful hachuring, with yellow wash shading into dark brown for steep slopes, dramatically lighted from the south. 'Sleevnaman mountain' is a mass of swirling black. Cultivation is depicted with a stylised patchwork of fields, while bogs, pasture and woodland are also named. Openings and hindrances to the safe passage of men are noted by the inclusion of mountain passes, roads, rivers, streams, fords, bridges, and coastal landing places. These are supplemented by additional notes such as 'horse pass' (on the top of 'Barnavayalavulla' in the Blackstairs mountains), 'tide ends here' (at 'Ennisteague'), 'fordable throughout its course' (King's river, Callan) and 'tide ends' (north of Carrick-on-Suir). The critical map-reader might question why landlord demesnes feature so conspicuously. But, as Andrews argues, the 'great house' was of military as well as of social and economic importance; here soldiers could be billeted in outbuildings, horses put to graze on the lawns, the woods would provide some cover, while the house itself could be expected to act as a focus of upper-class loyalty and co-operation.[7] There was also the fear that local disaffected Catholic families would shelter an invading force; that is probably the reason behind the attention to 'priest's house' and 'Mass house' at 'Rathgormuc' (west of Carrick on Suir). Waterford was described by Vallancey as 'a well-peopled part of the country will conveniently canton or quarter 12,000 men [French forces] and this may be done a few hours after the landing is done'. And only 'a very trifling defence' could be offered by Duncannon fort, which would be easily silenced in three hours. These original top-secret military maps were never published; however, Vallancey's work is well known through the published maps of his assistant Alexander Taylor (fig. 8), and those of a London-based cartographer, Aaron Arrowsmith.[8]

Despite the certainty among government advisers that there could be no invasion along the treacherous western seaboard, 1,100 French troops landed in Killala in August 1798, under the command of General Humbert. Map publishers were quick to present this disturbing event to the public, which placed Connacht,

7 Andrews, *Shapes of Ireland*, p. 258. **8** Ibid., pp 248–74; see also idem, 'Charles Vallancey and the map of Ireland' in *Geographical Journal*, cxxxii (1966), pp 48–61.

a most neglected European outpost, briefly but alarmingly centre-stage in Anglo-French relations. The routes taken by 'the Loutenant' (*sic*), General Lake and 'the French' were marked in colour on existing county maps and road maps.[9] *Wilson's modern pocket travelling map of the roads of Ireland*, which was based on Taylor and Skinner's 1778 *Maps of the roads of Ireland* (fig. 10), was overwritten by hand: 'divided into 5 coloured sections to whom the districts under generals Lake, Dundas, Crosbie, Smith and Dalrymple, 1796–97'.[10] Coverage of the 1798 rebellion provides excellent examples of maps being used for the most blatant propagandist purposes, most notably in Richard Musgrave, *Memoirs of the different rebellions in Ireland …* (Dublin, 1801), for which the military cartographer Alexander Taylor produced maps of Enniscorthy and of Vinegar Hill. In the case of County Wexford, cartography allows the rebel camps, hide-outs, 'crafty' plans, roadblocks (piquets) and barbarities, the 'hostile' landscape of the south-east as exploited by the natives, the road and river networks, and the spread of disaffection as the rebels move into adjoining counties, to be exactly located.

This introduction to military mapping has dwelt on topographical surveys, small-scale regional or all-Ireland maps, and the use of maps for propaganda purposes. But local military chiefs were not the only officers interested in Ireland's topography from a military planning perspective. The British army continued to interest itself in the mapping of Ireland after 1922, producing maps at headquarters in London at the scale of 1:500,000 (approximately eight miles to one inch); these are headed GSGS (Geographical Section, General Staff) and show railways, main roads, navigable rivers and canals with insets at a larger scale of selected towns (on GSGS sheet 4366, produced 1942, insets of Londonderry, Dublin and Belfast). There is also a GSGS series at the scale 1:253.440, 2nd edition published 1942,[11] and a GSGS 1:25,000 series titled 'Eire, Provisional Edition'.[12] The numbered series of the War Office, GSGS continues through changes in name to the Ministry of Defence, Military Survey; these maps are part of a major Ministry of Defence collection (of over 200,000 maps), dated 1881–1968, which has been deposited at the Map Library of the British Library. The Irish content is minor relative to the full collection which covers much more than the British Empire. At least some of the GSGS topographical maps are really photographic reductions of maps from larger scales. The 1:25,000 is at an ideal scale for troops moving across country or planning artillery attacks, showing field boundaries, contour lines, trigonometrical stations, roads and railways; close examination betrays that this is a reduction of the six-inch sheets (the parish totals, in acres, roods and perches, can be discerned) but with the addition of contour lines, heavily inscribed. The GSGS maps are valuable

9 Richard Musgrave, *Memoirs of the different rebellions in Ireland, from the arrival of the English: with a particular detail of that which broke out the 23 of May, 1798; the history of the conspiracy which preceded it, and the characters of the principal actors in it. Compiled from original affidavits and other authentic documents and illustrated with maps and plates* (Dublin, 1801), plate 10; see also map in *An impartial relation of the military operations which took place in Ireland in August 1798* (1799), (NLI, P.129). **10** NLI, MS 809. **11** 16 maps held by Yale University Library, 327 1942.

to the local historian seeking detailed information over an extensive area. They also represent the technical expertise and preoccupations of the military mapmaker over time, placing the local area within a larger world of power struggles, strategic alliances, defence planning and preparations for invasion. The German military (*Militargeographische*) was also interested in producing maps of Ireland during World War II. A map of telegraph services and the electricity network in Ireland produced in 1940 gives a hint of the preparations being made to invade the island as a stepping stone to Britain.[13]

The investment of the military authorities in large-scale mapping of their own properties is of most interest to the local historian. This occurred most spectacularly after 1823 when control of barracks in Ireland was transferred to His Majesty's Ordnance (replacing the Irish Barrack Board). The Ordnance Office itself was abolished in 1855, when its functions were transferred to the Office of the Secretary of State for War, but by 1857 this section, popularly known as the War Department, was integrated into the War Office (WO), with its headquarters in Pall Mall, London. The Ordnance Survey of Ireland from 1824 (discussed below) had military-style command structures, and was markedly reluctant in the founding years to employ civilians in any significant superintending or surveying role.[14] The War Office is best described as the third major British mapping agency, after the Ordnance Survey and the Admiralty, but because there is no published record or catalogue of its products (as there was for the OS and for the Admiralty), researchers may inadvertently overlook this source. There are occasional returns to parliament of 'all the maps, charts, plans, tables and diagrams published by the topographical and statistical department of the War Office'.[15] The term 'published' is the key, as the vast bulk of the War Office's products exist only as MS originals and hand-drawn copies, and were never intended for public consumption. Returns for WO maps published between 1855 and 1861 record how this centralised agency served British imperial interests throughout the world, making 'expedition maps' and maps for colonial offices. Irish maps are a very small item in the overall workload of this important office; an 'ancient map of the siege of Enniskillen' and the more significant all-Ireland barracks survey explained below are the only published items listed for the period 1855–61.

It must be emphasised that the extent and breadth of the War Office's work goes well beyond those maps which were produced for public or at least official circulation. Large-scale maps in particular were central to the record keeping and property management of the army (fig. 14). Its chief value to the local historian lies in the fact that practically every town in Ireland, and many rural districts, are represented in this archive. For an introduction to the all-Ireland barrack network,

12 A full set held by the Geography Department, UCD. **13** See map of Ireland's electricity network, 1:750,000 (Berlin, 1940), CUL Maps *c.*170.94.24. **14** Andrews, *A paper landscape*, pp 36–9. **15** *A return of all the maps, charts, plans, tables and diagrams published by the topographical and statistical department of the War Office for the year 1855 to the present time* [16 July 1861], signed Thomas George Baring, 1862 (80) xxii.779.

there are a succession of published maps which provide useful starting points. Herman Moll's *New map of Ireland*, 1714 distinguishes between 'Barracks' and 'Redoubts or small barrac' (*sic*), with an additional table, titled 'Catalogue of Towns and Places where Barracks are erected, for quartering the standing army, with the number of troops and companys which each of them are to contain thro. out the Kingdom' (*sic*).[16] William Duncan's map of Ireland dated 1823 shows how extensive was the geographical coverage when barrack administration was centralised in London while a map entitled 'Barrack and ordnance stations, Ireland 1854', gives an update of the situation at mid-century, when some long-delayed rationalisation was forced on the military by the Treasury.[17] Written lists or returns were also published by parliament,[18] as were several landmark surveys, such as the 1854 report on barrack accommodation[19] and the follow-up 1861 report on the sanitary condition of barracks and hospitals.[20]

Major holdings of large-scale barrack maps, and associated documentation (including correspondence, memoranda, reports, financial records) will be found in the War Office papers in the National Archives, Kew, catalogue online at http://www.pro.gov.uk/. The series WO 78 lists the maps held by the WO in its library or topographical section (and now in the National Archives, Kew), but this covers only a fraction of the maps produced and/or held by the military which can be consulted at Kew. The OPW archive in the National Archives of Ireland, Bishop Street, is also a major collection because so many military properties were taken over as constabulary barracks (Royal Irish Constabulary and Garda Síochána) as explained more fully in chapter 3. The OPW catalogue is available as an on-line database [http://www.nationalarchives.ie/]. The National Library of Ireland holds the Kilmainham papers, covering the Curragh camp. The Department of Defence Property Management Section, Dublin; the Military Archives (Rathmines); the Military College Library, London; the Royal Engineers' Corps Library, Chatham, and (in some cases) the barracks themselves, where still used as such, hold property maps. The catalogue *Maps for empire: the first 2000 numbered War Office maps, 1881–1905* (London, 1992), compiled by A. Crispin Jewitt, lists the published map output of the various sections of the WO (including the Intelligence Department [ID], the Geographical Section, General Staff [GSGS]), over a 24-year period, and features notes on where these maps can be consulted (the BL, the National Archives, Kew; the Mapping and Chart Establishment of the Royal Engineers, and the Ministry of Defence, Whitehall). The Irish content listed here consists largely of maps showing military districts (at a scale of 1:633,600), and the maps produced to accompany the

16 Herman Moll, *A new map of Ireland … according to the newest and most exact observations*, 1714, Yale, Maps *327 1714. **17** *Barrack and ordnance stations, Ireland 1854*, TNA, MF Q/1/549. **18** *A return from each barrack in the United Kingdom …* 1847 (169) xxxvi.321; *Returns of the names of all military stations in Ireland …* 1860 (36) xli.593; *A return of all military stations in the United Kingdom, including the Channel Islands …* 1862 (305) xxxii.577; *A return of all military stations in the United Kingdom, including the Channel Islands …* 1867 (305) xli.577. **19** *Report from an official committee on Barrack accommodation for the army, with the minutes of evidence, appendix and index*, 17 July 1855, 1854–5 (405) xxxii.37. **20** *General report of the Commission appointed for improving the sanitary condition of barracks and hospitals*, 1861.

annual reports on military manoeuvres. There is also a 1:253,440 six-sheet map titled *Map to accompany the general scheme for the defence of Ireland* (War Office, 1885–6); the note 'strictly confidential' explains why these are not as well known among local historians as they deserve to be.

The local historian needs to explore the purposes for which the army produced individual maps (Appendix 1) if they are to be 'read' to maximum advantage. In the case of the large-scale property records, it is important to know something of the key surveys upon which later revisions were made. Some information will appear on later maps simply because it was inherited from an earlier survey, rather than being newly entered. Large-scale plans were produced routinely from within the army itself in the planning, expansion and refurbishment of barracks, when leases were renewed, or when property was to be disposed of. Continued and secure access to military lands and buildings, preventing encroachments and ensuring that neighbours did not build against the walls or other defences of barracks were further reasons for regular map-making. Adjoining property owners, private companies, other state departments, local authorities and charities all initiated contacts with the military concerning barrack property; these discussions often revolved around maps.[21] The upgrading of living conditions in line with health and sanitary regulations was another reason for mapping. Some understanding of the management regime within which barrack maps were produced can be gleaned from a study of published barrack regulations. *Rules, orders, powers and directions for the good government and preservation of the barracks and redoubts for quartering the army in Ireland* (Dublin, editions of 1701, 1711 and 1726) was superseded by *Warrant for the regulation of the barracks in Great Britain and Ireland and the colonies* (London, 1824).[22] Huge efforts were made to ensure that each room was used for its authorised purpose and no other, an obsession which lay behind the close labelling on barrack maps of each store-room, privy and stable. Sanitary engineering was also a major force behind barrack mapping, as royal engineers, medical officers and senior management sought to prevent outbreaks of infectious disease among the soldiers.

While many maps were produced in response to individual or local needs (such as the renewal of a lease), there were also a number of landmark all-Ireland barrack surveys. Following on the transfer of responsibilities for Irish barracks to London, a 'Register of Ordnance and Barrack Property in Ireland' was submitted to the Board of Ordnance by surveyor John Dowdall in October 1827,[23] this was followed by a major cartographic project, undertaken by the Royal Engineer Willam Tyer.[24]

21 For example, see extensive correspondence between the barrack board and landowners such as the Marquis of Donegall (Belfast), the Callaghan family (Cork), Lord Tullamore now earl of Charleville (Tullamore) and the earl of Rosse (Parsonstown), TNA, WO 44/120. **22** *Rules, orders, powers and directions for the good government and preservation of the barracks and redoubts for quartering the army in Ireland*. The following editions have been identified: 1701 (BL): 1705 (CUL); 1711 (BL); 1726 (BL, TCD). See also copies of printed regulations in TNA, WO 44/22, 31 & 32, and in WO 43/215. **23** Annotated correspondence of John Dowdall, 6 October 1827, TNA, WO 44/110. **24** Memoranda of William Tyler, 17 October 1827, TNA, WO 44/110.

Its primary purpose was the delineation of the boundaries of all Ordnance property, barrack, fort, redoubt or other type of fortification, most usually at the scale of 40 feet to one inch. It does not appear that these manuscript maps were bound as a set or otherwise separated from the general map archive built up by the Board of Ordnance; they were certainly available to the officers of the Corps of Royal Engineers when preparing plans on the disposal or refurbishment of individual stations, and copies will be found throughout the WO papers (search under Barracks, Ireland), at the National Archives, Kew. In most cases, barrack sketch maps are filed along with the correspondence, memoranda and other lists that were part of the story, making the WO papers an exceptionally rich resource for the local historian.[25]

A major map-based country-wide survey of 'all the landed property vested in the Ordnance' was undertaken between July and October 1848, in this case 'describing the quantity without the Boundary Walls and the portion most likely to be wanted for the public service as shewn in the accompanying sketches, with the probable value of the same'. These consist of large-scale manuscript maps on tracing paper, but the scale varies, from 'one inch showing 100 feet' to a scale of one inch to 10 feet. These maps are folded and stored as a single file in the National Archives, Kew (WO 44 / 581). Some are accompanied by notes expanding on the information mapped, such as reference to public drainage works through Ballinrobe (September 1849). The written list most closely dated to this map collection is published 1847; with its interest in the dimensions, ventilation and occupancy of sleeping rooms and the facilities afforded the men for washing and cooking, it marks the move towards data collection and map-making for sanitary purposes, a theme which is taken up with a lot more vigour in the 1850s.[26]

The best known country-wide and large-scale barrack survey is titled *Plans of the barracks of England and Ireland* (London, 1858–67) and was directed by Colonel Henry James. The survey was 'reduced and lithographed at the Topographical Department of the War Office', and published in four volumes.[27] James' survey is the basis for many of the barrack maps held by the OPW and the WO. A pocket-book version titled *Plans of the barracks in Ireland, 1861–7*, 'designed for use in the field', was also published.[28] Each volume of James' atlas opens with an index map (one inch to one mile) showing the military divisions and locating each station with a red spot, while the location and extent of each site is also coloured on the relevant six-inch OS plan, and the appropriate OS one-inch plan is indicated. As with the 1820s 'register of barrack property', the full survey consisted both of lists

25 For example, see 'A return of the several barracks in Ireland the leases of which will expire within ten years from the 1st January 1827 prepared in obedience to the orders of the Honourable Board of the 23rd August last and in deference to the letter from the respective officers of 2d. August 1826', TNA, WO 44/574. **26** *A return from each barrack in the United Kingdom* ... 1847 (169) xxxvi.321. **27** BL, Maps 150 d. 8. **28** BL, Maps C. 21.a.13; the scales used here also vary, for example one inch to 500 feet or 1:6,000 (Athlone), one inch to 350 feet or 1:4,200 (Royal Barracks, Dublin), one inch to 100 feet or 1:1,200 (Sandymount tower).

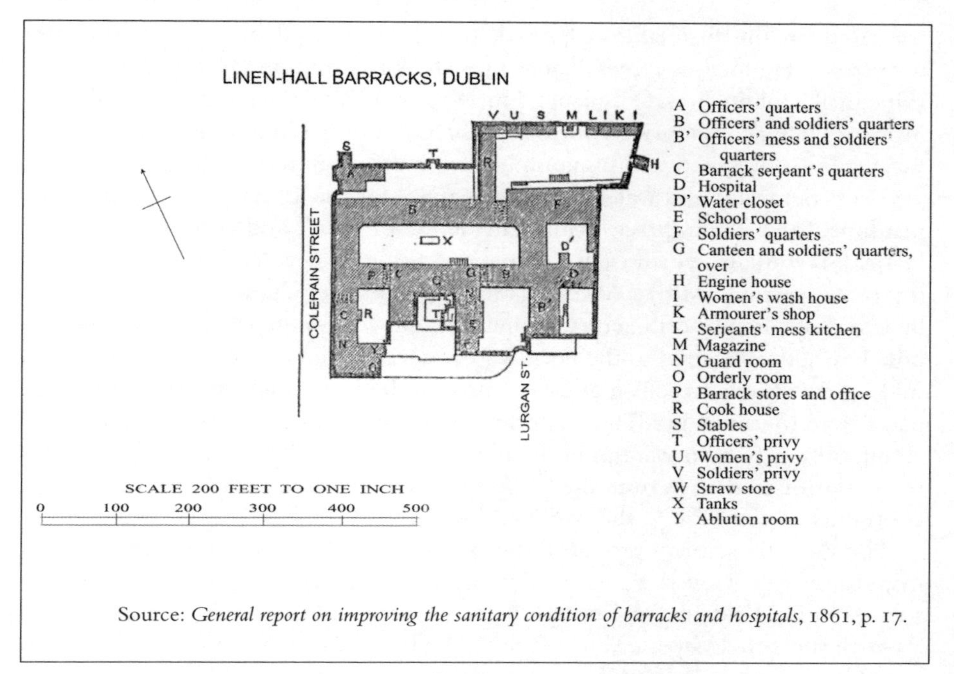

Source: *General report on improving the sanitary condition of barracks and hospitals*, 1861, p. 17.

14 Linenhall Barracks, Dublin a) Outline plan 1861

and large-scale maps. Tables of the accommodation available to officers, men and horses in each barracks were included on the maps, distinguishing between cavalry, artillery and infantry. However, there were also fuller lists, published as 'returns of all military stations'.[29] By 31 December 1866, James could report to parliament that barracks in Belfast and Dublin districts in Ireland, and most of England, had been mapped, 'and have been bound into atlases for the use of the military authorities'.[30] Common to all the James' survey maps are the following: breakdown of accommodation as noted above, classification of ground surface ('gravelled', 'pitcher pavement', 'syssel asphalte paving', 'flag paving' 'grass plots'), 'fall in drains' and drying posts, number of stories, means of lighting (candles, gas) both internally and externally, WD (war department) boundary, and the labelling of each unit on the plan. The scale of most of the maps is fifty feet to one inch (1:600), though some of the smaller stations, as in the batteries overlooking Lough Swilly (1863), are at a scale of 10 feet to one mile (1:528). It is this large-scale, uncluttered layout and complete coverage – for Ireland and England – that ensured this was a landmark survey in its day but also the basis for barrack management well into the next century. Colour

29 See footnote 18, p. 90. **30** *Report on the Progress of the Ordnance Survey and Topographical Depot, to the 31 December 1866*, 1867 (18301) xii.693, p. 709.

was used on the first edition barrack maps, and indeed on most of the later revisions and numerous copies. Figure 14 is an extract from the Henry James survey (Linenhall, Dublin city), first mapped in 1859, revised in 1884 and corrected again in 1891. Drainage is a priority, with the direction of fall indicated by arrows, and varying line styles used to distinguish between surface gutters, underground drains, pipes for rainwater and water supply, and pipes for gas. Every trough, pump, tap, manhole, brick drain, privy, urinal, latrine, water closet and cess pit is carefully marked. The use to which each basement and story is devoted is noted, including (by 1891) married ('Md.') soldiers' quarters women's wash-house, fumigating room, billiard room, library, cells, sentry promenade and coal vaults. Manuscript additions note how a small section to the north-east was transferred in 1906 to the OPW, and sold in 1907 (signed 1908), a good example of the uses to which these large-scale maps were routinely put. There appears to be a concerted investment in revision about 1883–4 (which were published), but some of the James' survey maps were revised prior to that, in response to immediate needs, and revision of these maps continued piecemeal into the twentieth century.

Alongside those maps produced specifically for military use, the army used OS maps to the fullest, overwriting them with military information. For very extensive sites, as in the Kildare and Curragh barracks, new buildings were marked on the six-inch sheets held by the War Office.[31] The large-scale OS town plans were also useful base maps, as in the OS Dublin five-foot sheet showing Aldborough House which was overwritten on 4 July 1866 and 23 May 1870 with notes such as guards-room, ablution room, ballcourts, bakery, hospital yard and drill ground adjoining a cattle yard, slaughter house and meat store ('to be divided by partition').[32] The 1:2,500 or 'twenty-five inch' sheets were to prove particularly useful,[33] though as this series was not sanctioned for all of Ireland until 1887, the military did not exploit it as fully as might have been the case had it been available earlier. The active interest taken by the military in the large-scale town surveys is clear from an atlas in the National Archives, Kew. Between 1874 and 1876, the War Office constructed a register of some of its property in Ireland by binding into a single volume selected sheets of the OS six-inch survey (first edition) which show military barracks, and numbering the sheets according to its own index. Thirty-four stations are included.[34] The notes connecting these six-inch sheets to the large-scale town plans then in preparation demonstrate how closely the barrack authorities followed developments in the topographical section. It lists towns that have been surveyed, at what scale, how many sheets each takes, and exactly where the sheet margins fall.

31 Six inch Kildare sheet 22, 1910 edition, barracks added in, TNA, WO 35/48 no. 28. **32** TNA, WO 78/ 2804 no. 1. **33** Cat Fort, City of Cork, 1919, TNA, WO 78/4742, no. 40, scale of twenty-five feet to one inch. **34** The stations covered are: Belturbet, Boyle, Ballyshannon, Castlebar, Cavan, Clonakilty, Carrick on Shannon, Cahir, Cashel, Dungarvan, Fethard, Fermoy, Gort, Granard, Kinsale, Longford, Mallow, Mitchelstown, Maryborough, Mullingar, Navan, Nenagh, Omagh, Parsonstown, Roscommon, Roscrea, Sligo, Trim, Tullamore, Tralee, Templemore, Tipperary, Westport, Wexford, Youghal. TNA, WO 78/23949 no. 1 (1874–76).

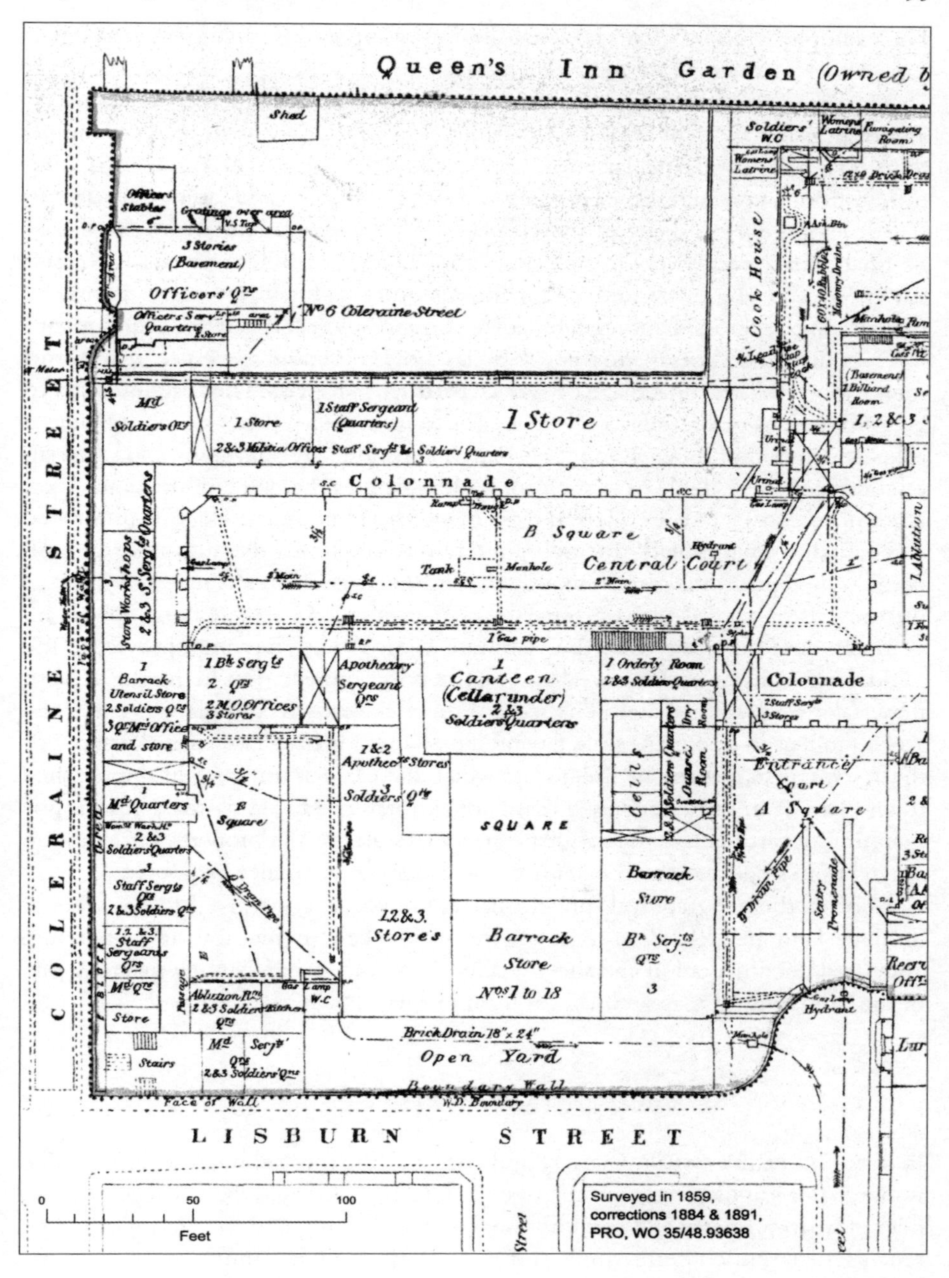

14 b) ground plan of 1859 corrected to 1891

For example, Mullingar barracks will be found on a single five-foot plan but is spread over 13 sheets at the scale of 1/500.

The military authorities did not want to redo the high-quality survey work already available to them from the Ordnance Survey, but they continued to produce their own maps for particular purposes. Between 1909 and 1919, barrack mapping moved from a concentration on accommodation, drainage and sanitary matters to the development of 'skeleton record plans'. They were published by the OS in Southampton (mostly in 1910 and 1919), annotated and further revised (signed and dated) by the engineers and commanding officers. These are more complete cartographic records, including tables of accommodation for 'staff and department', 'NCOs and men', horses, guns and vehicles, notes on cubic space per man, use of each room, and overall acreage of War Department property. They are intended to be comprehensive records of all new construction and repair, as each includes a brief description of the building itself, date of erection and extensions, and materials of construction. A 'Table of cost etc. of building' notes the name of the contractor, amount of contract, dates of commencement and of completion, costs (estimated and actual), and authority (including reference number of the authorisation). All floors were represented simultaneously in the James survey and its revisions (fig. 14); on these later record plans a separate plan is allowed to each floor. These are complemented in many cases by elaborate sections and drawings, providing the builder with all the architectural and engineering information he might need to undertake work immediately. The scale of these 'skeleton record plans' is either the 1:600 (50 feet to one inch) scale favoured by James for his 1860s barrack survey, or the 1:500 (10.56 feet to one mile) scale which the OS was using for its unpublished town plans. It must be reiterated that barrack maps produced by the army's topographical department were at a great variety of scales, as the properties in question ranged from single towers to extensive rifle-ranges and complex campuses.

In brief, the geographical spread, quantity of property involved (3,851 acres of military land in Ireland in 1862),[35] the care taken in revision and updating, associated documentation and sheer quality of product makes barrack mapping one of the most fruitful map archives for local history.

ESTATE MAPPING

In terms of quantity, quality, variety and coverage, the most important cartographic resource to local historians outside of the Ordnance Survey is the heritage of estate maps. Urban property and rural holdings, demesne lands and the lands farmed by tenants; quarries, collieries, mills and river courses, all and much more were the subject of privately commissioned maps, in manuscript form (on the whole) and

35 *A return of all military stations in the United Kingdom, including the Channel Islands, and all lands, tenements and appurtenances … held by the Military or Ordnance departments*, H.C. 1862 (305) xxxii.664–665.

now to be found in a large number of state and private repositories, as well as in private hands (explained below).

The professional surveying of estates in Ireland, for the purposes of management and forward planning (as opposed to surveying for the purposes of confiscation and redistribution of lands) dates at least from the 1590s.[36] An unpublished defence of 'the arte of surveying', written in 1682 expressly for an Irish audience, argues that the practice of estate mapping protects both landlord and tenants, as 'all parties are made to understand what they have'. The shift from the traditional practice of valuing land 'by the quantity of corn that is sowed in any farm and the number of cattel such farm is sufficient to grase' to accurate, paper surveys, had major implications for long-term planning:

> And by the mappe of an estate may the situation and distance of all the parts of the same be fully known, and how the profitable and unprofitable land is situated in respect of one another and what lands do bound uppon or joyn to an estate, with the highways and bridges or where highways and bridges may most conveniently be made all which may be plainly demonstrated and made at one view apparent to the eye for the satisfaction of the minde.[37]

Estate mapping was presented as a rational and enlightened practice which would enable the 'improvement' or commercial exploitation of the country's greatest resource, its land. And the benefits of this new practice accrued to small as well as to larger landholders. In the valuation of massive landbanks and modest holdings, by both absentee and resident owners, in urban, surburban, village and rural settings, the estate survey was to become a potent tool. In the private sector 'rationalisation' of land holdings, the planning of aristocratic (and profitable) new residential sectors from the 1730s onwards, and the transformation of rather nondescript villages into grand annexes to the 'big house', the estate map was a central document. For the transfer of property ownership, the practice of attaching a map to the lease probably dates from the late 1500s, following the example of the official surveys through which the planters had acquired these estates in the first place.[38] Not all landlords judged it necessary to include a map, relying still on verbal descriptions, but where they were used, early lease maps are typically simple, showing the leasehold boundary, naming the adjoining lands, and noting the area, in acres, roods and perches or, in urban sites, the length and breadth of the unit in feet and inches. There is rarely any interior detail. Under the Registration of Title Act (1891), property owners were (and are) required to deposit descriptions of the property unit, including a sketch map showing its bounds and dimensions, with the Land Registry of Ireland, Henrietta Street, Dublin 1.[39] Since 1922 properties held

36 Kissane (ed.), *Treasures from the National Library*, p. 185. **37** Thomas Knox, A survey of the lands of Termond Magraagh, 1682, NLI, MS 19,786, Leslie papers. **38** Andrews, *Plantation acres*, p. 116. **39** The Registry of Deeds (est. 1707), Henrietta Street, Dublin 1; the Land Registry (est. 1892), Dublin (Chancery Street, Setanta and Irish Life centres) and Waterford (Cork Road).

in Northern Ireland are registered with the Land Registry Office of Northern Ireland (Belfast). Instructions on how to access records in both registry offices are provided online at their respective websites [http://www.landregistry.ie and http://www.lrni.gov.uk]. Lease maps are very often found among family papers deposited in solicitors' offices, or held by the local bank as collateral for loans. Where the early estate office continues in existence, copies may still be held there, or may have been deposited in a local or national repository.

For a full discussion on the making of estates in Ireland, the wide variety of records they generated, and how to go about accessing these papers, the reader is directed to *Sources for the history of landed estates in Ireland* by T. Dooley (Dublin, 2000), which is published in this series of research guides for local history. Terence Dooley's *Decline of the big house in Ireland* (Dublin, 2001) is the major study of this world and its demise. A. Eríksson and C. Ó Gráda's *Estate records of the Irish famine* (Dublin, 1995), is organised on a county-by-county basis for the period from 1840 to 1855, with landlords listed alphabetically, followed by dates, type of records, location (principally NLI, NAI, TCD, PRONI), and reference as assigned by the repository. The code number 4 refers to maps, the addition LT where tenants' names are included. Estate descriptions are also specified: 3S, surveys and reports; 3V, valuation. This small guide, produced by the Irish Famine Network, provides a simple but authoritative introduction to the location of estate records, although the dispersed nature of these papers means that no single union list can ever be regarded as definitive. P. Collins' *County Monaghan sources in the Public Record Office of Northern Ireland* (Belfast, 1998) is designed to accompany rather than supplant the dedicated PRONI calendars or listings. It provides a lively introduction to the range of sources generated by estates, including maps, that are of interest well beyond the bounds of County Monaghan. Nolan and Simms' *Irish towns, a guide to sources* features sample illustrations, in colour, and discussion of estate maps.[40] *Treasures of the National Library of Ireland* also includes sample extracts from some of the finest estate maps in the NLI. These include the eight-volume survey of the earl of Kildare's estate by Rocque (1755–60), and the maps of the Domville estate, in the counties of Meath, Dublin and Wexford (1773). Andrews' *Plantation acres* is the classic account of the profession of land surveying in Ireland. It covers the complex development of estate mapping in Ireland, the immense variety in styles, units of measurement and classification, the technical processes of surveying, and places the work of individual practitioners within the larger history of estate cartography. It is generously illustrated with samples of their work.

Most fascicles of the *Irish Historic Towns Atlas* series published to date include a selection of estate maps of the town in question; the Maynooth fascicle by Arnold Horner, for example, reproduces maps showing the layout of the town on the brink of landlord redevelopment (1750s), along with the maps which the earl of Kildare commissioned in his efforts to 'improve' the town, not every element of which was implemented. The *Quarterly Journal of the Irish Georgian Society* published many

40 See Andrews, 'Estate maps', pp 31–5; Jacinta Prunty, 'Estate records', pp 121–36.

papers on estate mapping, and can be used as an introduction to the select world of the 'great estates', including Carton.[41] *A catalogue of the maps of the estates of the archbishops of Dublin, 1654–1850*, by Refaussé and Clark, reproduces maps of Church land in the county of Dublin and is a reminder that the Church was a major landholder and estate manager in its own right. There was massive investment in estate mapping prior to disestablishment on 1 January 1870, as the Church Temporalities Commissioners sought to assess the Church's land and property holdings. The observations column of each carries a note that the corresponding terrier and map for each parish or parish union was lodged 'in the diocesan registry'. Alas, all the maps, terriers and accompanying documentation formerly lodged in the registry were destroyed in the Four Courts conflagration in 1922.[42] Other major institutional estate owners were Trinity College, Dublin which held disparate but substantial land units from Donegal to Kerry,[43] and the City of London in County Londonderry.[44] The Corporation of Dublin found itself in possession of lands in counties Dublin, Kildare and Wexford following on the dissolution of the monasteries *c.*1539; the 'Book of maps of the Dublin City Surveyors 1695–1827' includes maps of these dispersed but valuable properties. Those with responsibility for military property in Ireland (the Irish Barrack Board to 1823, the War Department to 1855, the War Office from 1857) also contributed to estate cartography. Military concerns included boundaries and access, and the prospect of expansion or contraction of the holding; tenure details and the names of adjoining landholders are generally included on these estate maps. For statistical information on landholding prior to the land acts which enabled the transformation of the Irish tenantry into a new class of small landowners, John Bateman, *The great landowners of Great Britain and Ireland* (London, 1883) is a standard reference. This guide to the 'great landlords' lists names, residence(s), county or counties in which land is held, acreage and valuation for estates of 500 acres or £500 valuation and upwards. *Return of owners of land of one acre and upwards ... in Ireland* [c.1492], H.C. 1876, lxxx–61 is the most comprehensive listing. Many landlords held land in more than one county. The local historian needs to ascertain the names of those who owned land locally, as well as the geographic spread of these holdings, as access to estate records will require both landlord (such as, Shirley estate, Domville family) and townland names. A significant number of landowners held property in both

41 Arnold Horner, 'Carton, County Kildare, a case study of the making of an Irish demesne', special issue, *Quarterly Journal of the Irish Georgian Society*, xviii, nos 2 and 3 (1975); see also Gordon St George Mark, 'Tyrone House, County Galway', special issue, ibid., xix, nos 3 and 4 (1976), pp 23–66. **42** Wood, *Guide to records*, p. 231. **43** F.H.A. Aalen and R.J. Hunter, 'The estate maps of Trinity College: an introduction and annotated catalogue' in *Hermathena*, 98 (1964); R.B. MacCarthy, *The Trinity College estates, 1800–1923, corporate management in an age of reform* (Dundalk, 1992). **44** James Stephens Curl, *The Londonderry plantation, 1609–1914, the history, architecture and planning of the estates of the City of London and its livery companies in Ulster* (Chichester, 1986); idem, *The honourable the Irish Society and the plantation of Ulster, 1608–2000, the city of London and the colonisation of Londonderry in the province of Ulster in Ireland, a history and critique* (Chichester, 2000).

Britain and Ireland; in such cases estate maps of Irish property may be found in UK repositories; the Royal Manuscripts Commission online gateway ARCHON [http://www.hmc.gov.uk/archon/ archon.htm] is the most direct entry point to these holdings. Further detail on accessing estate records is given in chapter 3 of this guide.

In approaching estate maps, the questions posed in Appendix 1 may be applied with particular rigour. The Rocque map of the manor of Maynooth (1757), which hangs in Rhetoric House, NUI Maynooth (fig. 15) is an outstanding example of investment in cartography for prestige as well as for administrative purposes. It displays the village in the early stages of its redevelopment by the Fitzgerald family, along with the Carton demesne and surrounding land units. It incorporates reference tables on a townland basis, noting in acres, roods and perches exactly how much land is held and by whom. It is an example of what is termed the 'French school' of surveying, a style of manuscript cartography introduced to Ireland by John Rocque, developed by his pupil and brother-in-law Bernard Scalé, further advanced by a third generation of surveyors including Thomas Sherrard, John Brownrigg, John Longfield and Richard Brassington, and widely imitated by other surveyors throughout Ireland.[45] As an example of this style, the manor of Maynooth (1757) is both scientific and artistic achievement, vividly representing a productive landscape, with tilled fields (furrows) bounded by neat hedges, and houses with their outbuildings shown clearly in plan form. Map and tables combine to impress the visitor and overawe the tenant or whoever else might call to the estate agent's office, the most likely original home of such a large wall map.

From the perspective of the local historian focused on a particular parish or district, early estate maps can be a rich but also frustrating source of information. Some of those who practised as surveyors were highly skilled professionals with access to good instruments, a devotion to accuracy for its own sake, and familiar with the best examples of estate map-making to guide and inspire them. Others were practically self-taught, and likely to have combined surveying with other occupations. The customer base varied from the tenant farmer who had his land resurveyed by a local expert to ensure his landlord had not left him a perch or two short, to the owner who was lord of many thousands of acres and could well afford to employ a top-class surveyor to produce detailed maps and terriers to a standard that was comparable to the best work internationally, as in the employment of John Rocque by the Kildares. In the latter case, the estate maps were part of a much larger modernising project, intended in the long term to generate further wealth in the form of increased rent flows, and at the very least to protect existing holdings from being devalued by encroachments and expired leases. The purposes for which estate maps were drawn up therefore vary enormously. When the individuality of the surveyors is taken into consideration, alongside the different demands and resources of clients, the size and disposition of estates (some are more cohesive, others widely dispersed) and the differences in topography in Ireland, the variety in scale, size, style, ornament and content found among estate maps will be better appreciated.

45 Andrews, *Plantation acres*, pp 162–70.

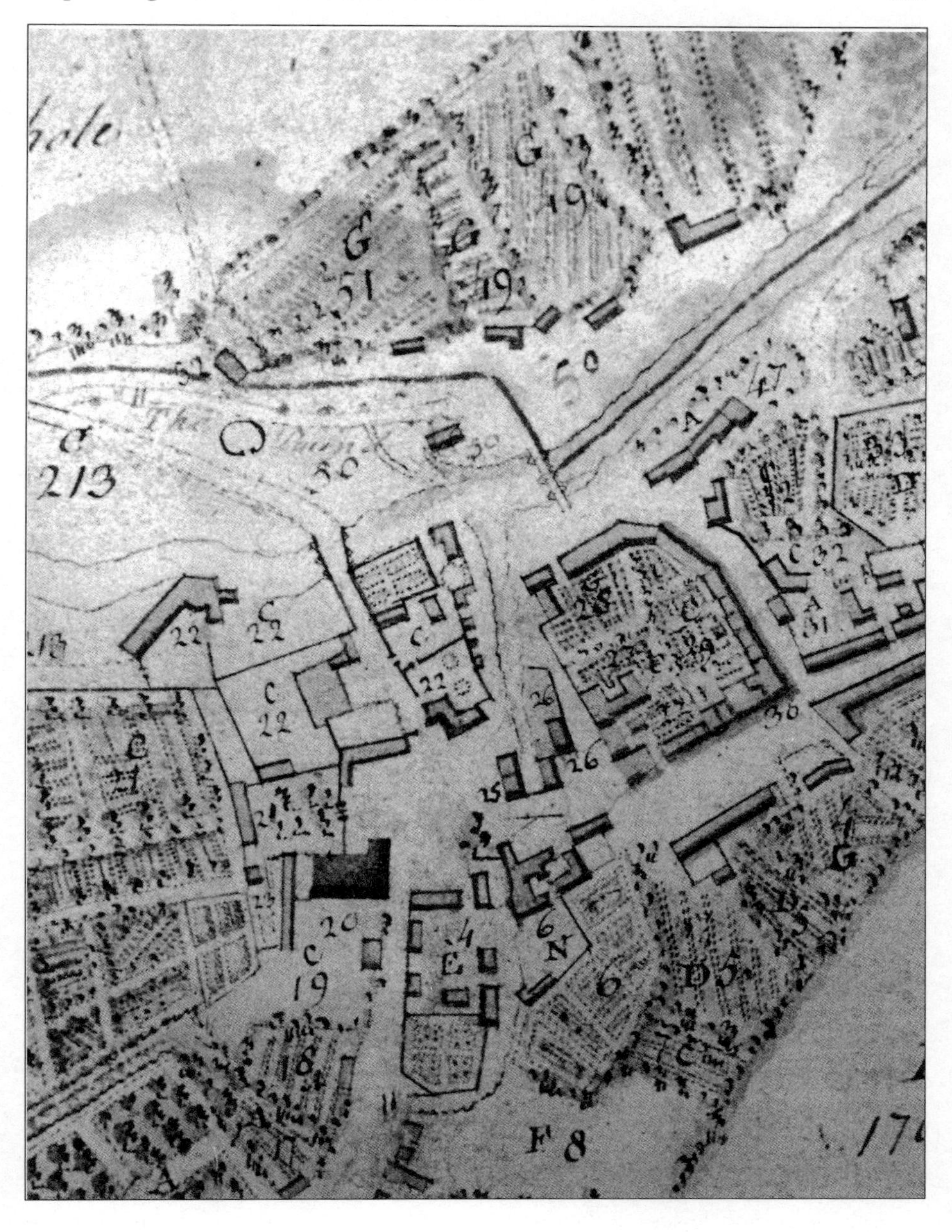

15 Map of the Manor of Maynooth, surveyed in 1758 by John Rocque, Rhetoric House, NUI Maynooth

And what appear on first viewing to be rather crude or 'empty' maps may turn out to be most useful to the local historian who can identify the boundaries on later OS maps. In terms of accuracy, even the humblest member of the profession, operating under intense local scrutiny, was skilled in the mathematics of surveying individual holdings.[46]

Figures 16 to 18 are taken from the papers of the Pembroke estate (known as the Fitzwilliam estate up to 1821) which are held in the NAI (no. 2011), and demonstrate something of the variety to be found under the heading 'estate maps', even within a single cohesive collection. This substantial and wealthy estate extended from Merrion Square in Dublin city south-eastwards to Blackrock, with parcels of land in Dundrum and Ballinteer, and a further unit known as the Bray estate in County Wicklow.[47] While some Irish materials are still held in Wiltshire, the NAI now holds the bulk of surviving records, including maps. The NAI catalogue divides its Pembroke records into six sections: deeds; maps 1692–1850; surveys and valuations 1762–1934; rent rolls and accounts 1751–1806; statements of accounts 1847–1916; and 'miscellaneous'. As always, the estate maps need to be seen as an integral part of a complex collection, and the map-making commissioned by the landowner understood against a larger background of developments in surveying and estate management. Perusal of the catalogue alone reveals the many uses for which this estate commissioned maps: laying out of new squares, streets, roads and 'avenues'; the development of 'a plot of waste ground'; the enclosure of strand; the clear depiction of disputed boundaries; leasing ground to the railway company and the canal company; submitting building plans for the approval of the Wide Streets Commissioners; preparing sites for school, house of refuge, church; showing leases granted in reversion, and leases about to lapse.

Figure 16 by Edward Cullen (1731) maps the course of the River Dodder from Ballsbridge to Ringsend. As an estate map it displays the owner's interest in a major engineering project, the construction of the 'Dodder canal' (as named on a map of 1806), which would allow a large tract of marsh to be drained and the land leased for high-class suburban development, as it was so advantageously sited relative to the city. Comparison of the 1731 map (fig. 16) with later estate and OS maps reveals which sections of the plan were implemented and which were abandoned, while later stages of this long-running saga are recorded in the candid estate correspondence of Mrs Barbara Verschoyle, agent to the estate (1776–1821). From the local history perspective, this map serves as a pre-OS record of the watercourses and associated mills in the Dodder valley. It shows the exact source of the raw materials ('fine sand' and 'brick clay') which were used in local glassworks and brickmaking, and demonstrates the disjointed nature of the road links with the city, a recurrent complaint among lessees and potential investors. It also shows the impact the construction of the new south wall (begun in 1707 under the Ballast Board) had on the geography of adjoining estates, opening up exciting possibilities where the landowner had the capital to invest in drainage.

46 Ibid., p. 136. **47** See map of the extent of the Pembroke estate in Prunty, 'Estate records', p. 122.

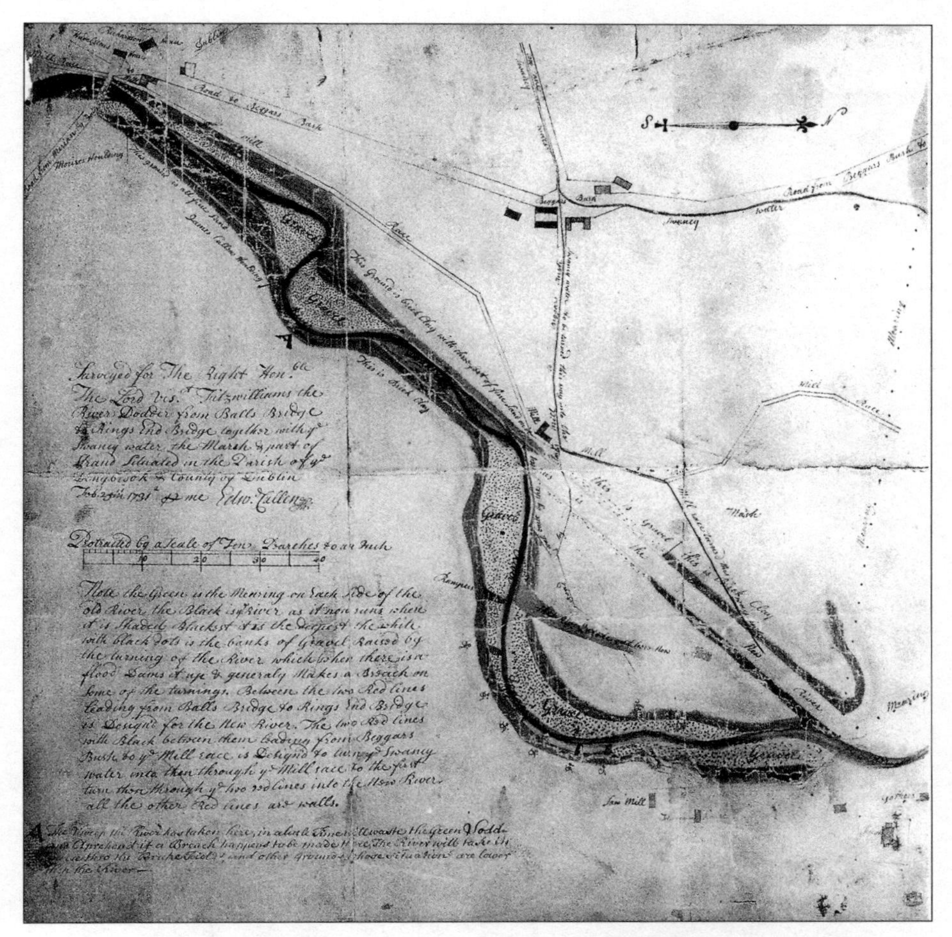

16 The River Dodder from Ballsbridge to Rings End Bridge together with ye Swany Water, the Marsh and part of the Strand situated in the Parish of ye Donnybrook, & county of Dublin, surveyed by Edward Cullen, 1731. NAI, 2011/2/1/6

Figure 17 is taken from the Jonathan Barker 'Book of maps and references to the estate of the Rt. Hon. Richard Lord Viscount Fitz William, 1762–64', which opens with an abstract of his entire holdings in Ireland. As an estate commission, only the land belonging to Fitzwilliam is mapped, although the newly-constructed and very grand town-house of the earl of Kildare is sketched in, presumably because his gardens were laid out on land leased from Fitzwilliam. The map extract reproduced here (fig. 17) is part of an overall guide, at a scale of 20 perches to one inch (approximately 1:5040)[48] to what was the medieval manor of Baggotrath (complete with castle), and is recorded as the townland name by the Ordnance Survey. Each unit is numbered, and the accompanying lists (separate from the map) give very full

48 Based on the Irish or plantation perch, where one perch equals twenty-one feet.

17 A map of Baggat-Rath and all its subdenominations, surveyed in 1762 by Jonathan Barker, scale of original: 20 perches to one inch. NAI, 2011/2/2/6

details: lessors' names, acreage, field boundaries, dimensions of plot, street widths, tidal/river levels, lease expiry dates, value of encroachment, yearly rental, and in the case of proposed developments, areas 'as designed and laid out for building'. The built-up areas are mapped at larger scales, including plans of Merrion Street and of Merrion Square 'with the intended new street' (100 feet to one inch) and of Ringsend and Irishtown (80 feet to one inch). The Dodder continues on its sinuous course (though somewhat attenuated by Barker where it traverses the strand), the note 'New River' showing that the core of the earlier plan has been implemented or is at least still an active ambition. As a local history source, the Barker survey allows the boundaries with adjoining landlords to be ascertained (Dublin Corporation, the Church, the earl of Milltown, the earl of Meath, Lords Palmerston, Trimleston and Carysfort). Some of the maps include miniature drawings or 'elevations' of noteworthy premises ('John Johnston's Linnen Printing Houses'; 'Miss Julian Blossett two new houses'). Because it is an estate-wide study, the status and fortunes of any single place can be compared with others and so its particular characteristics or difficulties better appreciated. In the disordered cluster of houses which made up the village of Irishtown (fig. 17), 60 of the 80 plots are returned by Barker in the survey notes as 'with encroachment' or 'part encroachment'. No other part of the estate can compare with Irishtown in this regard, testifying both to the irregular origins of this settlement (an unplanned fishing village) and the ambitions of the estate management, which will be to bring this village quite literally into line.

Periodical surveys or valuations were among the most useful management tools of the estate system as they provided an overview of how leases stood, and could be used as a basis for forward planning. The 1762 Jonathan Barker survey (fig. 17) is of that genre. The 1831 survey of the same estate by the firm of Sherrard, Brassington and Gale runs to sixteen maps; figure 18 is a section of the sheet showing Irishtown.[49] The physical growth of the village can be determined, and by comparison with earlier maps the extent to which order was imposed on the morphology of the village (an estate ambition) judged. The 1831 map gives the plot dimensions, with the name of the original lessee entered in red ink, and that of the occupying tenant(s) in black ink. Each plot is numbered and further details are summarised in the accompanying survey book. The use to which each holding was devoted is noted ('an old house in a ruinous state occupied as a forge'), allowing industrial or non-residential building to be separated out from dwelling houses. The notes are sufficiently detailed – and linked in each case to a numbered plot – to allow maps showing landuse and condition of housing to be constructed. Thus Irishtown is characterised by one and two story dwelling houses with over half 'in good repair', the remainder either 'tolerable' or 'in ruins'. Nearby Ringsend, by contrast, has mostly old three-story dwellings the bulk of which are 'in poor repair' or 'ruinous'. The non-residential use in Irishtown is limited to sea baths, RC chapel and St Matthew's

49 'Valuation and maps of certain parts of the estates of the Honourable Sidney Herbert in Ireland, out of lease or the leases of which will expire within fifteen years of the date of the valuation, 1829, 1830, 1831'; eight perches to one inch, NAI, 2011/2/5.

church, and some scattered 'offices', while Ringsend is overwhelmingly industrial with its warehouses, factories, salt works, glass works, shipyards and foundry, many in 1831 described as 'in ruins'. An estate survey such as this 1831 example (fig. 18) allows the socio-economic geography of a district to be reconstructed, with other local history sources (both inside and outside the estate), allowing the patterns established by the map evidence to be more fully explored. It must be noted that it was the largest, wealthiest, and best-managed estates, and those which were the subject of intensive and/or high-quality development, that produced the greatest volume of maps and associated records; the Pembroke estate is in that category. But wherever estate maps have survived the ravages of time and fortune they can be a goldmine for the local researcher.

The 'gentleman's seat' and associated demesne have long featured on maps at the county, regional and all-Ireland scale, as discussed already in the case of the military maps by Charles Vallancey, maps produced by the landlord body the grand jury, and commercial road maps. In the planning of their new recruiting barracks at Beggars Bush, Dublin, the military authorities were very grateful to have a Pembroke estate map by Arthur Neville (1813) to hand.[50] Private estate mapping directly informed the work of the 1820s Boundary Commission, which paralleled the Ordnance Survey townland mapping. Richard Griffith boasted that because of his 'personal acquaintance' with 'the greater number of the principal landed proprietors in Ireland, I have rarely been refused permission to make the necessary traces from their private maps, which are so essential to the accuracy as well as the rapidity of our progress'.[51] The standing of the 'big house' is reflected cartographically in the distinction made on the OS six-inch sheets between the 'ornament' used to separate demesnes from other land, the care invested in engraving the avenues, pleasure walks and names of each house, and the delineation of many demesnes as 'townlands' in their own right. Rather ironically, it was the OS six-inch townland maps, and large-scale town plans, which smoothed the way for the massive changes in land-ownership that characterised the second half of the nineteenth century.

In 1849 a Court of Commissioners for the Sale of Incumbered Estates was established with the duty of selling some lands with clear title, the property of an 'incumbered' near-bankrupt landlord, which would rapidly generate funds allowing the landlord to discharge his debts, and the new owner to occupy land with security as parliament conferred an indefeasible title on the new owner. Jurisdiction for encumbered estates passed to the Landed Estates Court in 1858, which in turn became part of the chancery division of the High Court (the Land Judges court) in 1877. From 1859 the court required the OS to provide it with countless six-inch extracts, but even before that date Ordnance Survey maps were essential to its work; many are now held by the National Archives of Ireland (as explained in

50 'Proposed site of the recruiting depot, Beggars Bush, 1826', TNA, WO 44/575.
51 Richard Griffith, 'Statement of the progress in the perambulation of the boundaries of baronies, parishes and townlands in Ireland, under the 6 Geo. 4.c.99, 28 May 1828', H.C. 1828 (23) vol. no. p. 349. See FN 182.

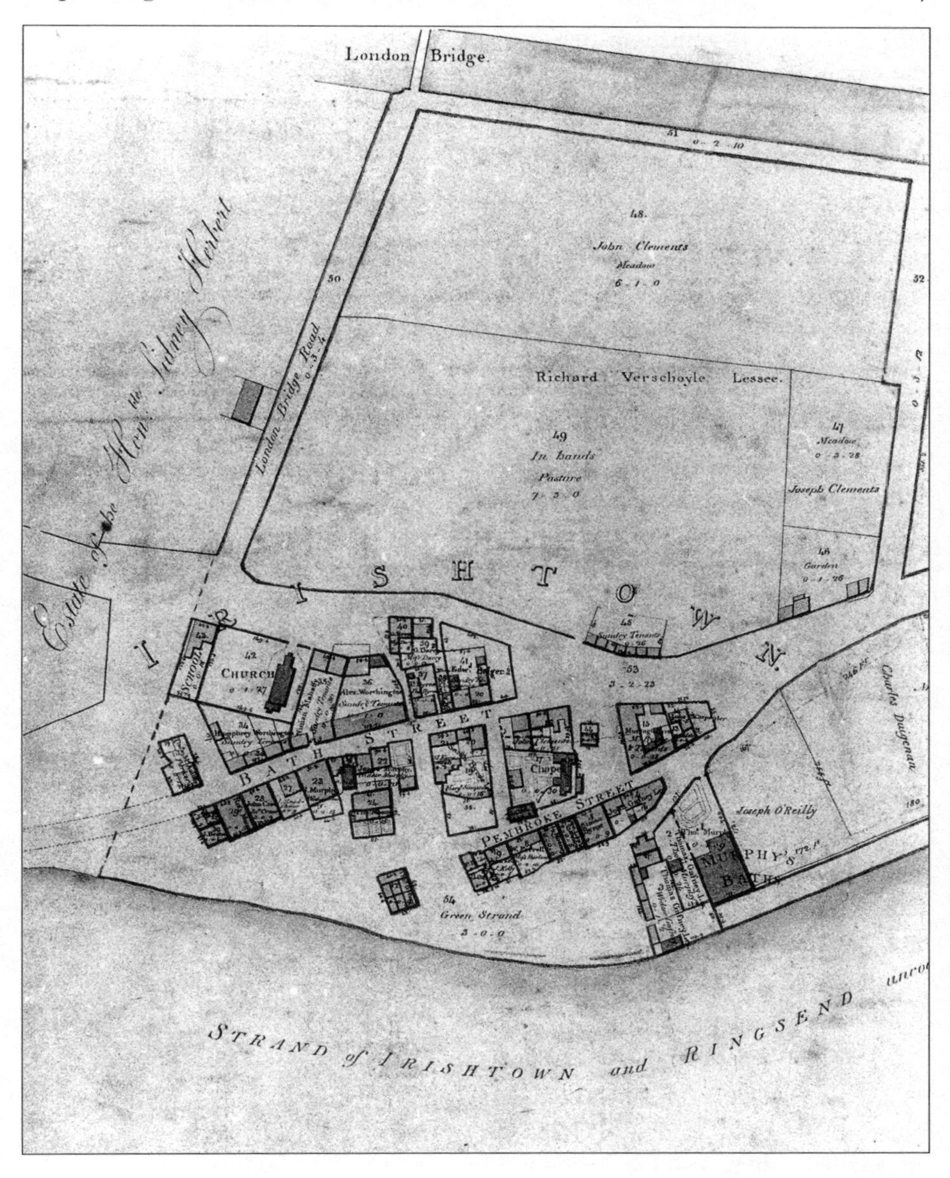

18 Survey of Ringsend and Irishtown by Sherrard, Brassington and Gale, 1831, NAI, 2011/2/5

chapter 3). The 1850 maps of Belfast (held by PRONI) produced for the Court of the Commissioners for Sale of Incumbered Estates in Ireland (14 Henrietta Street, Dublin), 'in the matter of the estate of Rt. Hon. George Hamilton, Marquis of Donegal', is a spectacular example of how large-scale state-funded mapping made possible the transfer of ownership, in this case on an urban estate. When the third marquis of Donegall succeeded to the title in 1844, he inherited debts of almost £400,000, many times the annual rental income; he had no choice but to hand over *c.*30,000 acres of urban and rural property to the courts to be sold. The Belfast maps number 27 sheets, at a scale of 160 statute perches to three inches (1:10,560, or six inches to one mile), complete with index map of wards and of individual areas. The sole concern is the exact location (numbered, bounded and dimensions noted) of each unit, and the status of the lease (present owner, yearly rent and duties, fine). Close examination allows the transfer of ownership of key sites to entrepreneurs such as the mill owner Arthur Mulholland to be traced, fuelling the industrialisation of what was for a short period the most populous city on the island.[52] The 'rental and particulars of sale' produced by the Incumbered Estates Court in respect of each property on its books, can be likened to an auctioneer's brochure, with its 'descriptive particulars and conditions of sale' aiming to present the property in as positive a light as possible. All of the lands were surveyed, maps, plans and a full rent roll drawn up. Some rentals are further enhanced with attractive line drawings. Mary Cecilia Lyons has reproduced a selection of these lithographs and extracts from the 'general description' in *Illustrated incumbered estates Ireland, 1850–1905* (Whitegate, County Clare, 1993). The way in which OS extracts were overwritten (on tailor-made extracts from 1862) for the purposes of establishing legal title and conveyancing is illustrated in figure 19 which shows a small farm in the townland of Killurin, on the estate of Lord Digby, King's County (Offaly), which is to be transferred to Eliza Dunne, under the 1923 Land Act. Land registry maps feature in many other legal situations, such as defending rights of way, or claiming turf-cutting privileges.

CANAL MAPPING AND INLAND NAVIGATION

Transport networks (post roads, rivers, canals and railways) each lend themselves most obviously to mapping. Grand jury maps of the eighteenth century have already been discussed. The boom in canal construction from the 1760s, and the wealth of maps, engineering drawings and public debate that arose from the development of inland navigation is a valuable resource for the local historian. While canal maps limit themselves geographically to a narrow strip of land on either side, they feature additional information such as cross sections that complement the longitudinal drawings, plans and sections of individual locks, bridges and

52 Town of Belfast, Incumbered Estates Commissioners, 1850–5 (PRONI, D/2122/4A–B and also D2/2122/2–3); see also the Town Parks and Malone estate volumes.

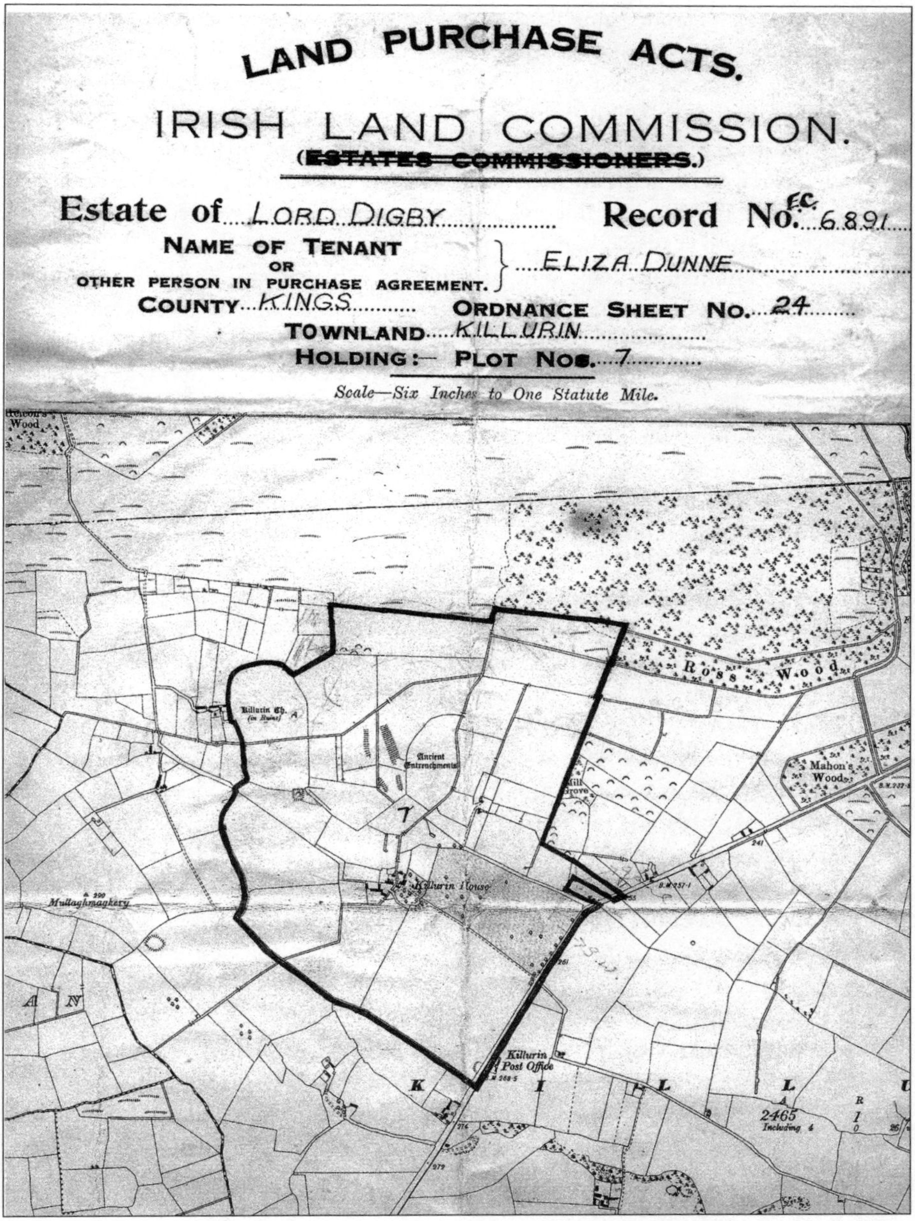

LAND PURCHASE ACTS.

IRISH LAND COMMISSION.

(~~ESTATES COMMISSIONERS.~~)

Estate of LORD DIGBY Record No. E.C. 6891

NAME OF TENANT OR OTHER PERSON IN PURCHASE AGREEMENT. } ELIZA DUNNE

COUNTY KINGS ORDNANCE SHEET No. 24

TOWNLAND KILLURIN

HOLDING:— PLOT No. 7

Scale—Six Inches to One Statute Mile.

Ross Wood

Mahon's Wood

Killurin House

Killurin Post Office

Mullaghmagkery

19 Killurin, estate of Lord Digby, plot to be sold under 1923 Land Act.

aqueducts, notes of underlying geology and water levels, and details of the overall fluvial geomorphology of the area. The construction of the Grand Canal for example, was a massive feat, with the Leinster aqueduct carrying the canal over the river Liffey just west of Sallins, and other lesser rivers and streams all needing to be integrated into the pattern, requiring the construction of further aqueducts, and additional canal branches. Construction began in 1756 and by 1779 the canal was open to Sallins; by 1785 the Barrow Line was completed to Monastereven, and in 1803 the Shannon was reached. The reports by Charles Vallancey and John Trail (both in 1771) include maps and critique of what is still today a marvel of engineering and vision. Where good county maps were already in existence, later cartographers and engineers were happy to make full use of them; the map of Kildare by Alexander Taylor (1783, figure 8) provided J. Brownrigg with an excellent base for his Grand Canal map of 1788.

Canal promoters maintained that economic prosperity would follow on a rational system of transporting goods by water, enabling Ireland to play a full part in international trade, finding markets for its own goods, and handling the goods of others.[53] Maps showed how the river Shannon was to be linked to Dublin, via both north (Royal) and south (Grand) lines; by the Grand to Waterford via the Barrow; to Drogheda via the Boyne; across the Shannon to Galway by the Suck; 'from the heads of the Shannon to the Heads of the river Earn [*sic*], and thence to the sea near Ballyshannon'. Further links from Lough Neagh to Newry and Belfast 'with the several junctions, communications and collateral branches' would complete an all-Ireland network.[54]

But despite the energy with which its advocates promoted it, the extension of the canal network was an immensely complicated and lengthy process. The construction of canals was only possible by compulsory purchase, on foot of an act of parliament; maps were part of the formal submission to parliament (as discussed below in reference to railways), although care must be taken to ascertain that the proposed route was in fact taken. In the search for canal maps, the *Journals of the Irish House of Commons*, are the first point of reference; for example, in the 1761 volume (xii, p. 292) will be found 'A plan and section of the River Nore from the city of Kilkenny to the town of Ennisteague' (*sic*).[55] The NLI map catalogue, under 'canals', will also direct the researcher to published papers on individual canal projects, many of which reproduce maps submitted to parliament. James Dawson's *Canal extensions in Ireland, recommended to the imperial legislature* (Dublin, 1819), and Charles W. Williams, *Observations for the inland navigation of Ireland* (2nd edn, London, 1833), illustrate how an entirely integrated inland navigation system was still being

53 Henry Brooke, *The interests of Ireland considered, stated and recommended, particularly with reference to Inland Navigation* (Dublin, 1759), p. 80. **54** Ibid., p. 131; see also 'A map of Ireland showing the number and extent of its rivers with the present and proposed lines of inland navigation, 1795' in Hely Dutton, *Observations on Mr Archer's statistical survey of the country (sic) of Dublin* (Dublin, 1802). **55** This map is reproduced in Patrick Watters, 'The History of the Kilkenny canal' in *Journal of the Royal Historical and Archaeological Society of Ireland*, 4th ser., ii (1872), p. 93.

promoted well into the nineteenth century. Both private and state inquiries into the condition of the Irish during the 1830s noted the potential of improved navigation systems to boost the rural economy and appended maps to their reports to illustrate how advantageous this could be.[56] The commissioners for the improvement of the navigation of the river Shannon produced many maps and engineers' plans for bridges and canals; their fourth report (1839) is valuable to researchers working on the midland countries on either side of the Shannon.[57] Kane's 1845 map representing 'the existing and proposed canals, railways and navigable rivers'[58] sees in the new technology, steam, but a further enhancement of an all-island transport network. The 1907 *Canals and waterways commission* (vol. ii, part III) provides a later reference point for the researcher who needs to know which canal branches and links were finally developed. It is essential to know the full title or company name associated with any particular section of canal before trying to access maps and associated records. From the eighteenth century onwards, the construction of canals was closely connected with that of railways, with some early railways built for the express purpose of bringing goods to the canal dock or quayside. The Midland Great Western Railway company was one of the major investors in rail in the 1840s, with several instances throughout the country of rail track laid parallel to or alongside canals. While separate railway companies are associated with separate lines of canal, charge of the canal network as a whole in the Republic was taken over by the state transport company Córas Iompair Éireann (CIE) under the Transport Act (1950). In 1986 the canals became the responsibility of Dúchas, and more recently of Waterways Ireland (Enniskillen, http://www.waterwaysireland.org. At the time of writing, the canal records formerly held by CIE at its headquarters in Heuston Station, Dublin had been transferred to the National Archives of Ireland but were not yet catalogued.

THE ORDNANCE SURVEY OF IRELAND (SIX INCHES) SERIES

The mapping of Ireland by the Ordnance Survey from 1824 onwards is by far the most important cartographic undertaking in this island's history, and ranks as the earliest large-scale mapping of any single country. The published map series – the six-inch sheets, town plans, one inch and 25-inch maps – are the principal fruits of this massively ambitious centralised project, but behind these is a rich store of related materials, including astronomical and trigonometrical readings, field-books, correspondence, accounts, 'fair plans', copper plates and scientific instruments. The National Archives of Ireland on-line catalogue of its Ordnance Survey holdings [http://www.nationalarchives.ie/] is in itself a useful introduction to the undertaking, the many steps involved, the huge numbers of persons employed in widely

56 *Report on the state of the poor in Ireland*, H.C. 1830 (667) vii.1; Charles Wye Williams, *Observations on the state of Ireland, the want of employment of its population* (Westminster, 1831). **57** *Fourth report of the Commissioners for the improvement of the navigation of the River Shannon*, H.C. 1839 (325) v.259. **58** Robert Kane, *Industrial resources of Ireland* (Dublin, 1845).

different capacities, the scale of the finances required, and the problems encountered as its originators 'trained and organized a completely new department', for the purposes of this 'minute survey', and devised methods 'by which large numbers [of surveyors] could be employed on it, at a moderate expense, with little risk of confusion or error'.[59] The National Archives of Ireland's introductory notes to its online catalogue are the basis for much of the following discussion; the series list is reproduced as Appendix III. As explained in chapter 3, the most important archive for OS records (excluding the published maps) is the NAI, but records are also held in PRONI, the NLI, the Valuation Office, the Royal Irish Academy, the Geological Survey, the National Archives (Kew) and elsewhere. In the discussion which follows, the prefix NAI identifies which materials will be found in Bishop Street, Dublin. Andrews' *A paper landscape*, is the major reference work, while his *History in the Ordnance map*[60] provides an authoritative account of the different map products of the early Ordnance Survey, specifying scales, dates of editions and revisions, and geographical coverage in each case. Publication and revision dates of the six-inch series, and dates for both unpublished and published town plans, at a variety of scales, are also appended to Andrews' *A paper landscape*. Charles Close's *The early years of the Ordnance Survey* (London, 1926) covers operations throughout the four countries of what was in 1800 the United Kingdom of Great Britain and Ireland. The study of Ordnance Survey maps is the specific focus of the Charles Close Society; its journal *Sheetlines* includes Irish-related material. The April 1991 edition (no. 30) was a special Irish issue. A joint publication between the Ordnance Survey of Ireland (OS*i*) and the Ordnance Survey of Northern Ireland (OSNI) titled *Ordnance Survey in Ireland, an illustrated record* (Dublin, 1991), is a succinct but generously-illustrated account of the island-wide survey from 1824. The Ordnance Survey presented its own annual report to parliament; from 1855 to 1921 these are published in the House of Commons series, full listings appended in Andrews' *A paper landscape*, p. 338. A brief overview of the Ordnance Survey's structures and materials is featured as a preface to the *Memoir* series, edited by Angélique Day and Patrick McWilliams (1990–1998, volumes i–xl, covering the Ulster counties), and also as an introduction to the *Ordnance Survey Letters* series, edited by Michael Herity (volumes on Donegal, Meath, Dublin, Kildare, Down and Kilkenny published to date).

Because of the importance of the Ordnance Survey six-inch maps to the local historian, it is worth pausing to consider some key questions (Appendix 1): what was the inspiration for this survey and the ends to which it was directed? What structures were created to enable to be carried out? What checks and counter-checks were put in place? An understanding of the instructions under which the officers, civilian assistants and labourers operated, and the many tasks behind the

59 Thomas Colby, *Ordnance Survey of the City of Londonderry*, i, 1837 (facsimile reprint, Limavady, 1990), preface. **60** *A paper landscape* was reprinted in 2002 by Four Courts while *History in the Ordnance map* was reprinted in 1993 by David Archer, The Pentre, Kerry, Newtown, Montgomeryshire, Wales SY16 4PD.

production of each published sheet, will assist the researcher in using the full range of OS records relating to any particular parish, to great personal satisfaction. Other questions the researcher might bring to OS maps are how did the OS map compare with previous output in that locality? What level of expertise and technology was available to the OS? What sets these maps apart from earlier surveys? Who held final editorial control? The Ordnance Survey of Ireland is a trigonometrical survey, that is, based on accurate measurement, verification and calculation (explained below); it was also a very orderly, centrally-controlled process, with distinct stages being reached along the line from first establishing a mathematical framework, through to the interior 'filling in' on a parish-by-parish basis of the countless smaller triangles that spread over this paper landscape, and the eventual knitting together into a county series of the material gathered so faithfully in the field in so many different notebooks. Its records therefore must be understood as tied into a particular stage of the process, and founded upon careful measurement. Determining boundaries was another essential prerequisite of the townland survey proper, as was deciding on placenames; both of these allied concerns are dealt with below, as is the general valuation of Ireland 1846–64, which relied upon the OS to produce the maps upon which this major cadastral survey was based.

The immediate background to the making of the six-inch or townland survey is well covered in a succession of parliamentary inquiries directly preceding the setting up of the Irish survey (1815, 1816, 1822 and 1824). The unsatisfactory state of Irish cartography is exposed, and the urgent need for an authoritative, comprehensive, up-to-date general survey that would define townland, parish and barony boundaries and their acreages is argued convincingly. The county cess was levied on occupiers of property according to its rateable valuation; without accurate, independent and up-to-date surveying, the basis upon which this local tax was built was notoriously inequitable. Very many different parties had an interest in a country-wide survey, including the census commissioners, whose ambition to calculate population densities could not be achieved without maps which included exact areas of parish and townland units.[61] The select committee on the survey and valuation of Ireland, reporting on 21 June 1824[62] exposed how variable in scale and reliability the existing cartographic record was, and the antiquated, obscure, irrational and unfair basis upon which taxes were computed. It provided the *raison d'être* for the new all-Ireland, scientific, state-sponsored survey, and a forum for current expert opinions on the ways in which the proposed new survey should be conducted. The 'great territorial survey or cadastre' of France, which commenced in earnest in 1808, was reviewed, along with large-scale surveys in Bavaria, Savoy and Piedmont. In the Irish case, it recommended a speedy, accurate, systematic survey, at the scale of six inches to the English (statute) mile, showing townland boundaries but not fields (a restriction that was later lifted), superseding 'all local

61 Andrews, *A paper landscape*, pp 13–14. **62** *Report from the select committee on the survey and valuation of Ireland*, H.C. 1824 (445) viii; Known as 'The Spring Rice report' after its chairman. The following quotations are taken from the reprint in Andrews, *A paper landscape*.

topographical proceedings whether under the authority of Grand Juries or otherwise'. The county of Kildare, for example, would require forty sheets at that scale, the county of Galway, 137 sheets. A 'central and effectual' control, along military lines, along with 'command of the best instruments' available was advocated, with the committee expressing the hope that the final product would be 'creditable to the nation, and to the scientific acquirements of the present age'.[63] Mountjoy House in the Phoenix Park, Dublin, which was used temporarily as a barracks, was handed over as the headquarters of the Irish survey. It was to here that the officer in charge of each department sent his monthly progress reports and field accounts, which provide invaluable detail on the huge numbers of staff employed in the Irish survey (peaking in 1839 at 2,139), turnover of personnel, and the range of skills required of these men.[64] It was to Lieutenant Thomas Larcom, at Mountjoy Barracks, that John O'Donovan and his colleagues who were employed to advise on the orthography of the placenames to be entered on the final maps, sent their field reports and letters 1834–43. There were four departments in all: Mountjoy, Geological, Trigonometrical, and Hill, while there were different 'divisions' working in the field. The 'Mountjoy returns' (NAI, OS 1) for example, include the monthly returns of officers (rank, names, stations, remarks); returns of civil assistants in receipt of 2*s*. or more per day (names, rates of pay, remarks including notes of transfer to Britain); returns of royal sappers and miners in receipt of 1*s*. per day (ranks, names, duties performed, rates of pay, remarks); equivalent listings for all those employed in Mountjoy House itself (royal sappers and miners, civil assistants, inferior assistants), and return of permanent articles of stores, that is, instruments and equipment. Each department had to account for its personnel, use of equipment, the progress (or 'state of forwardness') of the work entrusted to it, and its preparedness for the next stage, in a meticulous manner. Endless orders, circulars and memoranda were issued in Dublin and the OS headquarters in London or (from 1841) Southampton (NAI, OS 13); these provide insights into both the complex administration of the survey, and the technical processes of map production. The registers of documents (NAI, OS 51) provide an archival listing of the number and description of each document on a parish-by-parish basis, evidence of the sophisticated set of checks maintained on the work in the field. The engraving journals (NAI, OS 12) provide a concise summary of the labour costs of each sheet throughout the three stages of 'outline' or 'engraving', 'writing' and 'ornament'. Larcom's own private papers, running to 136 volumes, are deposited in the National Library of Ireland, while major files of correspondence between Dublin and London concerning the Ordnance Survey are preserved in the War Office archive in the National Archives, Kew.[65]

At the heart of the 'Colby system' (called after the founding director, Thomas Colby), was the division of labour, whereby each specialised in a particular skill, completed his task to an exact standard, and presented the results in a pre-

63 Ibid., pp 305–6, 308. **64** Ibid., pp 90–1. **65** For example, see Correspondence on the Ordnance Survey of Ireland, 1839–49, vol. 2, TNA, WO 44/703.

determined and closely regulated form. The manual drawn up to instruct officers in the surveying methods already employed by the OS (3rd edn, Dublin, 1862) explains each step in the procedure, emphasising that 'on an extensive survey one general system must of necessity be vigorously enforced, to insure uniformity in all the detached portions of detail'.[66] The teams in the field who produced the trigonometrical framework, recorded levels (or heights), plotted boundaries and noted endless detail on 'all communications by land and water, manufactories, geology, antiquities, or other matters connected with the Survey' (art. 68) entered their findings in numerous carefully-coded field notebooks, each ruled for a distinct purpose, following instructions. Other teams at headquarters processed countless columns of figures recording location, height, distance, area and direction. As these employees pursued their work at quite a remove in space and time from the original surveying, it was imperative that the information recorded in the field be exact and presented unambiguously. The first and most important of the many circulars and memoranda issued by the Ordnance Survey of Ireland is Thomas Colby's 'Instructions for the interior survey of Ireland, 1825' (NAI, OS 13, reprinted in Andrews, *A paper landscape*, Appendix B). The safe keeping and daily checking of the surveying instruments was also a matter of concern.

Each 'fair plan' (held by the National Archives of Ireland) covers a single parish, at a scale of six inches to one statute mile, on a sheet of double-elephant paper measuring 40 inches by 27 inches (some larger parishes take several sheets, and a number of smaller ones are found on the one sheet). This is the final manuscript drawing (1825–42), the culmination of the work of surveying, plotting and checking; it is aligned with north at the top, and includes placenames (NAI, OS 105). There can be some differences between these 'fair plans' and the final printed maps, including changes in orthography. For example, the 'fair plan' of Maynooth town (fig. 20, NAI, OS 105 A 56), includes the name and code for each trigonometrical station (T[1] Mariaville 207, R[1] Maynooth tree 204); and a number of minor names (the students' walk, slate quarry, pump, sheep paddock, watch house, lodge, wall, post office, new road) which were dropped at this last stage, while 'Bond bridge' on the fair plan became 'Pond bridge' on the published map of 1839 (fig. 21, six-inch, sheet 5, County Kildare, 1839). And while name-placing may be more aesthetically pleasing and legible on the published map (fig. 21), the more condensed lettering on the fair plan can often be of more assistance to the local historian seeking the exact location of an individual place, such as the 'old chapel' and police station on the fair plan of Maynooth (fig. 20). The ornament was also subject to change, with the grounds of Maynooth college for example being represented in a much more elaborate (and expensive) manner on the fair plan than on the final printed sheet. Additions can also be made, such as the note 'Library' on the first published edition of Kildare sheet 5 (fig. 21), and the insertion of townland

66 Frome, *Outline of the method of conducting a trigonometrical survey*, p. 32.

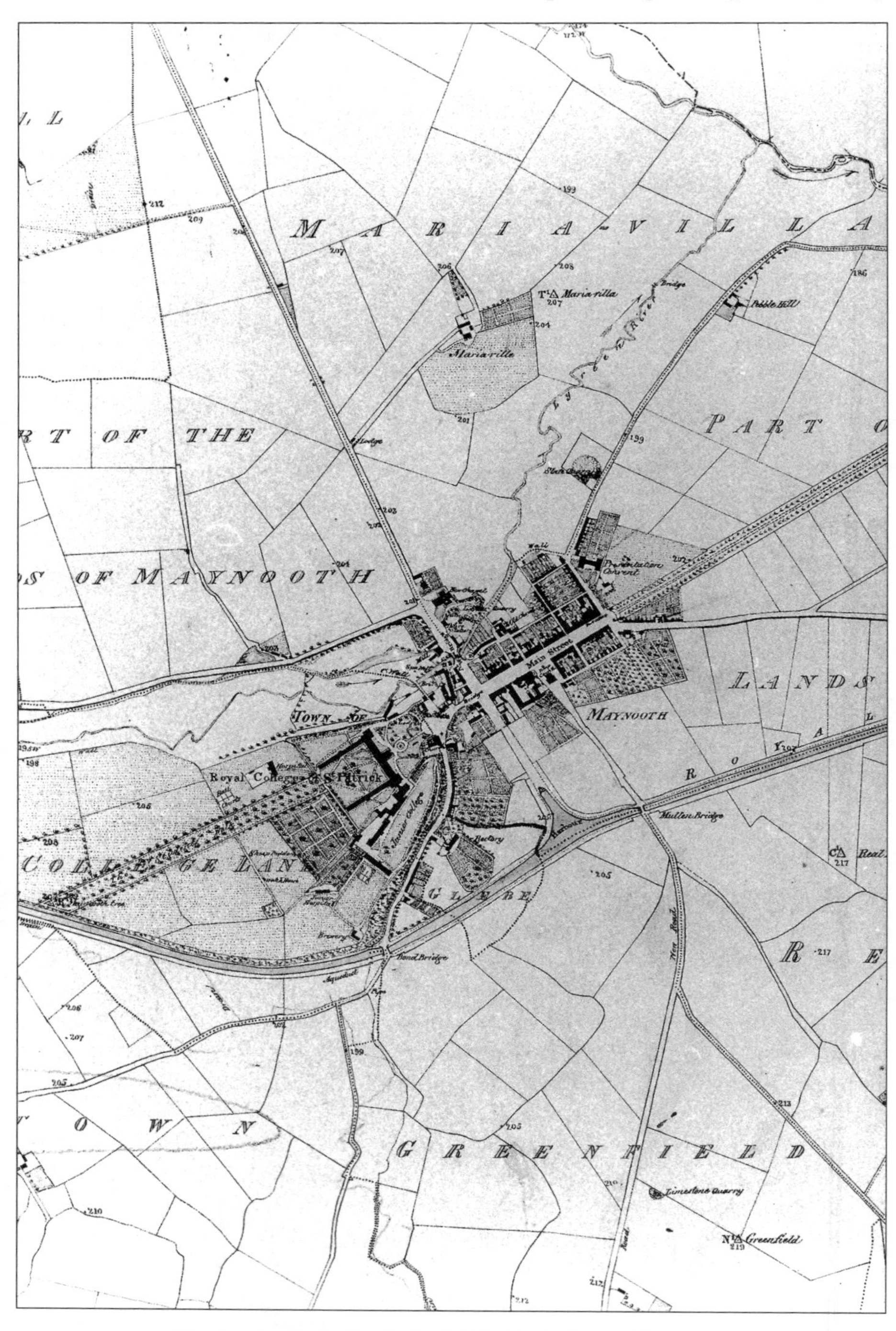

20 Maynooth town, OS six-inch 'fair plan' of Laraghbrien parish, County Kildare, 1838, NAI, OS 105 A 56

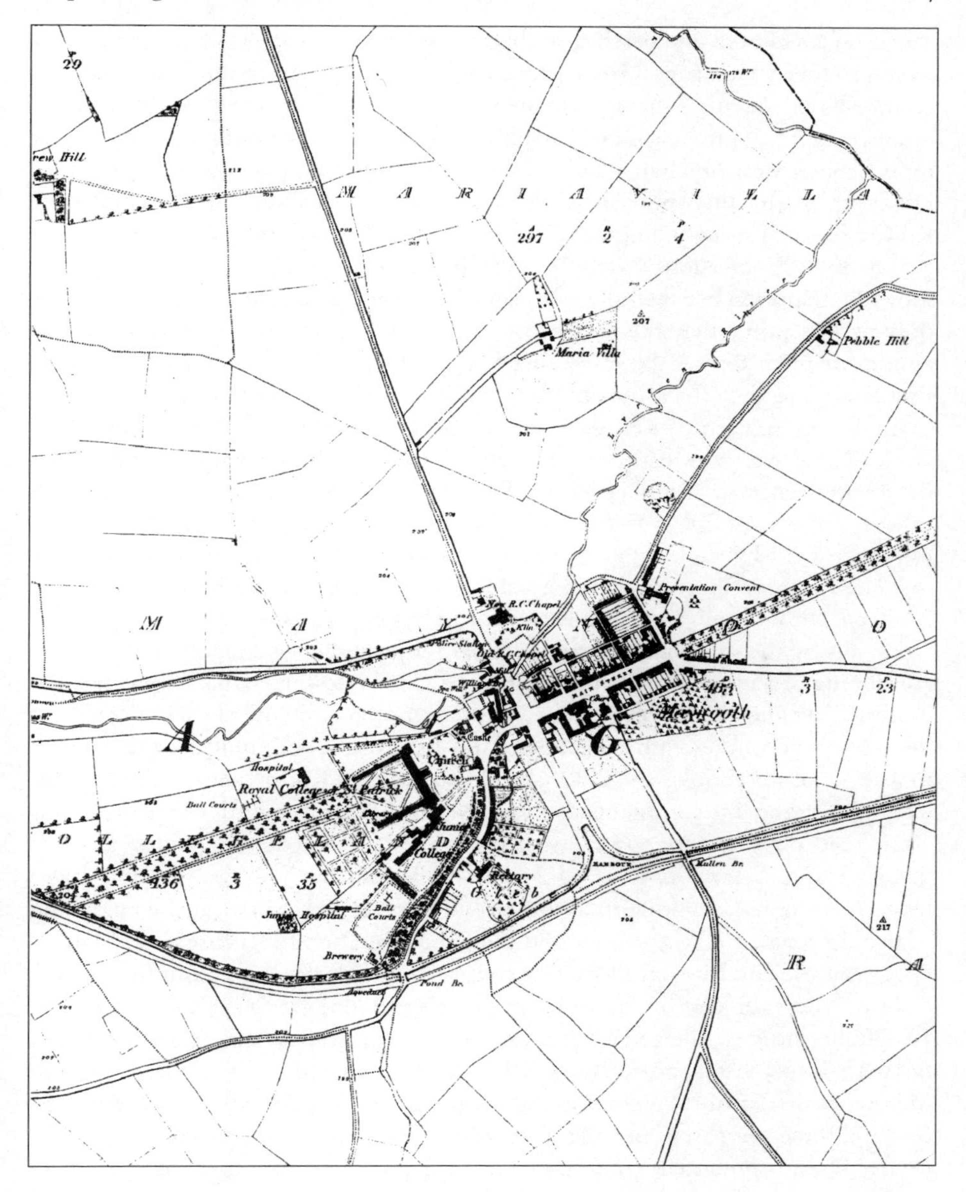

21 Maynooth town, OS six-inch, County Kildare, sheet 5, 1st edn, 1839

acreages (acres, roods and perches) at this final point. As noted already, the 'fair plans' were produced on a parish basis; these are combined to make the final 'seamless' county-based six-inch sheets, which were given their own distinctive 'county' numbering. The entire parish of Laraghbrien, of which Maynooth is a part, will be found centred on one 'fair plan', while the surrounding parishes are left blank. However, in the final published series the parish of Laraghbrien extends over Kildare six-inch sheets 5 and 6, and there is no break in coverage between parishes.

The proof impressions (NAI, OS 107) the first printing from the copper plates, show the transition between the 'fair plan' and the engraved and printed map. It was only at this point that the system of sheet numbering on a county basis was employed; prior to that the researcher follows the parish name in their inquiries. Proofs went to the officer (or officers) who had carried out the original survey for examination, and corrections were made directly onto the copper plate. When the work of alteration was made, the correction sheet was countersigned and dated by the various engravers concerned to show the changes were indeed made. Each officer was responsible for the area within his own triangles. They show the influence held by each supervising OS officer, and how even within a tightly-regulated system there is always a subjective element in the representation of landscape. Proof impressions are extant for only fourteen counties.[67]

Engraving the six-inch maps onto copper for printing was a highly skilled, time-consuming process; the engraving itself was divided into three separate processes of 'outline', 'writing' and 'ornament', with the duration and cost of each element of the work noted along with the name of the engraver(s), both military and civilian, in a series of notebooks. (NAI, OS 12). In the case of Kildare sheet 5 (fig. 21) for example, engraving commenced on 29 November 1837 and the ornament was completed on 5 October 1838, with the wage rates for the six named workers involved ranging from 4*s.* to 8*s.* 6*d.*[68] The number of days, hours and minutes spent by each are noted, resulting in a total labour cost of £14. 3*s.*1*d.* for the engraving of this sheet alone. The 40 sheets and index map of County Kildare cost £779 4*s.* 6*d.* to engrave, but this was modest in comparison with the larger counties, such as Galway, where the cost of engraving 137 sheets and index map came to £3,461 6*s.* 1*d.* Dublin sheet 18, the single most expensive sheet in the entire series (extract as fig. 68) cost £300 11*s.* 2*d.* to engrave. The final product, the published maps, were eagerly awaited; from the completion of the survey to publication took between two and three years in most cases. Revision of the first edition Northern counties (where the survey started) commenced in 1844, and proceeded on a sheet-by-sheet (rather than county) basis up to 1891, by which time 21 of the 32 counties had been resurveyed, stretching from Donegal to Carlow, along with part of County Roscommon. Revisions to the six-inch sheets after 1891 were carried out by reference to the fieldwork undertaken for the larger scale maps (the 1:2,500 series,

67 Proof impressions (NAI, OS 107), counties for which they are available: Cork, Carlow, Clare, Dublin, Galway, Kerry, Kilkenny, King's County, Limerick, Tipperary, Queen's County, Waterford, Wexford, Wicklow. **68** Engraving journal, Kildare 1838–40 (NAI, OS 12 no. 2).

commenced in 1887). The first edition six-inch maps of the 10 counties west of a line from Mayo to Wexford were revised through photographic reduction from the larger scales.[69] The end of the townland survey proper therefore can be dated to *c.*1891, although activity in the later decades cannot be compared to the scale of work in the peak decades of the 1830s and 1840s.

THE TRIANGULATION OF IRELAND

The first step in accurate topographical surveying over a large area is to decide on a number of points or stations whose relative position can be accurately established. The network is independent of the scale of the maps to be produced. In the case of the OS, a network of points, each exactly located, and the distances between each, exactly computed, was created over the entire island, and linked at several points to the British mainland. The geographical position of one of the fixed points of the triangulation must be determined in order that the survey as a whole may be correctly located on the earth's surface; this must be done by astronomical observations. A point outside Athlone, at latitude 53° 30' North, and longitude 8° 00' West, was chosen as the 'national origin'; the Ordnance Survey of Ireland was thus tied in to that of Britain in a single geodetic system (fig. 22). From the perspective of the local historian, if the place under study can be correctly situated within this fixed network or triangulation the full range of materials produced by the Ordnance Survey relating to that district (Appendix III) can be utilised.

The 'Principal triangulation for the Ordnance Survey of Great Britain and Ireland' originated with the task given to General Roy in 1783 to connect the Observatories of Greenwich and Paris, and was extended gradually under a succession of Ordnance Survey directors, to be completed during the term of Colonel Henry James.[70] Captain William Yolland, who took part in this enterprise, left a technical account of the Lough Foyle base measurement (1847),[71] while the Ordnance Survey published several other volumes listing the 'observations and calculations of the principal triangulation' and 'principal lines of spirit levelling'. A military handbook by Colonel Edward Frome titled *Outline of the method of conducting a trigonometrical survey for the formation of geographical maps and plans …* (London, 1862) explains the mathematics upon which triangulation rests, including worked examples. The *Admiralty manual* (pp 1–6, 332–6) also provides an accessible introduction.

For those approaching the OS records for the first time, it should be noted that the calculation and adjustment of triangulation relies upon the principle that a

69 Andrews, *History in the Ordnance map*, pp 18, 20, 24. **70** Alexander Ross Clarke, Henry James, *Ordnance trigonometrical survey of Great Britain and Ireland, account of the observations and calculations of the principal triangulation; and of the figure, dimensions and mean specific gravity of the earth, as derived therefrom* (Southampton, 1858), preface. **71** William Yolland, *An account of the measurement of the Lough Foyle base in Ireland with its verification and extension by triangulation together with the various methods of computation, following on the Ordnance Survey and the requisite tables* (London, 1847).

round of angles observed at one point must sum to 360°. In a triangle, the length of any one side can be calculated once the angles of the figure are known. If the length of any one side is known exactly (that is, the distance AB is carefully measured on the ground), and the value of any one angle is also known, the remaining angles and distances can be calculated. The first step of this ambitious project therefore required measuring a single base line along a level plain, the first 'leg' in the triangle and crucial as 'upon its accuracy that of every subsequent proceeding depends'.[72] The Tower Museum in Derry city vividly re-creates the making of what was termed the 'Lough Foyle base', stretching 7.89 statute miles in length along nearby Magilligan strand and crossing the river Roe (fig. 22). From this base line in County Londonderry the entire triangulation of Ireland was developed.

Triangulation proceeded in two stages: the primary or principal triangulation, through which the island-wide network was created, established lines of visibility between selected high points (fig. 22), linking in along the east and north coasts with points in Britain to create a single British Isles reference framework, for example from Divis in Belfast to Merrick in Scotland.[73] The OS published an alphabetical description of each of these observation stations titled *Account of the observations and calculations of the principal triangulation and of the figure, dimensions and mean specific gravity of the earth as derived thereon*, 'drawn up' by Captain Alexander Ross Clarke under the direction of his superior Lt. Col. Henry James (2 vols, London, 1858). This text also includes a lengthy exposition of the mathematics upon which the triangulation is based, the readings for each station, a description and illustrations of the instruments utilised, and the names and ranks of the officers, NCOs and others who were personally employed in the measurement of these base lines or in taking the trigonometrical or astronomical observations, and who were assisted in their work by 'a great number of computers'.[74] Most of the plates are in volume ii, which gives each section of the principal triangulation as well as the completed British Isles network (fig. 22). From this text the local historian can find a detailed description of the trigonometrical station which ties his or her local place into the British Isles framework:

> The station is on the western head of a large hill or mountain, about eight miles south-west of Tralee, and on the left side of the road leading from Tralee to Dingle through Blenerville. Curraheen Chapel is on the Dingle and Tralee road; and a person intending to visit Baurtregaum station from Tralee should leave the Dingle road at this chapel, and pursue the course of a stream upwards for about 2.5 or 3.0 miles. This stream leads to Caum Brack Lough, which is about 1.5 miles east of the station; the passage from Curraheen Chapel to Caum Brack Lough is of easy ascent, the remainder is difficult. The station is marked by a large pile of stones, within which will be found the centre stone, an irregular piece of sandstone, level with the general surface, and marked with a hole.[75]

72 Ibid., p. 6. **73** Clarke and James, *Ordnance trigonometrical survey of Great Britain and Ireland*, p. xiv. **74** Ibid., p. ii. **75** Ibid., p. 3.

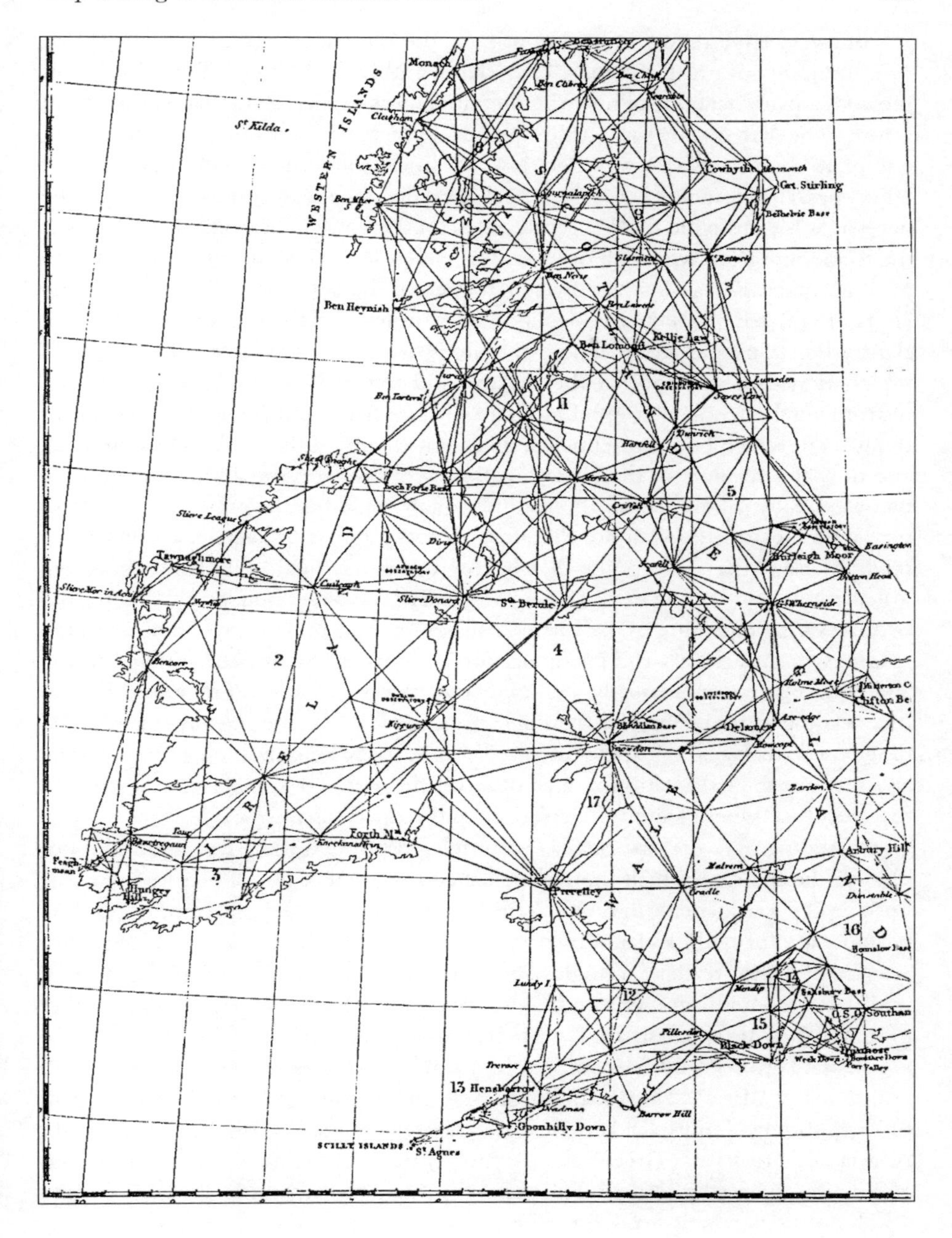

22 The principal triangulation of Ireland and Great Britain

Four very large theodolites were used by the Ordnance Survey to take angular measurements of great accuracy. The Clarke and James *Account of the observations* includes a lively account of how the theodolite was carried from place to place, 'in a four-wheeled spring-van', accompanied by a party of from six to ten men, complete with portable wooden houses, small theodolites, heliostats, pocket telescopes, chronometer, barometer, pocket compasses, measuring tapes, and 'all the necessities for camp life'.[76] The 'centre-mark' of the station, the point directly below the theodolite from which all observations were to be taken, was determined with great accuracy. This point was always recorded in the landscape, as in Hungry Hill (1832, 1843) parish of Ardrigole, about five miles north-east from Castletown, Bantry Bay, county of Cork, where 'the site is marked by a hole bored in the rock, which is within one foot of the surface, and the pickets which supported the instrument; the whole is covered by a pile'.[77] The Clarke and James *Account* explains the ways in which both horizontal and vertical circles on the theodolite were read, and how the verniers or sliding scales allow what are very small differences to be recorded. Tiny differences in the readings of an angle translate into differences of perhaps miles on the ground, therefore it is the mean or average of successive readings from the same point that is entered in the final column, back in the 'calculation room' of the field station. Triangulation is based on plane geometry, but because of the curvature of the earth, small adjustments must always be made to the readings to ensure that the triangulation is consistent with itself. Where all the angles have been each equally well observed (that is, there is no reason to believe one reading is more reliable than another), the 'excess', 'deficiency' or 'error' is distributed among all the angles involved. Complex astronomical observations were also made at the field station, and all observations, entered in ink, were returned to the OS office where lengthy calculations and reductions were made.

By August 1832 the first stage was completed (though the government did not publish the figures until all the six-inch sheets were on sale, lest private surveyors upstage the Ordnance Survey).[78] Within this primary triangulation, 'secondary triangles are formed and laid down in like manner by calculation, and if necessary a series of minor tertiary triangles between them'.[79] This stage of the work took until 1841. Each officer commanding was responsible for the surveying of a number of trapezia within parish units; this had the advantage that 'the effect of any accidental error should be confined to a single parish' rather than throwing the entire survey askew.[80] This further stage of triangulation is illustrated for the parish of Laraghbrien, County Kildare (fig. 23), where the trapezium labelled T[1] C[1] N[1] R[1] (Mariavilla, Real Park, Greenfield, Maynooth tree) covers the town of Maynooth and its immediate hinterland. The general principle of working always 'from the whole to the part' was stressed repeatedly to all officers and civilian assistants, as it was one of the chief scientific strengths of the survey.

76 Ibid., pp 44–6. **77** Ibid., p. 23. **78** Andrews, *Shapes of Ireland*, pp 280, 285. **79** Frome, *Outline of the method of conducting a trigonometrical survey*, p. 1. **80** Colby, *Instructions*, art. 58, reprinted in Andrews, *A paper landscape*, appendix B, pp 309–21.

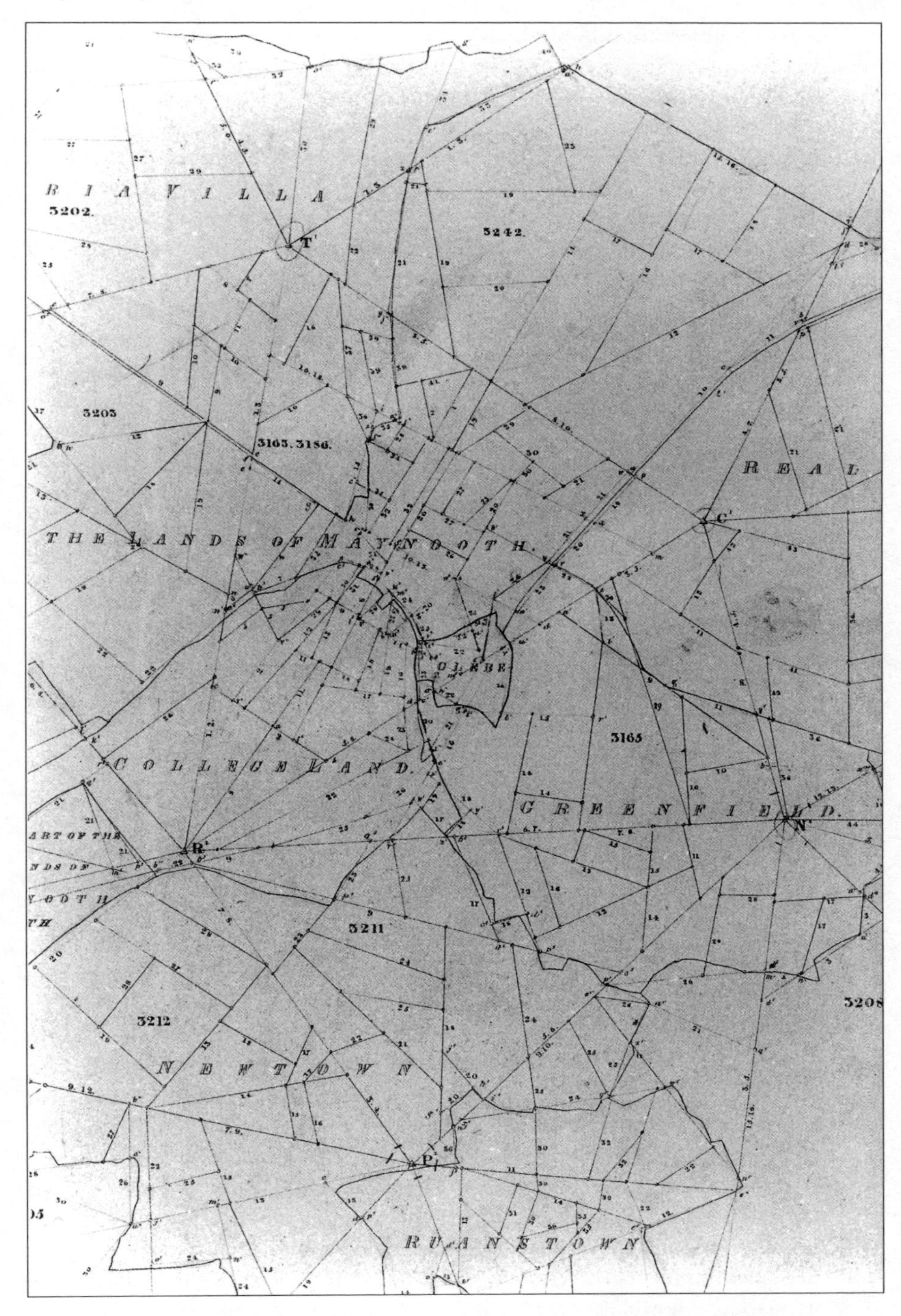

23 Extract from index diagram or outline plot for parish of Laraghbrien, County Kildare, NAI, OS 104 A Plots 56.1

The parish observation books (NAI, OS 43) contain a succession of theodolite observations, recording horizontal arcs (distance) and vertical arcs (height), in degrees, minutes and seconds, with the name of the observation station, and the figures of the observations to named points. They also include remarks on the instrument used, and notes on weather conditions. The levelling registers (NAI, OS 65 & 66), currently stored in the Four Courts, are also highly technical, but worthy of consultation if only to impress upon the researcher the enormous weight of resources behind this enterprise, and the great diligence shown by its officers. Of more assistance to the local historian is the series of notebooks titled 'Descriptions of trigonometrical stations' (NAI, OS 54). Notes relating to the 1850s and 1860s are often searches for stations used in earlier surveys, many having been disturbed or destroyed in the intervening years, including even major sites such as Knockmealdown. Some of the verbal descriptions are enhanced by sketch maps, as in Stacumney (fig. 24) in the townland of Elm Hall, County Kildare (sheet 11), which was successfully 'rediscovered' by William Burke on 18 April 1866 with the assistance of P. Murray and E. Murray, labourers:

> Situated on top of fence about 4 links from quick. The old centre was found with a hole formed (?) in the centre and is sunk 2 feet in the centre of fence. The site was found from copies from original main line books. (NAI, OS 54/M/3)

The Stacumney point was marked by a square of blue limestone, 6 inches square and 2½ inches thick ('for future finding'), and its location relative to five other points noted (fig. 24), including the 'Wonderful Barn' at Barn Hall and 'Maynooth monument'.

The choice of observation points at the local level depended on what was available; old castles, forts, church towers, factory chimneys, large houses and bridges were all used. The OS plots for the six-inch survey (variously called 'line', 'outline' and 'skeleton' plots) are single parish-based sheets recording these local decisions; Laraghbrien, County Kildare, for example (fig. 23), was plotted by W. Warren, and signed by Henry James (later director of the OS) on 19 July 1837 (NAI, OS 104 A Plots 56.1). While the OS plots appear on first examination to be an inexplicable network of numbered triangles, without any topographical information, they are in fact an excellent index to the content field books (figs. 25–26) where more intimate detail is recorded. The outline plot for the parish of Laraghbrien (fig. 23) gives the field book numbers for each of the triangles covering this parish (Maynooth town and its immediate hinterland will be found in numbers 3165, 3163, 3186, 3203, 3211, 3212, 3242). A number of real-world elements can be identified (fig. 23) as in the boundaries of the Glebe townland, the curved road (Parson Street) which follows the line of the college wall, and the curved road which crosses the canal at Mullen Bridge (labelled 'new road' on the fair plan, fig. 20). The skeleton plots in the content field books (fig. 25) are not to scale, nor are they oriented grid north; they are working notes which will assist the Dublin-based colleagues, who need to situate the local observation points within the larger

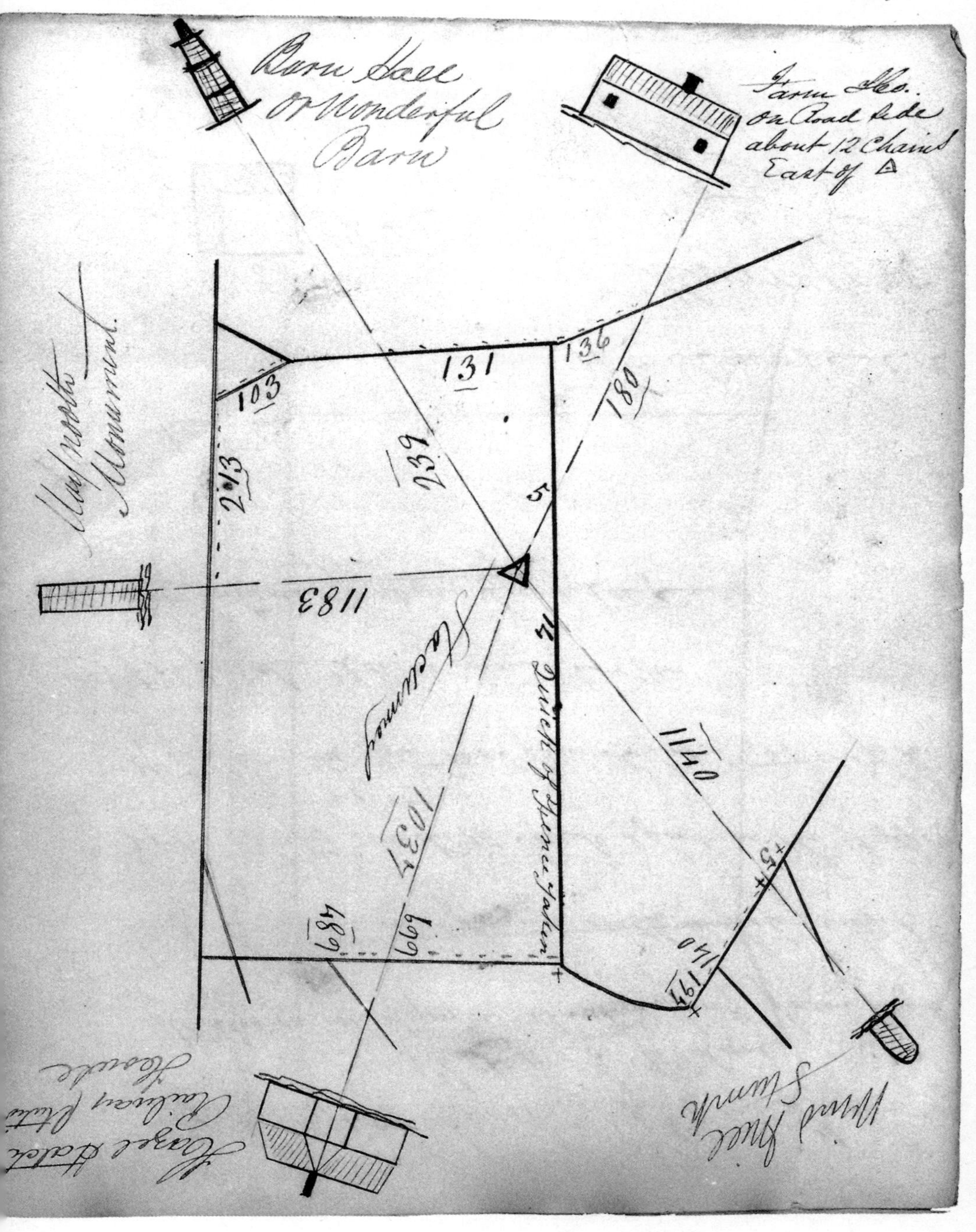

24 Stacumney trigonometrical station, Townland of Elm Hall, sheet 11, County Kildare, 18 April 1866, Descriptions of trigonometrical stations, NAI, OS 54/13/1, loose sheet

triangulation. The index diagram or outline plot for a complete parish (as in fig. 23, Laraghbrien) marks later stages of the same process but these are still working documents and are not yet oriented grid north. In the case of Laraghbrien, some of the local observation points decided upon by Corporal Andrew Bay, RE and C.A.R. Robert Harris in November 1836 were 'Maria villa' (T^1), 'Real Park' (C^1), 'Maynooth tree' (R^1), 'Greenfield' (N^1), and 'Newtown' (P^1) (fig. 23). While the skeleton plots (fig. 23) show the mesh of ever smaller triangles that was thrown over the country parish by parish, it is the OS six inch sheet, whether as a 'fair plan' (fig. 20) or as a published map (fig. 21) that places these points within their topographical context.

The first set of written instructions, issued in 1824, required the local surveyors to record the bearings and distances required to 'lay down the work', and also 'such other bearings and distances as may afford the best and most convenient checks for ascertaining and ensuring the accuracy of the work itself, and for connecting it with the surrounding portions of the survey'.[81] The measured detail of the six-inch survey was entered in 'content field books' and 'road field books' (NAI, OS 58 & 59);[82] these illustrate how at the most intimate level the theodolite triangles were subdivided into chain triangles, while the roads and all other topographical features (buildings, rivers, antiquities) were measured by chained offsets, that is, carefully-measured perpendicular lines drawn or 'offset' from the sides of the triangle (the main line) to the outlying point or feature. Figure 25, the content field book numbered 3186 (NAI, OS 58 A 56) illustrates the irregular plot pattern with which the surveyor was faced at the entrance to Maynooth castle and the college, and the immense investment of time and energy that was required on the ground. The rough diagrams in the content field books are not exactly to scale, 'but merely bearing some sort of resemblance to the lines measured on the ground for the purpose of showing at any period of the work their direction and how they are to be connected and also of eventually assisting in laying down the diagram and content plot'.[83] It is the measurements that matter most to the employees in the district or divisional offices who are plotting the returns. The content field books are carefully indexed and cross-referenced; the small area covered by figure 25 (from book 3186) can be placed in context by referring to other pages in this notebook ('to page 24') or in another notebook in the series ('page 12. 3163').

The work advanced on a parish-by-parish basis. The common or shared boundary between parishes was the point at which the information relating to each parish, and recorded in separate 'content field books' (such as fig. 25), had to be reconciled. This is accomplished in the OS 'common plots' (NAI, OS), sheets on which the same boundary is plotted from different books, and any necessary reconciliations are made. Small discrepancies did appear but were dealt with in a transparent manner. In the common boundary plot of Whitechurch and Bodenstown with

81 Colby, *Instructions*, art. 25. **82** Two separate books were used up to 1831, after which both boundaries and topographical measurements were entered in the same book. **83** Frome, *Outline of the method of conducting a trigonometrical survey*, p. 37.

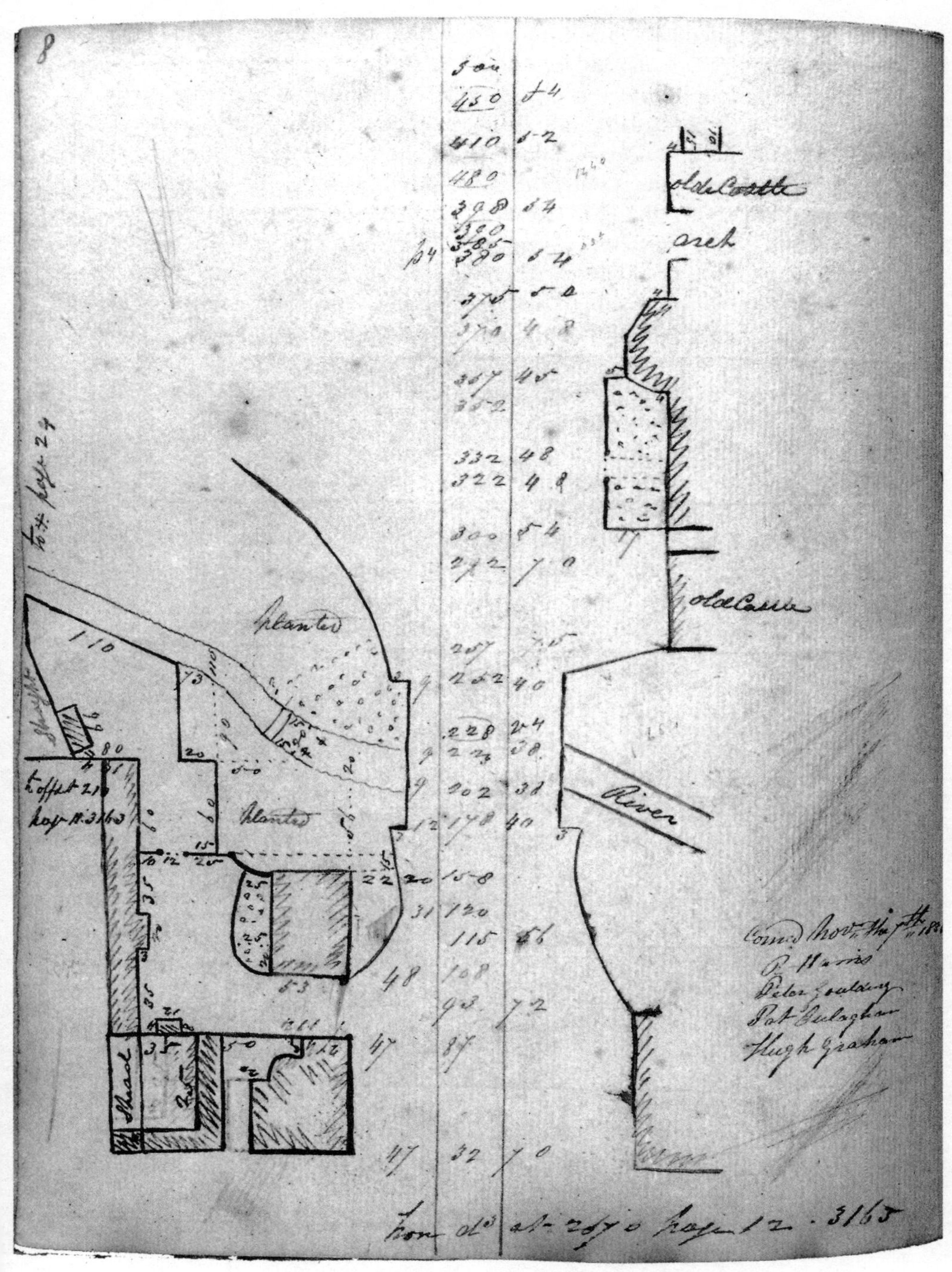

25 Parish of Laraghbrien, county of Kildare, content field book, 7 Nov. 1836, NAI, OS 58A/56, book no. 3186, p. 8

Straffan and Castledillon, for example (NAI, OS 103 15/21) where the boundary runs along the river Liffey, the resolution of one such difficulty in June-August 1837 can be followed from a succession of dated annotations from the two officers involved, George A. Bennett and Henry James, both with the rank of lieutenant, Royal Engineers: 'Both [lines] correct, one is the edge of quarry the other water, G.A.B.' 'Both correct one is water the other bank'. 'I have had this examined on the ground and mine is right, H.J.' 'Correct, G.A.B.' 'Red is correct, G.A.B.' 'This is a different bdy. Mine agrees with sketch and remarks, H.J.' 'Difference referred to Mr Griffith. Copy of letter of reference herewith sent, G.A. Bennet, 2 Aug 1836'. Such notes bear testimony to the high standards to which the Ordnance Survey aspired at all stages of the project, while they are also a reminder that for all its technical brilliance the OS was primarily reliant on the diligence and good judgment of its officers whether in the field or desk-based in the Phoenix Park.

The founding purpose of the six-inch survey was to establish the exact square area or 'content' of each townland (later extended to each field), which were of course highly irregular in shape and size. As explained to the trainee officers, 'the more minutely the triangulation is carried on, the easier and more correct will be the interior filling in'. The system was continued down to the local level (the tertiary triangulation), with all manner of 'conspicuous permanent objects' such as churches, mills, big houses, canal bridges, used as points, and the sides of the triangles measured with chains. Above all, 'the old vicious system of measuring field after field and then patching these separate little pieces together should be most carefully avoided'.[84] Once the additional triangles were drawn, this left only the awkward area between the sides of the triangles and the townland boundary to be calculated, which was done on paper. As a further check, the area of the entire parish was computed in a similar manner but using larger triangles. If the surveying and mathematics were correct then the sum of the areas or content of the separate townlands in that parish would exactly match the area of the parish computed independently as a whole. If the figures did not add up, then one had to return once more to the figures supplied by the field surveyors, and see what might be astray, or require to be surveyed again. Each commanding officer was responsible for the area within his own triangles, and could be called on at any stage (as each document is signed and dated) to explain any errors or contradictions. There were therefore a number of checks and counter-checks built into the system which acted as a control on quality, while as an organisation the Survey maintained a passionate commitment to accuracy for its own sake, aiming always to render this 'great national work' 'as perfect as possible, before giving it to the public'.[85]

84 Ibid., p. 31. **85** Clarke and James, *Ordnance trigonometrical survey of Great Britain and Ireland*, p. xiv.

THE BOUNDARY SURVEY

The Ordnance Survey of Ireland, launched in 1824, was to be a topographical survey; however, for the new maps to be useful for valuation purposes, the determining and marking on the ground, and on the maps, of townland boundaries was central to the project. The Spring Rice report of 1824, which set up the townland survey, could proffer only the vaguest 'leading general principles' in this respect, conscious that the 'survey must necessarily take precedence, the basis of the valuation being obviously the proposed maps of counties, baronies and parishes, divided into their respective townlands'. As the 'filling up of the triangulation proceeds', it was hoped that 'tracings of these skeleton maps may be furnished', to prevent the boundary work delaying the survey proper, or vice versa. However, three important principles 'to be adhered to and combined' were established in 1824:

1) A fixed and uniform principle of valuation applicable throughout the whole work, and enabling the valuation not only of townlands, but that of counties, to be compared by one common measure.

2) A central authority, under the appointment of government, for direction and superintendence, and for the generalization of the returns made in detail.

3) Local assistance, regularly organized, furnishing information on the spot, and forming a check for the protection of private rights.

Under the Boundary Act, 1825, the first task, that of demarcating the boundaries (or meres) of townlands, parishes, baronies and counties was the responsibility of a new body, the Boundary Commission, which commenced work on 27 August 1825. This statutory body, under the direction of Richard Griffith, had the duty of establishing the key boundaries through a lengthy process of public consultation, marking them on the ground, and pointing out this information to the officer in charge of the Ordnance Survey party. A handbook was produced by Richard Griffith, titled *General instructions for the guidance of the district and assistant boundary surveyors, in the performance of their respective duties* (Dublin, 1832); this contained lengthy verbal instructions as well as drawings of the abbreviations and symbols to be employed on the sketch maps (pl. 2, reproduced as fig. 26). Special care was to be taken at 'each angle or principal bend in the mearing', and 'at each change of feature in the boundary'. Under the 1825 act, the surveyors were empowered 'to enter upon any land or grounds through which any such Surveyor shall deem it necessary and proper to carry any Boundary line for the purposes of this Act, at any time or time whatsoever, until the marking out of any reputed Boundary Line shall be completed'. They also enjoyed sweeping powers in setting up ground markers, compelling the collectors of the county cess and grand jury rates to place their local

86 *An act to repeal an act of the last session of parliament relative to the forming tables of manors, parishes and townlands in Ireland, and to make provision for ascertaining the boundaries of the same,* 6 Geo. IV, c. 49 (5 July 1825).

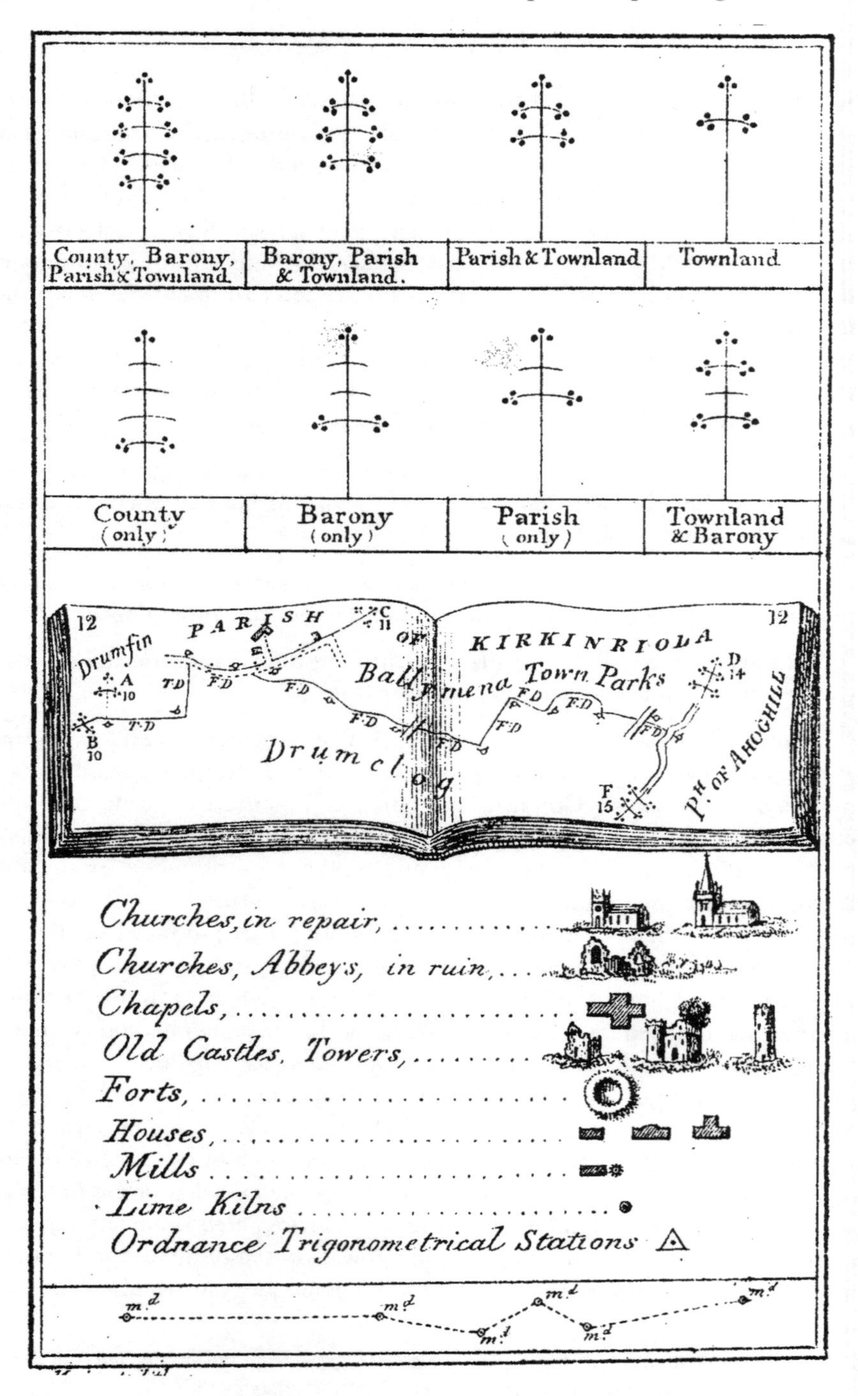

26 Town of Maynooth, parish of Laraghbrien, boundary remark book, 21 Sept. 1836, NAI, OS 55A/56, no. 400, p. 25

knowledge at the service of the survey, and in short to require the fullest co-operation, under pain of heavy fine, of all subjects of His Majesty's realm.[86] The boundary commissioners had the task of determining the boundaries, which would then be recognised in law; the job of the Ordnance Survey was merely to record them, in words and with diagrams, and to enter them on their six-inch maps. Records were thus kept by both authorities (the Boundary Commission and the Ordnance Survey). Figure 27, the boundary remark book for the town of Maynooth, for example (NAI, OS 55A, no. 400, p. 25) was kept by Corporal Andrew Bay, RE, who, with his meres-man, John Hewitt, and labourers Michael O'Hare and James Creegan, at 8a.m. on 21 September 1836 perambulated the boundary between the 'lands of Maynooth' (to the north of the college) and 'College lands', the newly-defined townland consisting of the demesne or grounds of the Royal College of St Patrick. The boundary runs along the centre of the river (CR) which is roughly parallel to the mail coach road; it then leaves the river to follow the line of the college wall (marked 'c.wall' on the fair plan, fig. 20). The enormous body of 'boundary remark books' (NAI, OS 55) which resulted were carefully indexed, cross-referenced and dated; regrettably many of these have not survived. These remark books formed the basis for the 'boundary registers', created by the Boundary Commission, and later part of the Valuation Office collection; the sketch maps (1826–41) created by the Boundary Commission are also part of the Valuation Office archive, and are deposited in the NAI.

Colby, in his evidence to the 1824 committee, asserted that the additional time required to include townland boundaries on the maps would not be very considerable, 'provided the boundaries were set out'. But therein lay the problem. In an 1828 statement 'on the progress in the perambulation of the boundaries', Richard Griffith bemoans the incompetence of the collectors of the county cess, who were the persons named as meresmen in the act and required to point out the boundaries or meerings of baronies, townlands and parishes to the officers of the Boundary Commission. The boundary men did not yet have the OS maps, but had to rely upon their own sketch maps, and the willingness of local proprietors to allow them 'to make traces of the boundaries from the maps of their estates, which traces being afterwards taken into the field by the Boundary Surveyors, would serve as a check on the meersmen, should they, from ignorance or design, point our a wrong boundary'.[87] It was a laborious and tedious job; however, by 1831 the Boundary Commission had covered half the country, and was far in advance of the Ordnance Survey.[88]

Standardising placenames was more especially the charge of the Ordnance Survey, but the boundary surveyors could not proceed at all without deciding what each townland they were to delimit was to be called. Griffith instructed each assistant in 1832 'to ascertain, as well as he can, the true pronunciation of such name, and its meaning, after which he is to adopt such orthography as will enable

87 Richard Griffith, *Statement of the progress in the perambulation of the boundaries of baronies, parishes and townlands in Ireland, under the 6 Geo. 4.c.99, 28 May 1828*, H.C. 1828 (420) v, pp 347–50. **88** *Report and statement of the expenditure incurred by the Boundary Department of the Ordnance Survey of Ireland*, H.C. 1830–31 (364), xiv, pp 3–24.

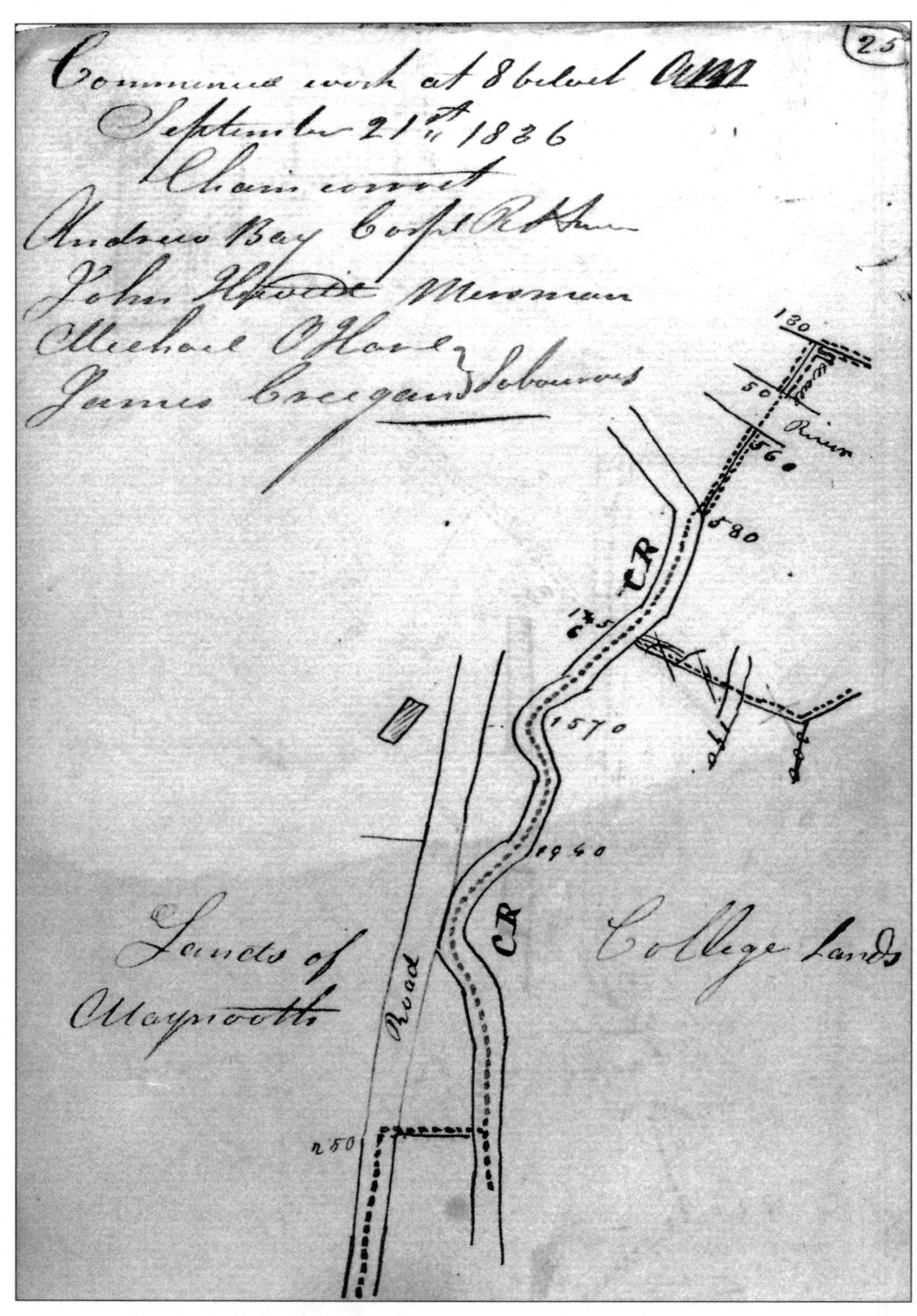

27 Legend for boundary sketch books, Richard Griffith, *General Instructions for the guidance of the district and assistant boundary surveyors, in the performance of their respective duties* (Dublin, 1832)

or oblige the reader to pronounce it correctly'.[89] Recognising the gross inadequacy of existing place-name lists, each assistant boundary surveyor was ordered to make out and forward to his superior two lists, one 'the same as that made out by the Barony Constable', but on the other 'the orthography is to be according to the manner in which the names are pronounced by the inhabitants of the Townland, or Parish'.[90]

The variety of local territorial units in existence in Ireland had been discussed by William Petty in his *Political anatomy of Ireland* (London, 1691), and again alluded to in the 1824 Spring Rice report setting up the townland survey. Griffith's boundary surveyors were now faced with difficulties that could no longer be evaded. The hierarchy of county, barony, parish and townland looks very tidy on paper, but there were countless local discrepancies. Some parishes stretched over two or more counties; 'in such case, the boundaries of such parish may be sketched in the same field book as circumstances shall require'. The problem with 'half barony' subdivisions was to be dealt with by treating each as a whole barony, 'and to be marked out as such'. Not every townland was part of a parish; those 'which pay neither tithe nor church cess are to be considered as extra parochial' and to be 'considered and marked out as belonging to that Parish by which they are wholly, or for the most part, surrounded'.[91] A single townland was sometimes divided by a parish boundary; this too could not be tolerated, but 'each of the parts into which it is divided is to be considered and marked out as a whole Townland in, and of, the Parish in which such part is situated'. There were also 'ancient demesnes' that have been 'returned by the High Constable as being assessed separately from, and independently of, any other division'; these are now to be marked out as townlands in their own right.[92] In each case the solution was pragmatic, and took immediate effect; the cases of 'College lands', 'Glebe' and 'Carton demesne' in Maynooth are typical examples of what happened.

There was nothing speculative or theoretical about the demarcation of boundaries. The surveyor's job was 'the marking out or defining of boundaries by marks, mounds, stakes, or posts of stone or wood, and by lockspitted lines'. They were to be marked literally on the ground, recorded in field sketch maps, described verbally in the field books, and, in due course, pointed out to the Ordnance Survey officer for inclusion in the six-inch sheets. In his published *General instructions* of 1832, Griffith explains in detail the shape, height, diameter and depth of these markers, and the materials of which they were to be constructed. Stone posts, for example, sunk two to three feet into the ground, were reserved for defining parish, barony or county boundaries. The lockspitted line was 'a narrow trench cut in the ground for the purpose of defining any undefined boundary which requires to be marked out'; it was to be at least 12 inches wide at the top, and nine inches deep.[93] Only lines that were already carved into the landscape, however temporarily, were to be transferred into the field sketch.

89 Richard Griffith, *General instructions for the guidance of the district and assistant boundary surveyors in the performance of their respective duties* (Dublin, 1832), p. 25. **90** Ibid., p. 72. **91** Ibid., pp 10, 11, 15, 24. **92** Ibid., pp 11, 15.

Local historians need to be alert to the code employed for registering each boundary mark, mound and post in the field books (fig. 26), and to the instructions regarding the production of sketch maps. In the tracings, and in the field book, 'the positions of all churches, chapels, castles, houses, mills, lime-kilns, bridges or other remarkable objects, situated on or near the boundaries, are to be represented'.[94] In the final work, the assistant surveyor 'prepares from his notes a general hand sketch map of the parish, on each of the divisions contained, in which he notes the nature of the boundary, whether it be a stream, drain, earth, fence, wall &c., and mentions the part of the fence to be measured to'.[95] The abbreviations are noted in the 1832 *General instructions* (pp 21–2):

CD	centre of drain
FR	face of rocks
CR	centre of river or road
LS	lockspitted
CS	centre of stream
RH	root of hedge
CW	centre of wall
TD	top of ditch
FB	face of bank
QB	quick bed
FD	face of ditch
Md	mound

Colour coding enhanced the sketch maps: 'tinging with yellow' for barony boundaries (which were not also parish boundaries); 'edged on both sides with a tinge of green' where the barony boundary was also a parish boundary; 'light Indian ink for glebe land, bishop land light tint of purple'. All additions were to be made in red, alterations to be cancelled by cross lines in red, and the surveyors warned by Griffith 'never to cut or make erasure, under any circumstances', a warning that greatly enhances the historical value of these documents.[96]

PLACE-NAME RESEARCH BY THE ORDNANCE SURVEY

> It is scarcely necessary to remark, that a map is in its nature but a part of a Survey, and that much of the information connected with it can only be advantageously embodied in a memoir, to which the map then serves as a graphical index.[97]

While the Ordnance Survey was referred to as the 'trigonometrical survey' and was firstly a mathematical achievement, the proposed maps were but one part of the

93 Ibid., p. 9. **94** Ibid., pp 11, 23. **95** Griffith, *Statement of the progress in the perambulation of the boundaries of baronies, parishes and townlands in Ireland*, p. 349. **96** Griffith, *General instructions for the guidance of the district and assistant boundary surveyors*, pp 29–31. **97** Colby, *Ordnance Survey of the county of Londonderry*, p. 7.

project, as conceived by Colonel Thomas Colby, its first director, based in London. Colby considered that the maps should be accompanied by written information, 'statistical additions' that would clarify the origins of the placenames and other distinctive features recorded on the maps of each parish. It was this ambition for a maps-cum-memoir project that led the OS management to direct their officers in the field to collect a wealth of additional local information, well beyond notes on placename variants, their spelling and possible origin, which were most obviously required for authoritative map-making. Despite the failure of the memoir project – only that for the parish of Templemore, Londonderry was published at the time – the structures put in place by the OS at the start of the survey for the collection and careful recording of information, the importance given to scholarly research into placenames (the vast bulk derived from Irish language words but already heavily anglicized), and the meticulous archival systems of the OS ensure that the researcher has a wealth of information at the level of the townland and parish. The principal sources, as explained below, are the memoir materials (for the northern counties), the name books (MS, on microfilm and typescript copies) and the correspondence between fieldworkers and headquarters (the 'O'Donovan letters'). The maps themselves, as fair plans and published maps (figs. 20–21), are a vital source, allowing the researcher to determine which variant(s) were under active discussion and which form was eventually decided upon. The plots and content field books are also worthy of attention, as these were the spellings used in the field, prior to decision-making at a higher level.

The Memoir *project*

In Colby's *1825 Instructions for the interior survey of Ireland*, the requirement to collect additional information is outlined in a very general way:

> Each officer, employed in the districts, is to keep a journal in which he is to insert all the local information he can obtain relative to communications by land and water, manufactories, geology, antiquities or other matters connected with the survey. [art. 68]

The following year, in the annual report to the Master General for 1826, Colby listed the immense amount of extra material that will be collected in the course of the survey. In the process he exposed the presumption – commonly held – that Ireland's future prosperity lay in the discovery of mineral wealth, new manufactures, and improved communications, rather than in tackling the agricultural and land-holding crises that were so shortly to result in the catastrophe of the 1840s famine.[98] In his preface to what he bravely titled 'volume the first', *Memoir of the city and north-western liberties of Londonderry, parish of Templemore* (1837), Colby explained how the 'elaborate search for books and records required to settle the orthography of names to be used on the maps', when combined with the requirement to make

98 Ibid., p. 145.

'a geological examination', and the unprecedented opportunities offered by the organisational machinery of the Survey, afforded 'means for collecting and methodizing facts, which were never likely to recur'. In brief, 'a sufficient extension of the original orthographic inquiries, to trace all the mutations of each name, would be, in fact, to pass in review the local history of the whole country'.[99] Colby credited his assistant, Lieutenant Thomas A. Larcom, who was in charge at Mountjoy (1828–46), with the inspiration for expanding the survey, at a 'small additional cost', to 'draw together a work embracing every species of local information relating to Ireland'.[1] But Colby also had a personal interest in seeing the Londonderry memoir into print, which local history reveals (Tower Museum, Derry): he had married a local girl, Elizabeth Boyd, and lived at 8 Shipquay Street, Londonderry, for a short while. Larcom produced instructions *c.*1834 (NLI, MS 7550) on the gathering of memoir material, and most valuably, a list of 'heads of inquiry' under which the material could be arranged. This is reproduced as Appendix IV.

The scale and ambitions of the memoir project are clear from the Londonderry volume, the only one to be printed by the Ordnance Survey itself. It runs to 335 pages of massively detailed material, followed by appendices and plates. The range of scholarship and quality of production is breathtaking, including archaeological, architectural and natural history drawings, facsimile reproductions of early maps of the city (1600, 1625, 1689 Neville, and 1788), landscape views ('Londonderry from the south west bastion', by George Petrie), statistical tables, accounts of the annual Apprentice Boys and Orange Lodge commemorations of the 1688–9 siege, and even the musical notation for an unpublished tune 'appropriated to their service' titled *No surrender*, a melody which is likely to be of great antiquity, but 'adopted for its pleasing and mirthful fitness'.[2] But what is also immediately striking is its shapelessness (the 'heads of inquiry' proposed by Larcom prove impractical), the indigestibility of the lengthy transcripts from historical documents and reports of local societies, the overflow of historical material into areas headed 'modern', and the political discussion which must in this divided city fire claim and counter claim of sectarianism and partisanship. In the mass of facts and opinions, compiled by an impressive but unwieldy number of expert contributors (as listed in the preface), there was much of 'utility' to be found, but extracting this material required assiduous effort on the reader's part.

The memoir project was first promoted as complementary but subsidiary to the map-making, merely taking advantage of the unprecedented opportunity presented by the Survey's fieldwork to compile valuable information. In practice the Londonderry memoir exposed this branch as consuming far more military time and state funding than had ever been intended. An inquiry was held in 1843 which recommended the complete separation of the geological survey from inquiries into historical, statistical and antiquarian matters.[3] Although the Londonderry volume

99 Ibid., p. 8. **1** Ibid., pp 5–6. **2** Ibid., p. 197. **3** Andrews, *A paper landscape*, pp 144–77; *Report of the commissioners appointed to inquire into the facts relating to the Ordnance memoir of Ireland*, 1844, Cd. 527, xxx.

remained the sole published example for many years, materials had been collected for much of the northern counties. Some enterprising local history societies published sections of interest to their membership, and a major publishing project was carried through by the Institute of Irish Studies, Queen's University Belfast, in association with the Royal Irish Academy, in a 40-volume series from 1990–98. Despite the lack of close controls on what the officers collected, most of the memoir material is of good quality. Some memoirs leave several headings blank, in others the zeal of the contributor reaches to countless additional pieces of information ranging from the nonsensical to the invaluable. For example, a brief extract from the parish of Kilmacteige (Kilmactigue) in County Sligo, submitted by Lance-Corporal Henry Trimble in December 1836 shows how fact is intermingled with the strong opinions or prejudices of the observer:

> *Farms and condition of the people*.... in general the inhabitants of this parish are extremely neglectful, thus complain in general of the lands being too high, rents etc. but they themselves in no shape or other are acquainted with farming. In general the farms are set by the bulk on the most moderate terms, particularly on the south east side of parish. Notwithstanding, they live in the most deplorable state for either clothing or raiment. This poverty principally originates from the making of poteen whiskey, which is the general practice with the entire of the farmers. The grain is destroyed. The wretches themselves drink as much of that liquor as would defray the expense of carhire in sending their oats to Sligo properly for sale...
>
> *Communications*: Bogs plenty in this parish. Roads in bad repair and many bridges required or pipes required. The civil engineer or rather county surveyor appears not to exert himself.[4]

An abrupt order from the master-general to Colby on 1 July 1840 brought the shutters down on the *Memoir* project: the Survey was 'to revert immediately to its original object under the valuation acts'. Existing materials could be arranged but no more collected, in effect ending the project.[5] Essential place-name research did of course continue – the Topographical department continued its work for a further two years – and by this route valuable observations on local society and economy, gathered in the field still found their way (via the 'name books', dealt with below) into the Mountjoy headquarters. Miscellaneous papers connected with the production and publication of the Templemore memoir are held in the National Archives of Ireland (NAI, OS 95), and filed under the following headings: source materials, graphic materials, manuscript drafts, and printer's proofs. Memoir material relating to places other than Templemore are also stored in the National

4 *Ordnance Survey memoirs*, xl, Counties of South Ulster, p. 190. **5** Letter from Ordnance to Colby, 1 July 1840, OS registered correspondence 4739, quoted in Andrews, *A paper landscape*, p. 165.

Archives of Ireland (NAI, OS 96); the documents are arranged within counties, under the headings 96/1 source material and 96/2 graphic materials.

OS parish name books

Concern for place-name derivation and orthography can be found in the very first *Instructions* issued by Colby in 1825, as he outlines the procedure to be followed in the field:

> The name of each place is to be inserted as it is commonly spelt, in the first column of the name book: and the various modes of spelling it used in books, writings &c. are to be inserted in the second column, with the authority placed in the third column opposite to each. [art. 34]
>
> The situation of the place is to be recorded in a popular manner in the fourth column of the name book. [art. 35]
>
> A short description of the place and any remarkable circumstances relating to it are to be inserted in the fifth column of the name book. [art. 36].

The exact way in which these instructions were followed is illustrated in figure 28, an extract from the field name book for the townland of Lehinch, parish of Kilmannaheen, barony of Inchiquin, County Clare. The process of approving Irish place names country-wide, and how each should be spelt, requires exceptional linguistic skills as well as a first-hand knowledge of the local topography, folklore and local history, an intimate familiarity with early Irish annals, martyrologies, literature and genealogies, and above all an ability to engage with local people of all classes, creeds and ages, using the various dialects of Irish as well as English. The field name book (fig. 28) in the first instance, provided the raw material on which a team of Irish language scholars set to work.

As with all Ordnance Survey records, the field name books are arranged on a parish/barony/county basis. Each name book opens with its own index (to the townlands within that parish), and is signed and dated by the officer responsible. These are the namebooks delivered to John O'Donovan as he and a small number of colleagues worked on the orthography of placenames throughout the length and breadth of the country. He annotated the field books, posted them back to the OS headquarters, and specified which volumes he now required. The paper is thin, and wherever the ink shows through it can be difficult to decipher in full. The original namebooks are held by the NAI (OS 88); due to their fragility and unique importance, the NAI has undertaken a major microfilming project. At the time of writing counties Antrim, Armagh, Carlow, Cavan, Clare and Cork are available on microfilm (series MFP 1/25), and so are the only volumes available for public consultation. There are typed transcripts which are useful but incomplete proxies, as they do not cover all 32 counties. From 1926, Fr Michael O'Flanagan produced a typescript of the OS letters in the Royal Irish Academy (explained below) and of the bulk of the parish name books which were then in the Phoenix Park.[6] The Hayes catalogue, under Ordnance Survey, lists the namebook typescripts held by the NLI; twenty counties only are included, and not all these counties are complete.[7] Nevertheless, pending the completion of the microfilm project in the NAI, the typescripts will continue to be of major importance. Sets or partial sets of the O'Flanagan typescripts have been located in the Royal Irish Academy, the British Library, PRONI and the Irish Folklore Commission collection in UCD, as well as in the NLI, and there may be further sets elsewhere.

The first Irish language scholar employed by the Ordnance Survey to research the form of Irish placenames was Edward O'Reilly (Aodhgáin Ó Rathaille). He was succeeded by John O'Donovan, to be joined by Patrick O'Keeffe, Thomas O'Connor, and Eugene and Anthony O'Curry, each an outstanding scholar of Irish (along with Latin and Greek). Thomas Larcom, later director of the OS, can be creditied with the placename policy adopted by the OS, whereby the name approved for final engraving should be that version which came nearest to the original Irish form of the name.[8] To this end, the Ordnance Survey invested remarkable care and resources in retaining, indeed rediscovering, early Irish placenames, as the parish name books, and associated Ordnance Survey letters (known as the O'Donovan letters, although he was not the author of all, as explained below) will illustrate.

In the case of the townland of Lehinch (fig. 28, NAI, MFP 1/29, Clare 89), it is in the parish of Kilmannaheen (coded B 331), and covered by the OS parish name books numbered 668 (the village of Laheensy, p. 55) and 669 (Lahench Lodge, p. 73). Under 'orthography' several variants of the name are listed (Lahensy, Lehinch, Leagh inchy, Lahinchy, Leahench, Lahench), and the authority for each. The research behind even minor names appears systematic and thorough, though some places quite simply had more extensive written source material, such as being named in the Down Survey, the inquisitions of the 1600s, and surviving Irish language manuscripts. Before any decision was made on the name of Lehinch, existing county and other maps were consulted (Pelham's county map), other written sources held by the local grand jury ('the county book', 'the barony book'), a resident clergyman was interviewed (Archdeacon Whitty), and some local officials (Mr Thomas Flanagan, assize [or asst.] collector, and A. Callinan, sheriff's bailiff), as well as a Mrs Mary Moran whose letter on the matter was dated 17 November 1840. Elsewhere, oral evidence was gathered from local landowners and their agents along with schoolteachers. In the case of Lehinch, there were so many contradictions between the names offered locally that the first column, 'received name', was left blank until John O'Donovan entered the spelling 'Lehinch' (signed JOD). He notes that the name is derived from the Irish '*Leith-inse*, half island or

6 Herity, *Ordnance Survey Letters: Donegal*, Introduction, pp xx–xxii. 7 Typesecripts are held for counties Armagh, Antrim, Donegal, Carlow, Cavan, Dublin (city and county), Kerry, King's County, Leitrim, Limerick, Longford, Londonderry, Mayo, Meath, Monaghan, Roscommon, Sligo, Tipperary, Westmeath, Wicklow; not all are complete. NLI, Ir 92942.
8 Andrews, *A paper landscape*, p. 122.

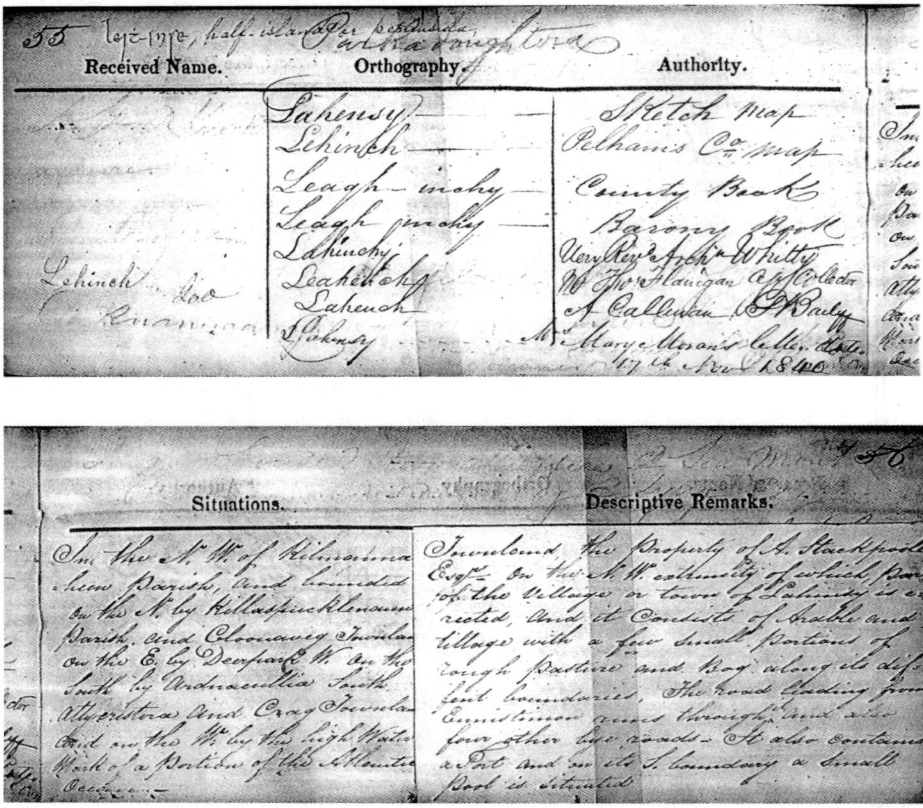

Received Name.	Orthography.	Authority.
	Lahensy	Sketch map
	Lehinch	Pelham's Co. map
	Leagh-inchy	County Book
	Leagh inchy	Barony Book
Lehinch	Lahinchy	Very Revd. Archd. Whitty
	Leahenchy	Mr. Thos. Flanigan Cy. Collector
	Lahench	A. Callenan & Daly
	Lahensy	Mrs. Mary Moran's letter dated 17th Nov 1840

Situations.	Descriptive Remarks.
In the N. W. of Kilmannaheen Parish, and bounded on the N. by Killaspucklenane Parish, and Cloonavery Townland on the E. by Deerpark W. on the South by Ardnacullia South Attyeristora and Crag Townland and on the W. by the high Water Mark of a Portion of the Atlantic Ocean.	Townland, the Property of A. Stackpoole Esqr. On the N. W. extremity of which part of the Village or town of Lahinch is erected, and it Consists of Arable and tillage with a few small portions of rough pasture and Bog along its different boundaries. The road leading from Ennistimon runs through it and also four other bye roads. It also contains a Fort and on its S. boundary a small Pool is situated.

28 Extract from OS Parish name book, townland of Lehinch, parish of Kilmannaheen, county of Clare, OS book 668, pp 55–6

peninsula', and made his decision accordingly. In the nearby townland of Lissatunna, 'a good [house] with offices, at present occupied by a herd', has the received names of 'Lahench Lodge' (on local authority) and 'Lahensy' (according to a letter of 16 November 1840, by A. Stacpoole Esq.); this premises was renamed 'Lehinch Lodge' by John O'Donovan. The OS decision in this case was not taken up locally, and the spelling 'Lahinch' was adopted as the standardised version locally and most importantly by the Post Office. Only the census authorities and the Valuation Office adopted the official OS orthography immediately; the failure by the Ordnance Survey to speedily publish its decisions in list or directory form led to the continued use of different forms of the same name by some official bodies and local authorities.[9] The continued discrepancy between maps and road signs in

9 Ibid., pp 125–6.

Lehinch/Lahinch is but one example of how the OS decision could be ignored locally.

Under 'situations' (fig. 28) the exact location of the place in question, relative to surrounding townlands and to natural features, is noted, so that there can be no error in ascribing the name so painstakingly investigated to the correct unit. Under 'descriptive remarks' valuable information may be gleaned, though this depends on whether or not the field officer made the effort, or indeed had the good fortune to meet with well-informed and co-operative residents. The townland of Lehinch (fig. 28) was the property of A. Stackpoole Esq. (elsewhere spelt Stacpoole) 'on the N.W extremity of which part of the village or town of Lahinsy is erected', hinting at the ambiguous standing of this settlement, not yet deserving of the name 'town' (the term adopted by the census commissioners in 1841 'for every assembly of contiguous houses', numbering at least twenty).[10] There is also a general note on the use to which the land is devoted (arable, tillage, 'some portions of rough pasture'), antiquities ('contains a fort') and the road network ('the road to Ennistimon runs through it, also four other bye roads'). Where the OS encountered particularly voluble informants, and chose to record their observations, the 'descriptive remarks' column can be intriguing. There were few to compete with the Clare landlord, James O'Brien Esq., who managed to have many additional observations made about the districts which he owned, as in his comments on the heavy taxes to which his townland of Kells was subjected (county cess and tithes), boasting of the newly-opened school with eight boys and six girls in attendance, and the abundance of trout, eel and perch in Lough Cullana on the NW boundary (NAI, MFP 1/25 reel 28).

The headings of the OS parish name books (fig. 28) are repeated in the case of street names in towns and cities, where the variant spellings, the authority for each (including commercial directories, Paving Board, name boards, registry records), derivation of name (occasionally) and exact location are noted. In the case of Dublin city (1837), for example, some of the brief notes are wonderfully illuminating. Mountjoy Square North, 'street wide and clean, macademized footways, gravelled and flagged, lighted with gas, inhabitants private families, members of the legal profession' contrasts with nearby Purden (or Purdon) street, the location of 'Institution for worm complaints for the poor, street narrow and dirty paved, middle footways lighted, houses of various qualities with dirty reres, inhabitants provision dealers, huxters, a great number of destitute poor, dissolute and depraved characters in both sex, laborers'.[11]

The consultation process had its limitations. The educated English-speaking clergy and minor gentry, who dominate the list of OS informants, were not necessarily the best authorities in the matter of placenames, and the interests of the OS were confined to administrative units and a very limited number of other features. Within urban areas, the Ordnance Survey confined its orthographic exertions

10 For definition see 'General introduction', *Report of the Census Commissioners (Ireland)*, 1841, vol. I: reports and tables, p. vii. **11** Dublin city namebook, no. 50, 1837; from typescript, Department of Irish Folklore, UCD.

largely to the streets proper. In the case of Dublin city, for example, the innumerable minor or back lanes, alleyways and inner courts are ignored in the name books. Nevertheless, considering that many of the officers of the Royal Engineers Corps were English-born and untrained in what was really historical research, they did at least produce a basic document, from which in due course a name and spelling was determined upon, and entered on the manuscript 'fair plans' at the Mountjoy headquarters, from 1828 onwards. And the details which they recorded under 'descriptive remarks' provide tantalising glimpses of a world, townland by townland and street by street, which was to be so soon refashioned under the catastrophic famine of 1845–47.

The O'Donovan letters

Etymology, which is the investigation of the derivation and original signification of words, required firstly that the local pronunciation of the name be heard, which obviously could only be done in the field. The first stage of this task was undertaken by the OS officers on the ground and recorded in the parish name books, as illustrated in figure 28. From 1834 to 1843, John O'Donovan, as did some fellow workers, travelled the length and breadth of Ireland, perambulating each county in turn, starting with County Down (counties Antrim and Tyrone were completed before this type of field work was determined upon), and finishing in County Kerry. As already explained, the relevant parish name books were forwarded to O'Donovan, annotated and returned to the OS. But alongside these name books the scholars employed by the OS corresponded regularly with headquarters on the matter, from practically every corner of Ireland. Twenty-nine of the thirty-two counties are covered in the Ordnance Survey or (more accurately) O'Donovan letter series; there are no letters from Counties Antrim, Tyrone or Cork. The absence of Cork letters is regretted, as the Ulster counties have rich memoir materials to compensate for that loss. The manuscript OS letters are deposited in the Royal Irish Academy; in their unpublished form they have provided historians with an incomparable entry point into the rich background to placename tradition in Ireland. They have also been a most important primary source for placename publications, include P.W. Joyce's three-volume work, the *Origin and history of Irish names of places* (Dublin, 1869). Local history scholars have made transcripts of sections of interest to their own constituency (for example, counties Westmeath and Wicklow),[12] but more importantly, from 1926 Fr Michael O'Flanagan produced a typescript of the letters in the Royal Irish Academy (29 counties), and of the bulk of the name-books as explained above.[13] As with the name books, several bound copy sets were made and can be consulted in the National Library of Ireland, the Royal Irish Academy, in

12 Paul Walsh (ed.), *The placenames of Westmeath, Part 1: the Ordnance Survey letters referring to the county, abridged and edited* (Dublin, 1915); Christian Corlett and John Medleycott (eds), *The Ordnance Survey letters: Wicklow, from the original letters of John O'Donovan, Eugene Curry and Thomas O'Connor 1838–40* (Wicklow, 2001). **13** Herity, *Ordnance Survey Letters: Donegal*, Introduction, pp xx–xxii.

the Irish Folklore Commission collection in UCD, in the British Library, in several universities, county libraries and other repositories, though some of the sets are incomplete. Michael Herity, MRIA, has recently edited the letterbooks relating to counties Donegal (Dublin, 2000), Dublin, Meath and Down (2001), Kildare (2002) and Kilkenny (2003). This scholarly series, which is to be continued, includes an introduction to the personnel and challenges of each county-based operation, and is comprehensively indexed, at least for family names and place names. It also reproduces some of the maps which were important to the field workers.[14]

The O'Donovan letters provide invaluable 'behind the scenes' insights into the work of the Topographical Department, which was headed by the archaeologist and artist, George Petrie. The letters provide a sense of the scale of the undertaking and the care invested, and an unrivalled introduction to Irish society in many of the least visited Irish-speaking places on the island before the great famine of the 1840s remodelled these places for ever. In Clonmany, County Donegal, for example, O'Donovan claims 'I never heard Irish better spoken, nor experienced more natural civility and innocence than in that very secluded and wild parish'.[15] Even as he pursued his place-name inquiries in the field, John O'Donovan was conscious of the role the Ordnance Survey was playing in the inexorable movement towards modernisation. Though himself a servant of that body, he had mixed feelings about the long-term effect of 'opening up' these remote areas, as he described in the case of Boylagh in west County Donegal, left alone 'until the red coated sapper dragged the chain along thy dreary glens and o'er thy azure peaks'.[16]

O'Donovan and his co-workers collected an immense amount of local folklore, history and other information that they expected to be incorporated into a *Memoir*, on the style (if less ambitious in scale) of the Templemore/Londonderry volume published in 1837. As already discussed, this material will be found in the parish name books, letter books, and also in the files of original *Memoir* materials held by the NAI (OS 95: Templemore memoir, OS 96: Memoir, places other than Templemore). But the primary purpose of these scholars was to listen carefully to how the locals pronounced the placenames of their districts, struggle with the etymology, and come up with a compromise solution in which the recurring elements (such as *kill, gleann, turas*) would be standardised but still maintain the sense of the original. There were some elements that could be standardised country-wide (*drum, bally, caher*) but the vagaries of the Irish language, in its several dialects, and differences in local pronunciation, meant that the most that could be hoped for was to standardise elements on a fairly local scale. In Inishowen, for example, O'Donovan grappled with whether to adopt the anglicisation *bin* or *ben*, before deciding that 'as the Latin is *pinna* and the Welsh *pinn*, the former seems the more analogical and I have accordingly adopted it'.[17] O'Donovan greatly respected many of his Irish-language informants, but was anxious above all to consult written authorities, the early Irish annals, histories, law tracts, genealogies and other manuscripts (few had

14 *Ordnance Survey Letters: Donegal*, 24 Oct. 1835, pp 111–13. **15** Ibid., 23 Aug. 1835, p. 9. **16** Ibid., 12 Oct. 1835, p. 90. **17** Ibid., 30 Aug. 1835, p. 18.

been published by the 1830s) for confirmation of names of persons and places and their varying orthography. Dublin-based members of the OS topographical department carried out specific searches on his behalf in the libraries of Trinity College, the Royal Irish Academy, the Dublin Society, Marsh's Library and elsewhere. For the townland of Kells in County Clare, for example (fig. 28), the existence of a pedigree of the O'Briens in Trinity College Dublin in manuscript form is noted, alerting the local historian to a source that may be of interest. The correspondence which survives allows the researcher to move from the scholarly place-name research of O'Donovan and his colleagues to the final published six-inch maps and town plans and see which variant was eventually adopted. The Valuation Office maps, with their manuscript revisions, showing place-name changes, can then be consulted, to bring the story a step further.

Critique of the OS placename research

The Ordnance Survey parish name books are the starting point for serious local history enquiries into place-name origins, and an essential complement to the maps (manuscript and published) of the Survey. The O'Donovan letters and memoir materials, for those counties fortunate enough to have both, provide the researcher with an understanding of the processes behind the name-collecting, and the prejudices and sympathies of individual officers and employees. However, the Ordnance Survey archive, despite its volume and the undoubted scholarship behind the making of this record, preserves only a fraction of Ireland's toponymic heritage. From its foundation to the early twentieth century, the OS has been, by virtue of its mandate, heavily biased towards making placenames pronounceable to an English-speaking public. The setting up of separate OS authorities in 1922 did not advance the cause of Irish language advocates appreciably. However, an official 'Placenames Branch' was established as part of the OS*i* in the Phoenix Park in 1956–8 (now at 17–19 Hatch Street, Dublin 2) while a voluntary association, the Ulster Placename Society, had already been set up in the Department of Celtic, Queen's University Belfast in 1952. Major research on Northern Ireland townland names, edited by G. Stockman, has been published as *Placenames of Northern Ireland* (7 vols, Belfast, 1992–7). An overview of the work of the placenames branch of the Ordnance Survey is included in *The placenames of Ireland in the third millennium* (Dublin, 1992) edited by Art Ó Maolfabhail. Research has been carried out at a local level throughout Ireland with the support of various statutory and charitable bodies, such as the Cork Place Names Survey, now extended to County Kerry. Patrick J. O'Connor's *Atlas of Irish placenames* (Newcastle West, 2001) includes an introduction to placename study and a critical review of the work of the OS and boundary commission. Based on the townland index of 1861 and the index maps to the six-inch survey, he identifies 72 root elements in Irish placenames – such as bally, kill, town, druim, cnoc, cluain, lios, rath, gort, cor, cúil, ard – and maps their distribution. The cause of bilingual mapping was advanced in 1978 when the new 1:2,500 metric series inserted settlement names in Irish as well as in English; however, as it excluded all minor names such as street names, the number of names in Irish per sheet was very limited. More intense placename work has been published by Tim Robinson, in his celebrated maps of Connemara, the Aran Islands and the Burren.[18] His Connemara maps record hundreds of minor placenames within the townland framework, making it clear how much more there is to be noted beyond the OS record. Irish language placename maps have been published by Eoghan O'Regan of Léaráid, Dún Laoghaire, covering the Cork district *Coise Laoi* (1986), and Gaeltacht areas under the heading *Léarscáileanna Gaeltachta*: de na Rosa (1986), Corca Dhuibhne (1986) and Léarscáil Shligigh (1986).

Comparative work on Dún Chaoin, County Kerry and Kilcock, County Kildare, undertaken by Irish language scholar Mícheál Ó Dubhshláine, displays the wealth of placenames which field research at the local level reveals. Fields and ditches, boithríns and small patches of garden, crossroads, mountain paths and streams, every headland, cliff, bay and beach, each with its own name and local associations are most richly preserved in the Gaeltacht districts but still accessible elsewhere in Ireland. Nollaig Ó Muraíle's placename research over several counties reveals 'an astonishingly rich substratum of microtoponymy still surviving' in forms that are 'utterly undisguised and undistorted'.[19] Work undertaken by Ó Catháin and O'Flanagan on the townland of Kilgalligan in north-west Mayo is an early example of how fruitful research into minor placenames can be. While exploiting the Ordnance Survey name books and associated documentation to the full, the local historian needs to go beyond this, drawing especially on the local communal memory, and asking more probing questions. And the OS personnel, after all, followed more limited lines of inquiry and were then ordered on to the next parish.

THE VALUATION OFFICE MAPS

The maps produced in connection with the rateable valuation of property, known as the Griffith's or 'tenement' valuation maps, and their revisions, are perhaps the single richest island-wide cartographic resource for local history (see figs. 29 and 32). They consist of six-inch maps and town plans (most at scales of *c.*1:500 and 1:1,056), covering both rural and urban areas, which were supplied by the OS but annotated by the Valuation Office (VO). Essential to the interpretation of the VO maps is the written information on each plot, collected in manuscript field books, and in the case of the first edition only (*c.*1846–64), published as the *General valuation of tenements in Ireland*. When working with these maps therefore, it is necessary to understand the legislative framework surrounding their creation, the instructions which the surveyors followed in filling up both maps and field-books (termed 'housebook' for rural areas, and 'town book' for urban centres), the codes

18 Tim Robinson, *Connemara* (Roundstone, 1990); *Oileáin Árann, the Aran Islands* (Cill Rónáin, Árain, 1980) *The Burren, a map of the uplands of north-west Clare* (Cill Rónáin, Árain, 1977). **19** Nollaig Ó Muraíle, 'Settlement and placenames in P.J. Duffy, E. FitzPatrick and D. Edwards (eds), *Gaelic Ireland:* c.1250–1650 *land, lordship and settlement* (Dublin, 2001) p. 225.

employed, and the very practical ways in which both maps and written record (in the 'cancellation books') were periodically updated, by hand, before being superseded by new editions. The questions posed in Appendix 1 can be usefully applied here. There can be some confusion over the period 1830–44, as there were two distinct valuations in existence, hence the explanation which follows. In addition, the position of the ubiquitous Richard Griffith as head both of the townland valuation and its successor, the tenement valuation, and his role as author of handbooks to support the valuators employed by each survey, may add to the confusion.

The first general valuation, variously known as the 'government', 'first Ordnance', and Townland valuation (as the townland was the unit of assessment), was begun in 1830, under the act of 1826, and an amending act of 1836.[20] Larger houses were also valued, but the vast bulk of ordinary residences were not distinguished separately, limiting its usefulness for local history. As already noted in the discussion of the Ordnance Survey six-inch series, the 1826 act envisaged that a new and equitable valuation 'should be proceeded upon, in the several counties, baronies, parishes and divisions, as soon as the same shall have been respectively surveyed'. The Ordnance Survey was ordered to hand over to the central authority (the lord lieutenant or chief secretary) the completed map or plan 'from time to time when and as such survey shall have been or shall be made and completed, or when and as any such map or plan shall be required by the chief secretary to be transmitted as aforesaid'. The townland valuation therefore was to advance closely on the heels of the mapping project. And as the new scientific survey had abruptly halted the grand jury mapping projects, so too it was ordered that the new valuation replace all previous listings made for the purposes of calculating county cess charges and grand jury rates. From 1826 these local taxes, required for the upkeep of roads, bridges, the county jail, and so on, were to be raised on the basis of an independent, uniform and closely regulated system. Land values were to be calculated with reference to 'fixed standards', 'certain general average prices' specified in the act (wheat, oats, barley, potatoes, butter, beef, mutton and pork), while houses were to be valued at two thirds of 'the sum or rent for which every such house respectively could be let by the year'. Entries, signed by the valuators, were to be entered in field books, based (as with the OS) on the parish.

The 1834 amendment clarifies exactly how the written and cartographic record were to complement each other, and the right of the public to inspect these records:

> The commissioners of valuation shall make out a copy of the field book of each parish, and shall make out a field map, showing the several portions of land which may have been separately valued, and distinguishing and numbering the same so as to correspond with the said field book and the said commissioners shall deliver such copies of the field books and maps, attested by his signature, to the Treasurer of the county, who shall keep the same in his

20 7 Geo. IV c.62 (26 May 1826); 6 & 7 Wm. IV c.84 (17 Aug. 1836).

> office, there to remain open to public inspection, and any owner or occupier of land within such parish shall be at liberty to make copies thereof or extracts therefrom without fee or reward.[21]

The Poor Relief (Ireland) Act of 1838, setting up the poor law unions, required that property be assessed to raise the poor rate, which went to support the union workhouse and allied concerns.[22] The poor law guardians appointed their own local valuators, to carry out this separate poor law valuation, rating each occupier of property separately on the basis of the 'net annual value of the land to let'. It was completed about 1842, and did not have access to the new OS maps. In fact, these valuators do not appear to have relied upon or produced any cartographic materials, but operated from lists produced at local or union level.

The confusion and contradictions that resulted in running two parallel systems – the townland valuation (which used the new OS maps) and the poor law valuation (which relied on lists alone) – led to the 1844 Select Committee, which recommended that there should be only one valuation for all the purposes of local taxation, that it should distinguish each tenement separately according to the 'net annual value of land', and that the townland valuation should be discontinued. Legislation followed in 1846, and in 1852 the act (15 & 16 Vict. c.63) authorised one uniform valuation of Ireland according to tenements. Both the townland valuation and the poor law guardians' valuations were effectively scrapped by the 1846 and 1852 legislation. The townland or first survey had not reached the counties of Cork, Dublin, Kerry, Limerick, Tipperary and Waterford at that stage, while the major urban centres of Cork, Dublin, Drogheda, Limerick and Waterford were similarly overtaken by the new developments.[23]

The new uniform, general valuation of Ireland, titled the 'Tenement Survey' or the 'Primary Valuation', popularly known as Griffith's Valuation, commenced in earnest in 1846 and was completed in 1864. It was to be an ongoing project, with funds earmarked for annual revisions, although it was hoped that 'revaluation costs' would be modest, considering the extent and quality of the work accomplished in the initial tenement valuation.[24] Much had been learned from the earlier government (or townland) survey, and the continued presence of Richard Griffith at the helm ensured that the experience gained in the earlier survey would be built upon in this much more ambitious project. *Instructions to the valuators and surveyors appointed under the 15th & 16th Vict. Cap.63 for the uniform valuation of lands and tenements in Ireland* were published by the General Valuation Office, Dublin in 1853 (and reprinted in the parliamentary papers for 1882).[25] The 1853 *Instructions* repeats

21 4 & 5 Geo IV, c.55 (13 Aug. 1834), p. 420 par. VIII. **22** 1 & 2 Vict. c.56 (1838). **23** Richard Griffith, *Return of the whole cost of the townland and of the tenement valuations of Ireland during the last three years, specifying the cost of the revision in each county and year separately*, 14 May 1866, H.C. 1866 (300) (300–1) lx. **24** Ibid. **25** Copies of the Instructions issued by the late Sir Richard Griffith in the year 1853, under the provisions of the 15th and 16th Vict. c.63, to the valuators and surveyors acting under him in making the Tenement

verbatim large sections of the guidebook produced by Griffith for the earlier (townland) survey (editions of 1833, 1836 and 1839);[26] by the 1850s he had developed a massively sophisticated but ultimately transparent and reasonably fair system of land valuation. The *Instructions* were widely distributed and copies of various editions have been located in the NLI, the RIA, the Russell Library and TCD. There were objections, but the overall system was widely, if grudgingly, accepted 'on the ground'.

In his comprehensive *Instructions to the valuators and surveyors* (1853), Griffith provides mathematical formulae and tables to be used in the complex task of estimating the value of houses and lands. He also provides specimen maps and forms (Appendix 5), and instructions on how exactly the maps are to be labelled. This manual is essential reading for scholars using the field books, published record, and Valuation Office maps in local studies. The introduction to the Valuation records held in PRONI by Trevor Parkhill (*Ulster Local Studies*, xvi, no. 2 (1994), pp 45–58) is well worth consulting by scholars working on VO materials for any place in Ireland, distinguishing clearly between the various valuations (and the Belfast revaluation of 1900–6), the legislation upon which they were founded, and the resulting records. The typescript catalogues to the Valuation Office materials held by the National Archives of Ireland are shelved in the reading room (alongside the OS catalogues).

The VO map adds in field subdivisions where necessary, and identifies the proprietor of each unit, annotating the six-inch OS first (or later) edition in coloured ink (as in fig. 29). Brackets show where very small or irregular plots were joined to a larger unit and valued with it. Each fieldbook was to be made out 'with the utmost precision', the notes 'should be clear and adequate' and the reference letters and numbers, in all cases, were to correspond exactly with those on the map (art. no. 21). Three major distinctions were made: land ('ground used for agricultural purposes only') was separated from 'house and offices' (used for residence only) and other buildings ('more particularly described as "brickfield", "brewery" etc.'). These distinctions were followed through by affixing reference numbers and letters to each and every unit marked on the map. An urban example can be seen in the VO map for Waterford city (fig. 31), where housebook no. 6, 'calculated 19 May 1848', records the dimensions of each numbered unit (length, breadth, height), then 'quality letter', followed by the calculation of valuation[27] and finally 'observations'.[28]

The legal definition of 'tenement', and its delineation on the map, was the first task. According to the *Instructions* of 1853, a tenement was 'any rateable hereditament' held for any term, tenure or agreement not less than from year to year. This meant that a single person could hold several tenements, or one tenement could be held by

Valuation of Ireland, *Return relating to Tenement Valuation (Ireland)*, H.C. 1882 (144) lv. **26** *Instructions to the valuators appointed under the 7 Geo. IV Cap.62, I and II Will. IV Cap.51 and 2&3 Will. IV, Cap. 73, for the uniform valuation of lands and tenements in Ireland* (Dublin, 1833). **27** Column headings: number of measures (that is, units of 10 square feet), rate per measure, amount of items in £ *s. d.*, gross amount, valuator's estimate, yearly rent, value of garden, lease rent. **28** Housebook no. 6, 'calculated 19 May, G. Williamson', and signed by James Keenan, NAI, VO 5.3404, shelf Z 6 4.

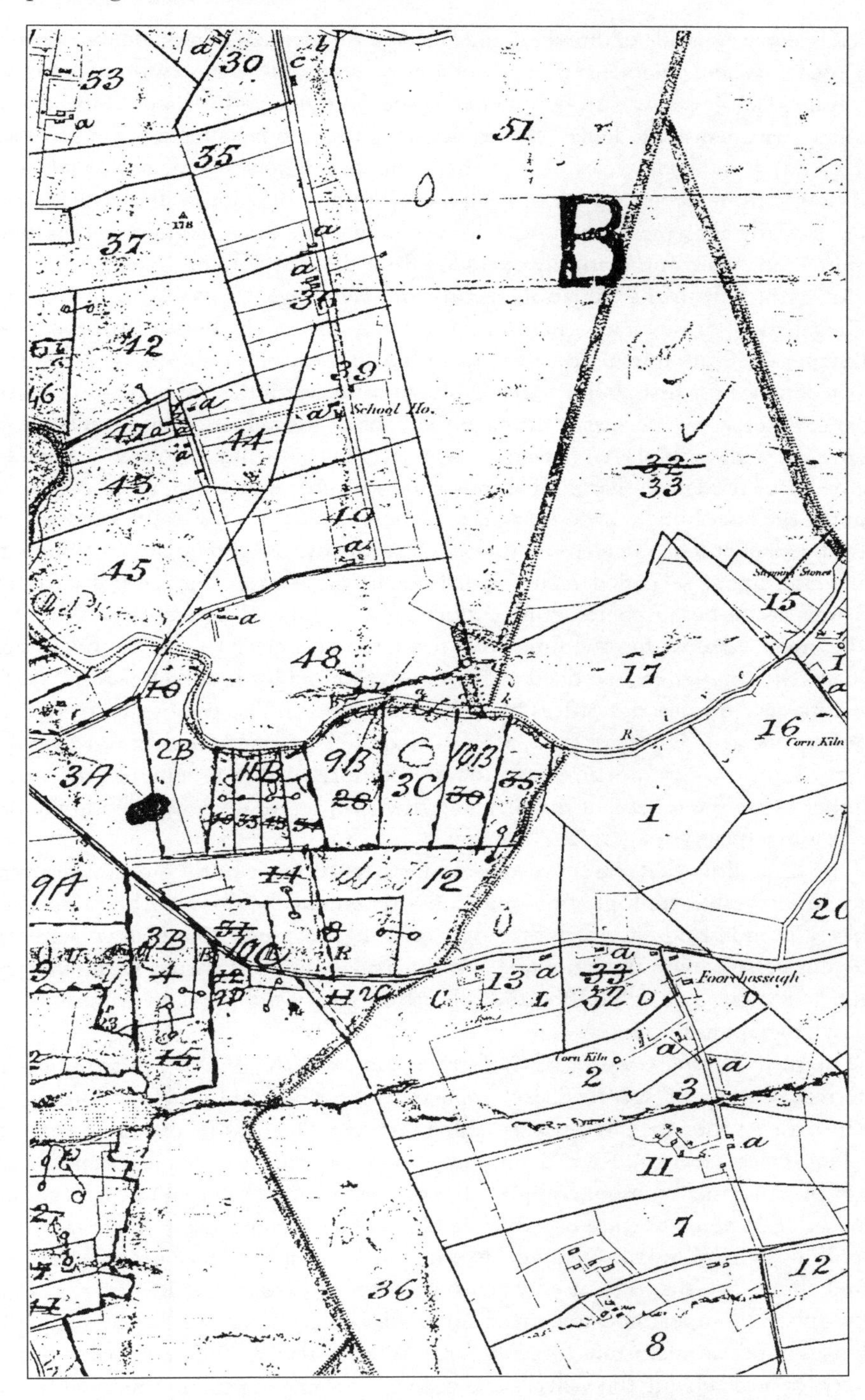

29 Rural valuation map, Beihy, VO six-inch, County Leitrim, sheet 20 (1842)

several people. As a rule of thumb, 'distinct receipts on payment of rent indicate distinct tenements'. Where tenements were let for a single crop, or to a 'succession of persons', the immediate lessor was to be entered as the occupier, and for any house 'let in separate apartments or lodgings', the immediate lessor is to be entered as the occupier (art. 12, 34). It is therefore clear that the huge numbers who occupied rooms, cottages and houses from week to week or from month to month, without the security of a lease or yearly agreement, or those who were living in a house conditional on their employment, were automatically excluded from the first column, 'occupier'.

An extract from the Valuation record for Glenveagh Castle, Gartan, County Donegal (fig. 30) illustrates the methodical way in which the valuator worked, following Griffith's instructions: 'the main house or dwelling should be measured and accounted for first, then its several returns, afterwards the offices' (art. 191). An enlarged sketch was necessary to make the basis of the valuation absolutely clear. Here each wing or block was given its own letter (a to m) and calculated separately, the better or residential wings calculated at 1*s.* per cubic foot, down to 3*d.* per cubic foot for the boat house, the 'estimated total' being adjusted only at the last stage, in this case 'deduct for difficulty and non-occupation' by 40%. Frequently a tenement was made up of several detached portions; in these cases 'the several parts or portions are to be numbered consecutively, as 5, 6, 7 &c.' Where cottagers' houses and gardens were within the limits of a farm, the farmer's house and offices were to have 'the *italic* letter *a* prefixed to the number of the lot in which it is situated', the cottagers' houses and offices by b, c, d, &c. (art. 26). The use of superscript or index letters (as 1^1, 1^2, 1^3, or 2^1, 2^2, 2^3) was advocated 'where more than one original quality lot, or where portions of such are comprised in a tenement', the main number being the tenement number, and the small figure to show 'the number of the original quality lot' (art. 27).

The difficulties that the surveyor could expect to meet in the poorest districts were innumerable: multiple tiny tenements, tenements held jointly but a portion sublet, common land for which there was no immediate lessor but many occupiers, different persons with the same first and last names, different parties with different rights to the same stretch of bogland, all matters which did not lend themselves to neat cartographic solutions.

The most complex aspect of Griffith's system was the division of agricultural land into 'quality' or 'sub-lots', each separately described and valued according to its 'intrinsic' rather than 'temporary' value (art. 48). This was recorded on the OS six-inch sheets (1:1056). Each 'quality lot', commonly less than 30 acres and made up of one or (usually) more complete fields, was intended to comprise land of the same value. Where a section of a field had land of a different quality that was to be noted on the map in words (rather than making it a separate lot); index figures, as noted above, could also be used. Farms were not individually valued; the second stage involved superimposing the farm boundaries on the quality lot maps to calculate a net annual value for each farm. Where farm boundaries changed, the process of recalculating the valuations would be a simple matter (in theory at least).

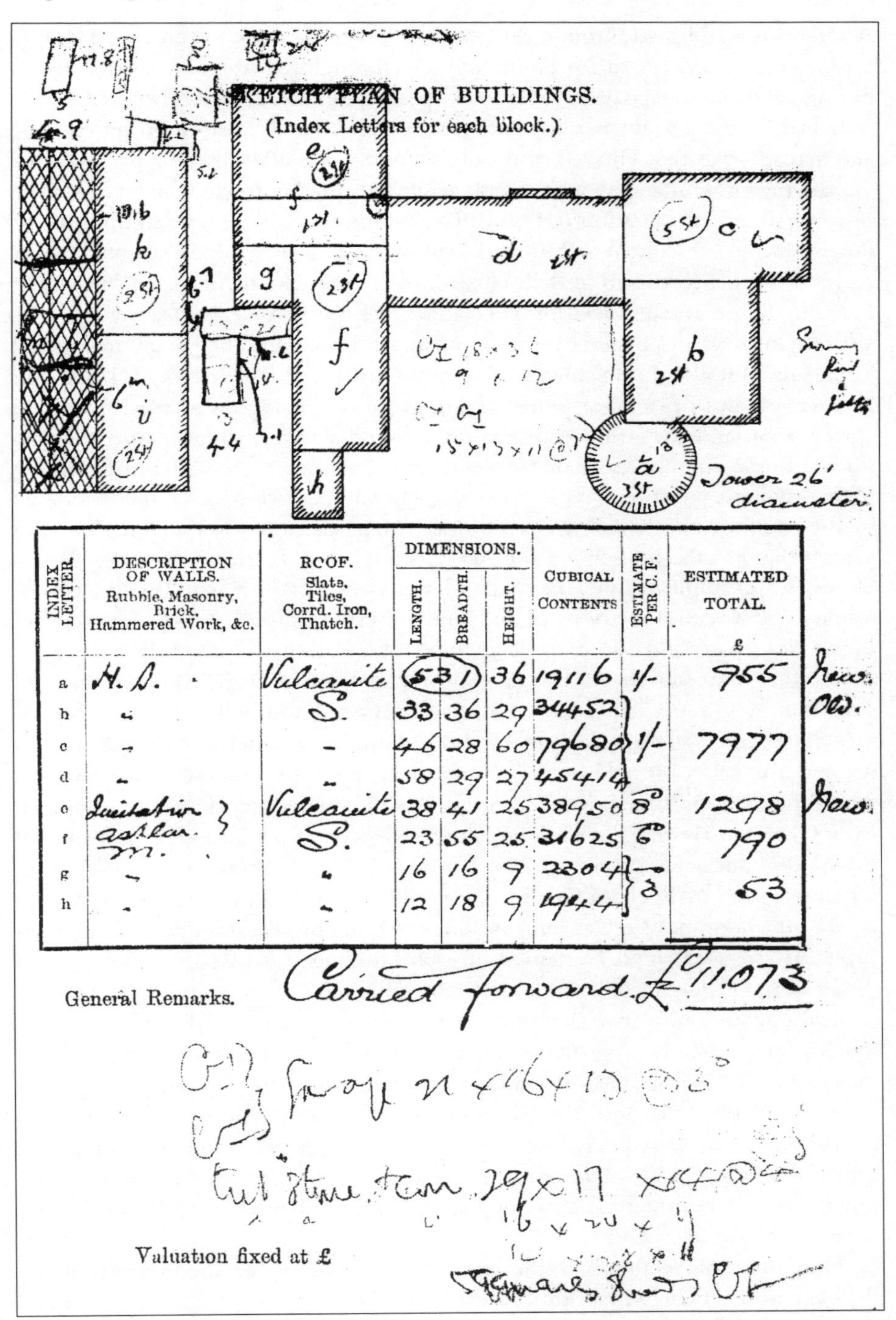

SKETCH PLAN OF BUILDINGS.
(Index Letters for each block.)

INDEX LETTER	DESCRIPTION OF WALLS. Rubble, Masonry, Brick, Hammered Work, &c.	ROOF. Slate, Tiles, Corrd. Iron, Thatch.	DIMENSIONS. LENGTH.	BREADTH.	HEIGHT.	CUBICAL CONTENTS	ESTIMATE PER C.F.	ESTIMATED TOTAL. £	
a	H. D.	Vulcanite	(531)		36	19116	1/-	955	New.
b	–	S.	33	36	29	34452			Old.
c	–	–	46	28	60	79680	1/-	7977	
d	–	–	58	29	27	45414			
e	Imitation ashlar M.	Vulcanite	38	41	25	38950	8d	1298	New.
f		S.	23	55	25	31625	6d	790	
g	–	–	16	16	9	2304	3	53	
h	–	–	12	18	9	1944			

General Remarks. Carried forward £11.073

Valuation fixed at £

30 Glenveagh Castle, Gartan, County Donegal, VO

A scale was established through this complex process, ranging from class 1, prime soils, average price: 30*s.* to 26*s.* per statute acre, through to class 12, cultivated moors or bogs, with an average price as low as 1*s.* per statute acre. The calculated value (the scale price, subject to further refinements), based in a rural situation on land quality and acreage, was first entered, and only secondly was allowance to be made for 'modifying circumstances', such as distance from a market town, which could raise or lower the final valuation (art. 113). These figures could be revised upwards or downwards in subsequent valuations, usually at 10-year intervals. In the manuscript revisions for the townland of Beihy (parish of Cloone, barony of Mohill, county of Leitrim), figure 29, the valuation per statute acre was revised downwards in 1869, with the new figures pasted over the outdated columns and the pencil note added 'there are no houses in this townland worth £5. 0. 0 a year' (NAI, VO OL.4.2698).

The system of placing a value on buildings was analogous to that of valuing agricultural land: an estimate of the intrinsic or absolute value of the building (linked to the amount spent on its construction, varying directly with the solidity of the structure, age, state of repair and cubic capacity), a price which is then modified by the circumstances which govern house lettings (art. 172). In the Waterford city extract (fig. 31), the housebook for numbers 2–4 along Patrick Street (near Broad Street), for example, records the quality letter 1C+ meaning first class in terms of solidity ('built with stone or brick and lime mortar'), but third class quality in terms of age and repair ('old but in repair'). Under 'observations' for no. 1 Chapel Lane is noted 'this is an old chapel but in good repair is situated in rather a backward situation', nos. 4–6 Chapel Lane Little are 'very old and bad' and 'set to several weekly tenants', while after nos. 5 and 6 (renumbered 7 and 8), 'schoolhouse', is entered 'this school is held by the monks and supported by subscription ~~national school~~ free school Patrick Murphy a monk is the owner', succinctly recording an early Christian Brothers' school, which was for a short while under the National Board of Education. Appendix V is a summary of the buildings' classification scheme created by Richard Griffith and used by the Valuation Office in both field books and housebooks. Capacity or 'cubical content', the number of measures (units of 10 square feet) contained in each building was based on the external dimensions of height, breadth and length. The 'tabular value' used for turning cubical contents or 'square measures' into rateable valuations depended on the 'quality letter' already assigned, so that a 2B house ('medium, slightly decayed, but in repair') of 50-square measures, and eight feet high, is valued at 18*s.* (unadjusted, town calculation). As with the valuation of land, the 'quality letter' assigned to buildings was calculated strictly in accordance with the rules, and any addition or subtraction in the value of the house (such as 'unusual solidity and finish', situation relative to the commercial core, or named 'deficiencies') was to be made in a separate column (art. 239).

The process through which the maps were updated in an ongoing manner by the Valuation Office, with additions made in different coloured inks, signed and dated (in the field books), ensure that each map is a unique record, including manuscript detail not available elsewhere. However, the maps must be used in conjunction with the written materials: the manuscript field books (the terms 'housebook' used for rural areas, and 'town book' for urban centres), the 'cancellation books', and the published valuation. In the process of keeping the valuation record up-to-date, the order that all errors were to be cancelled, not erased (art. 336) is especially useful, ensuring that matters such as the renumbering of houses, and changes in street naming (and spelling) can be traced, important evidence for which there is otherwise no record. As with the Ordnance Survey, the valuation mapping of Ireland was a milestone in the cartographic history of Ireland, a country-wide map-based property record which could at last supersede the Down Survey. While the quality of the maps and field notes is widely accepted by local historians and historical geographers, it is well to caution that there are documented instances of valuations which were vehemently disputed (as reported to parliament), and the field workers employed by the Valuation Office worked to punishing schedules for low wages, as contemporaries observed.[29] The possibility of appealing a valuation, and the periodical revision of all valuations however ensure that any serious errors in the first tenement valuation are likely to be picked up, a regulatory mechanism that applies to very few historical sources, cartographic or otherwise.

LARGE SCALE OS AND OTHER PLANS

The six-inch scale worked well on a country-wide basis, but a larger scale was needed for towns and cities to allow individual properties to be distinguished. The starting point in ascertaining what large-scale OS plans are available for any particular town at what scale, whether published or not, and date of edition and revision, is the OS town plans index, part of the online 'OS series list' in the National Archives of Ireland (Appendix III). This index and explanatory notes also acts as an excellent entry point into the processes behind large-scale OS mapping. An accessible and authoritative list (see online databases, http://www.nationalarchives.ie/) it clarifies for researchers what maps were published for their town, at what scale, covering how many sheets, and their OS reference numbers; armed with this knowledge, the published maps may then be followed up in the local or county library. Only the pre-publication materials such as 'fair plans' and content plots may be consulted in the NAI, the final published six-inch sheets are not available to researchers here. Researchers working on Northern Ireland towns should also consult Gertrude Hamilton, *A catalogue of large scale town plans prepared by the OS, 1828–1966 and deposited in PRONI* (Belfast, 1981). This is a union list, with towns in alphabetical order, and includes valuation ID, reference, date, town, scale and number of sheets.

The Valuation Office was the driving force behind the creation of large-scale OS maps of urban areas. From 1833, the scale of five feet to one mile (1:1056) was adopted for valuation town plans, after a period of experimentation; the many

29 *Civil Service Gazette*, 29 Sept. and 27 Oct. 1855.

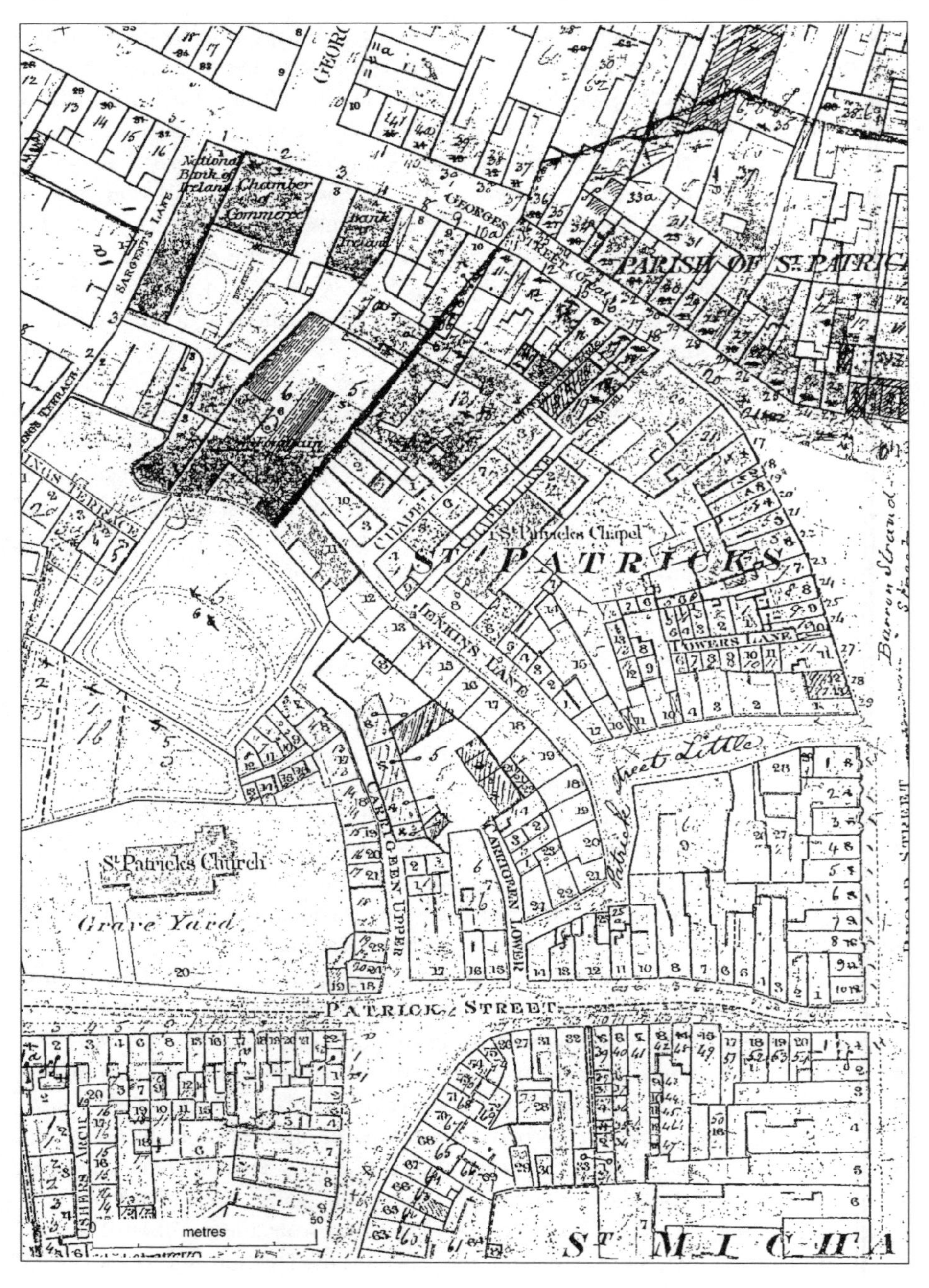

31 Urban valuation map, VO, 1:1056 (five foot), Waterford, (1851).

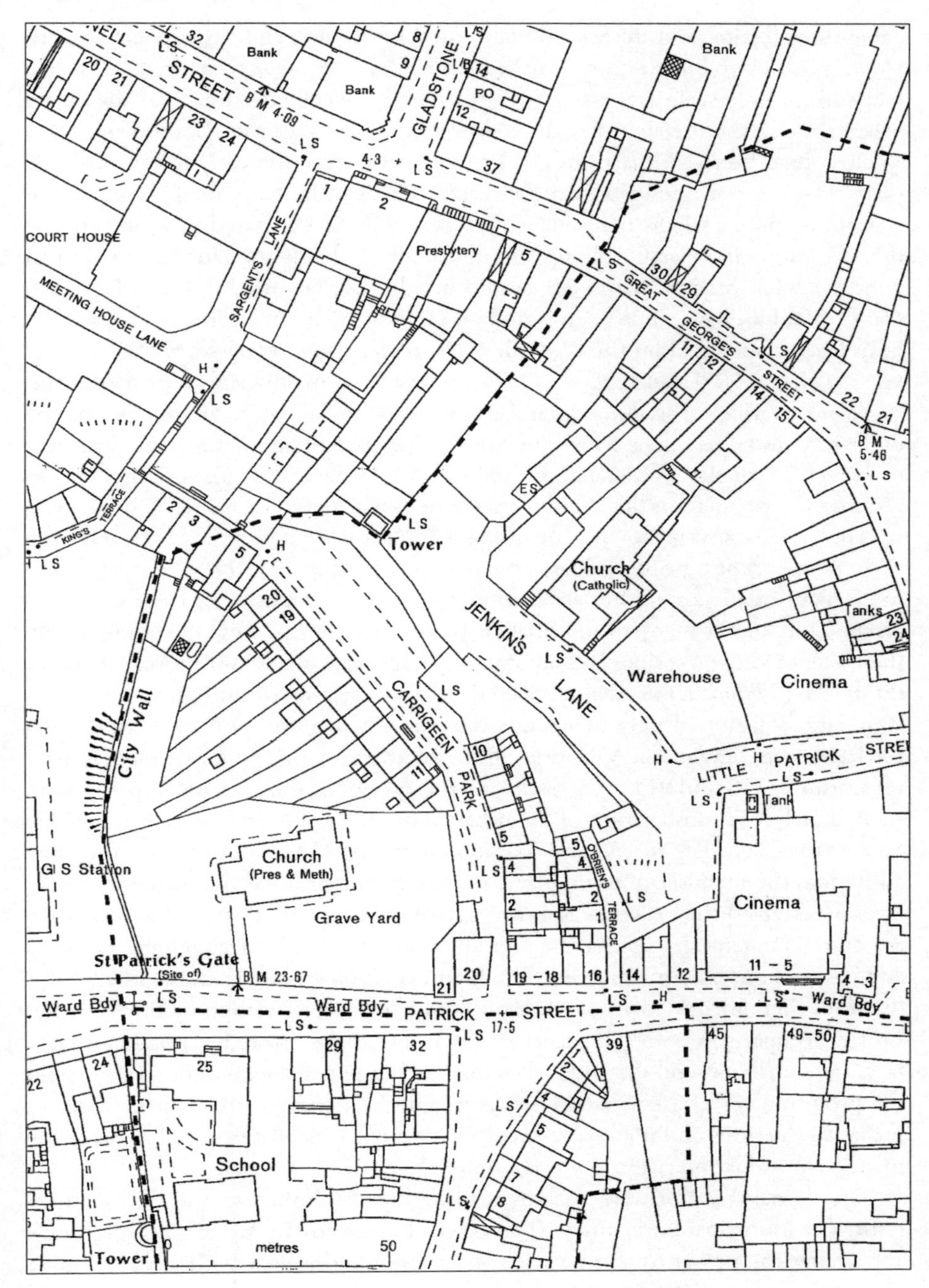

32 Waterford city, OS, 1:1000, Waterford (1985)
Mapping by Ordnance Survey Ireland, Permit no. APL0000304 © Government of Ireland

exceptions to this, and the varied scales used for manuscript town plans 1830 to 1848, are displayed in Andrews' *History in the Ordnance map* (1974), p. 27. The five feet to one mile scale (1:1056, fig. 31) proved an excellent choice for the larger urban centres; the metric scale adopted by its modern successor (1:1000, fig. 32) is so close that direct comparisons can be made. However, the early five-foot sheets (1:1056) were not intended for publication. The 'Castle sheet' for Dublin, which was put on public sale in 1840 out of professional pride, generated such interest that the OS authorities were obliged to publish all 32 sheets for Dublin city and to consider what other urban areas should be likewise favoured.[30] The 1847 town plans of Dublin have since been issued on CD-ROM by Eneclann Ltd., edited by Seán Magee, included on the disk titled *The 1851 Dublin city census, Chart's index of heads of households* (Dublin, 2001). The OS move from manuscript to printed form, and from a variety of scales to a standard of 1:1056 and 1:500, is complicated by the budgetary restraints under which the Survey operated, obliging it in some instances to survey at smaller ('older') scales than it would have wished, and, in 1872, to increase the population threshold for separate town surveys from 1,000 to 4,000.[31]

The OS manuscript town plans (NAI, OS 140) were completed between 1830 and 1842; excepting only Belfast city, which is held in PRONI, the manuscript town maps of all 32 counties are held in the National Archives of Ireland, Bishop Street, Dublin. For many of the smaller towns, whose maps were never published, these coloured manuscript plans are the only large-scale maps ever produced. Work on the *Irish Historic Towns Atlas* revealed that many early Valuation Office town plans were largely if not entirely independent of the OS manuscript ones, even though the latter were drawn with Valuation Office needs in mind. The *IHTA* relies heavily on both the OS and VO town plans for the production of the principal map in each fascicle, a reconstruction of the urban landscape as it stood as close as possible to the year 1840. The town plan examination traces (NAI, OS 142) have the same content as the manuscript town plans, and where the original OS manuscript plan has not survived they can act as a surrogate. However, due to their fragility (they are on tracing paper), they may be consulted only by special arrangement with the archivist responsible for this collection. Figure 33 is an extract from the manuscript five-foot sheet for the town of Maynooth which was overwritten by the Valuation Office; comparison with the earlier six-inch series (figs. 20 and 21) makes it clear how much more useful the large-scale map is for the management of property and the provision of infrastructure. The complementary information, in the form of printed valuations and manuscript revisions (in the cancellation books), place the manuscript valuation maps in a class of their own.

The demand for printed OS town plans may be linked to the mid-century reform of municipal government. The Towns Improvement Act of 1847 gave local authorities the power to name streets and to number houses, while in 1855 these authorities were invested with authority to change street names with the approval of four-sevenths of the occupiers of that street. Burdened with new responsibilities

30 Andrews, *History in the Ordnance map*, p. 26. **31** Ibid.

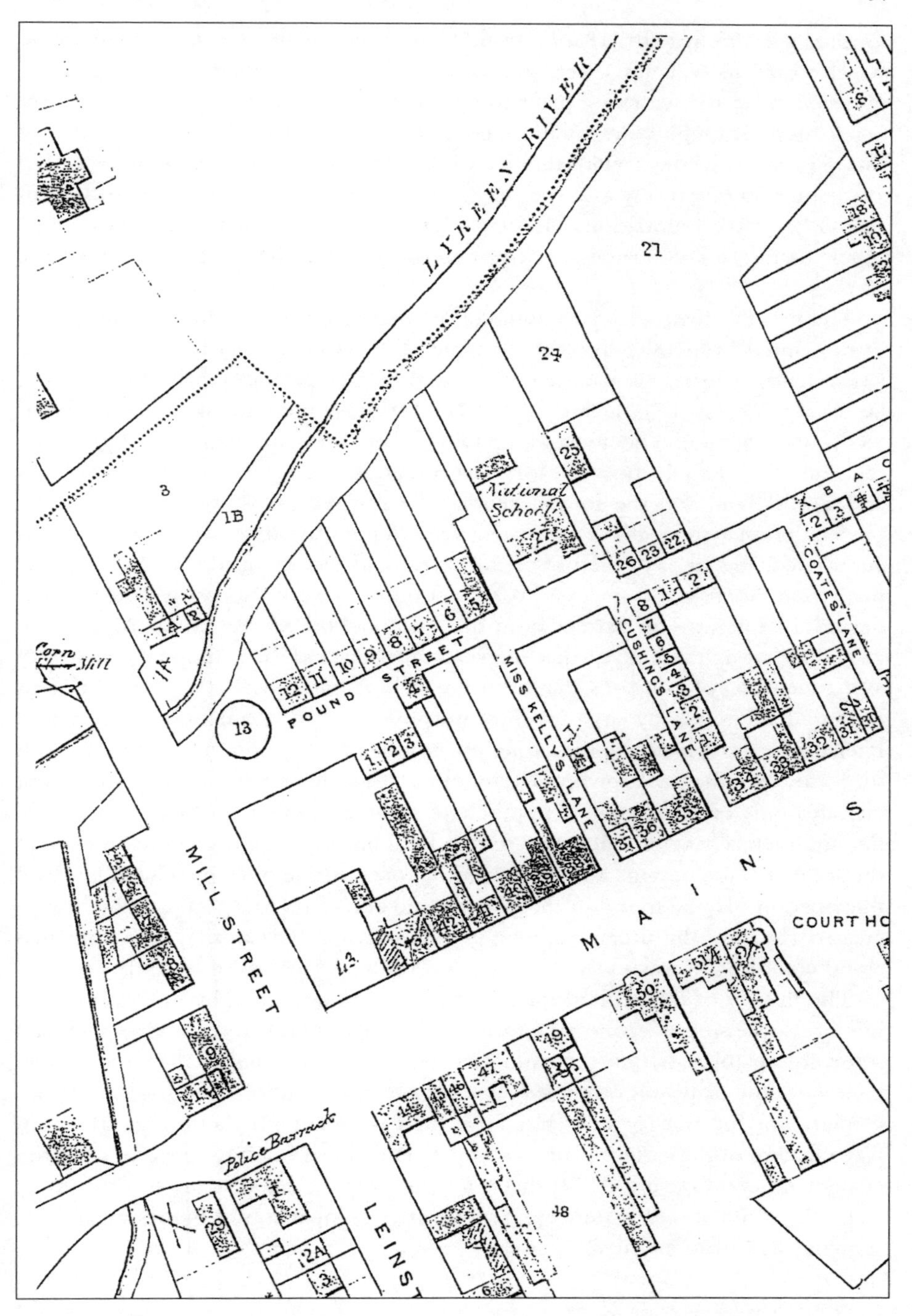

33 Valuation Office, Maynooth, 1:1056 (five foot), (1840)

for public health from the 1840s onwards,[32] local authorities needed maps at a scale which would allow drains, sewers and water mains, artisan housing and new abattoirs, libraries and street widening plans to be laid out and carried through. The water department of Dublin Corporation, for example, was quick to overwrite a set of the 1847 five-foot sheets with the line of the city sewers as they were inherited in 1850, most leading neatly into the river Liffey, and to plan their new network from this rather stark foundation. Municipal authorities were in fact empowered to supply themselves with large-scale maps under the Towns Improvement Clauses Act of 1854.[33]

The first OS town plans (excluding the special case of the Dublin castle sheet) were printed from 1847–1908 at the scale of 1:1056 (five feet to one mile). The larger urban centres were also revised within that period: Cork (1870, 1892); Waterford (1871, 1908); Belfast (1861, 1873, 1884–95) and Dublin city (1847, 1864, 1888 and 1907–9). The expanding Dublin suburbs also benefited by several revisions, as did a selection of Northern towns, including Ballymena, Portadown and Enniskillen, and the tourist centre of Killarney (1885, 1895). In 1855 the preparation of 1:500 plans of urban centres with populations greater than 4,000 was authorised, and in 1857 the population threshold was reduced to 1,000.[34] These maps were published from 1859 to 1894; a number were also revised during that period. Figure 34 is an extract from the 1:500 published sheet 4 for Maynooth, 1873, illustrating the high quality of ornamentation, and the additional detail, such as the inclusion of footpaths, that this large scale allows. Indeed, the 1:500 scale was criticised as excessively large for most purposes, with the Valuation Office under Richard Griffith preferring the earlier urban scale of 1:1056 (fig. 33).[35] It took seven sheets to cover the small town of Maynooth at the scale of 1:500; to map the central city area only of Londonderry required 36 sheets. Mapping at this large scale was not commercially viable. However, for the local historian the printed town plans at the scale of 1:500 provide a layer of information available nowhere else. Additional information may be found in the 'town name books' (NAI, OS 144) which were created as part of this urban mapping project; as is appropriate, they are catalogued along with the OS town plan index in the National Archives of Ireland.

The history of the printed town plans has been described by J.H. Andrews as one of a succession of economy measures. The early published town plans included some colour (blue, sienna, grey and carmine); this was ended in the 1870s.[36] The content of the plans was simplified after 1872 by the omission of minor detail from gardens. The interior layout of buildings, such as Maynooth Castle (fig. 34) which was a feature of the earliest plans was largely ended in 1881.[37] In terms of scale, the current 1:1,000 OS*i* and OSNI sheets for the larger urban areas are the successors to the five-foot sheets, manuscript and printed; comparison of figures 31 and 32, showing Waterford city, make this clear.

32 Prunty, *Dublin slums*, pp 69–71. **33** 10 & 11 Vic., c.34, sections 13–16. **34** Andrews, *A paper landscape*, p. 249. **35** Ibid., pp 250–1. **36** Andrews, *History in the Ordnance map*, p. 32. **37** Ibid., p. 32.

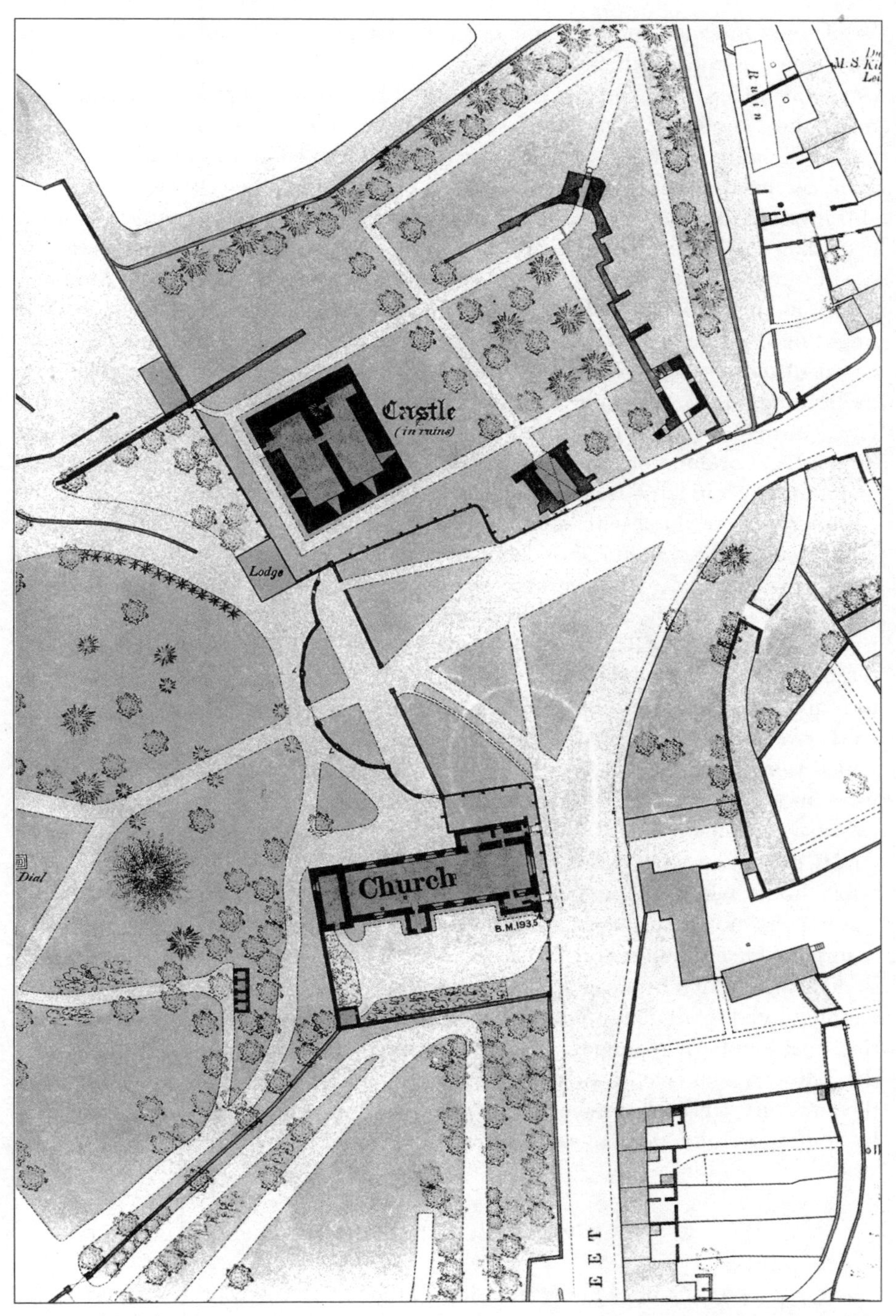

34 OS 1:500 Town of Maynooth, parish of Laraghbryan, barony of North Salt; drawn by G. Stuart, 1873, sheet 4 (previously numbered as V.16.14), NAI

While the Ordnance Survey dominated the town plan market in an overwhelming way from the mid-nineteenth century, at least a small number of other producers can be identified. Many private estates invested heavily in mapping, with large-scale maps showing individual holdings, boundaries and encroachments being of immense utility in taking stock of properties about to fall out of lease; this type of map has already been introduced under estate mapping. Fire insurance companies also invested in large scale mapping for the specific reasons that can be identified from their maps: ascertaining the quality of the party walls and what materials were used for roofing and stairways; identifying passage ways which might help or hinder access; and noting water hydrants, sprinklers and other facilities that the fire services might need in the event of an emergency. From such details the risk of fire could be calculated, and some assessment made of the likely losses, all important elements in deciding on the insurance premium to be paid. Figure 35 is an extract from the Goad insurance plan for Dublin 1893, centred on the Olympia Theatre, Dame Street. The London company of Charles E. Goad Ltd produced similarly detailed plans for the cities of Belfast and Cork; in each case the area covered is the commercial core along with warehouses and industrial property such as breweries which were insured with them. These maps include residential property only in so far as it is enclosed within commercial districts, and were updated periodically. Copies are available in a range of institutions, including the Linen Hall Library in Belfast (the Belfast maps), the NLI (Dublin maps), UCD Geography Department (Dublin and Cork), and the Cork City Archives. The practice was to lease the maps to insurance companies, calling in the early edition when a revised sheet was made. And coverage is confined to the commercial heart of the towns, those districts which were likely to provide the insurer with custom. The 'Explanation of symbols used on insurance plans of towns and cities', produced as a separate sheet, is an essential accompaniment to these maps. Hydrants (fig. 35) are indicated with open circles, private ones with a smaller circle and dot. D represents dwelling, S shop, PH public house, while the number of storeys is also given, with ½ for an attic. Roofing materials are noted, such as small circles for slate, P for patent (felt), ASB for asbestos materials. Fuel supplies and fire halts, door materials (single iron doors, SID), skylights (indicated by small angles), the widths of streets, archways and lanes, and a sophisticated three-digit numbering system to identify every premises supplements the street numbering, providing the insurer with a graphic summary of the hazards this property will face in the event of a fire or other disaster occurring. Likewise the customer, who was provided with an extract covering his own immediate area, has a most useful document to assist in managing his property.

THE OS TWENTY-FIVE INCH SERIES AND SUCCESSOR MAPS

There was a pressing need in the latter half of the nineteenth century for rural maps at a larger scale than the six-inch series due largely to the massive land transfers that were carried out in Ireland under the Encumbered Estates Court (from 1849), the

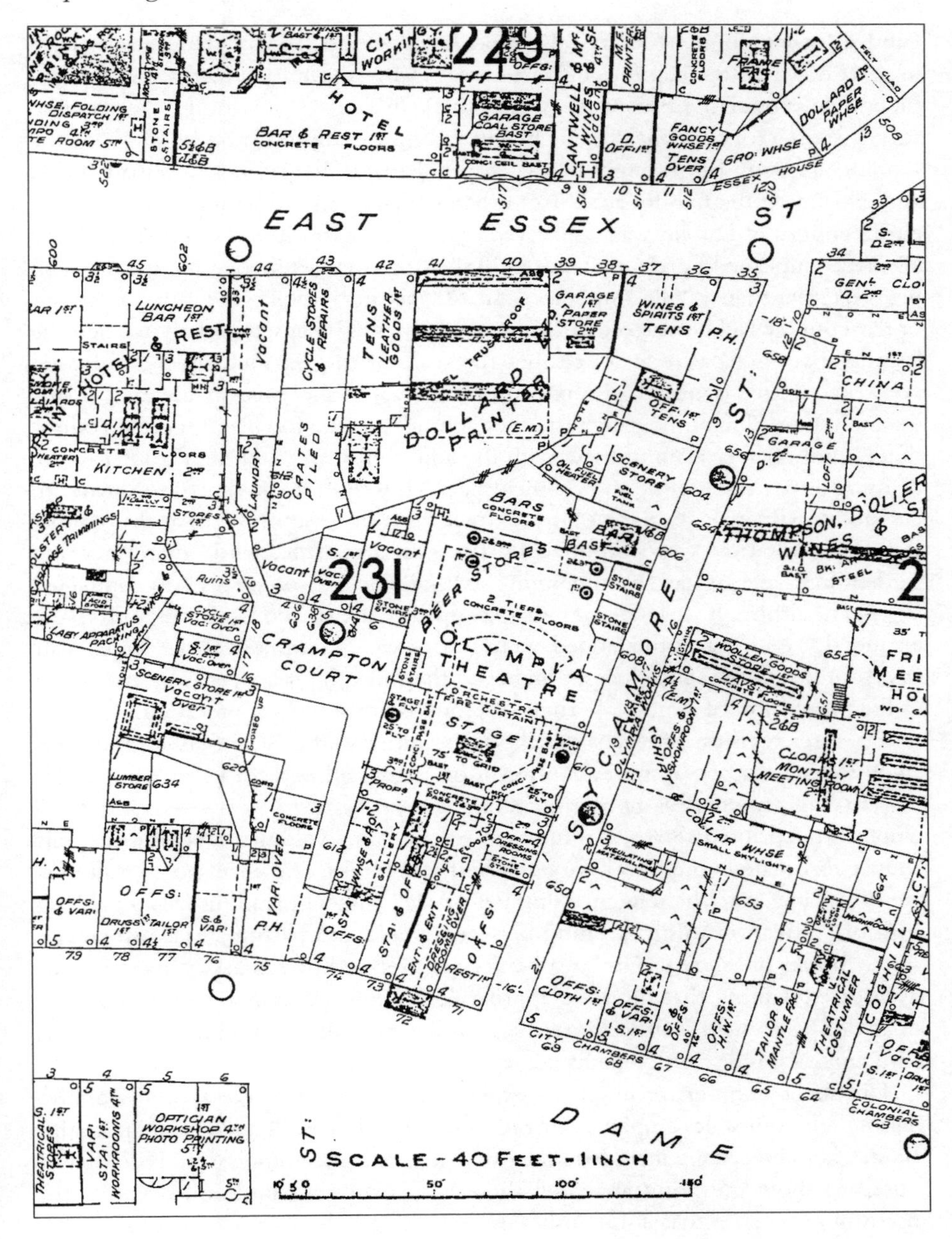

35 Dame Street, fire insurance plan, Charles E. Goad (London, 1893)

Landed Estates Court (1858), and the Land-Judges Court (1877).[38] The authorisation in 1863 of a 1:2,500 map for cultivated areas in Britain not yet mapped at the six-inch scale provided a precedent. While the Valuation Office found the six-inch survey satisfactory for most of its purposes, the expanding suburbs south of the Dublin city boundary brought particular problems, as the six-inch was inadequate, and expanding the five-foot (1:1056) urban coverage would be impracticable. The entire county of Dublin was resurveyed at this new scale of 1:2,500 in 1864 (with all sheets published by 1867); it was published by parishes (hence the term 'parish plans'), rather than as 'filled-up' sheets. In 1887 authorisation was given for the rest of the country to be surveyed at this scale.[39] The new maps, popularly termed the 'twenty-five inch' were at four times the scale of the six-inch sheets, and were constructed on a county basis; sixteen of the 1:2,500 sheets cover the area of one six-inch sheet. The numbering system established for the townland survey (county, sheet number) was again utilised, with the addition of an extra digit from 1 to 16, defining its position within the six-inch sheet. Thus the 1:2,500 sheet for Maynooth is County Kildare V.16 (fig. 36); this extract from the edition of 1977 shows how the ornamentation was simplified by the OS over time, while demesnes and gardens in particular have become more stylized. The 1:2,500 survey was completed by 1913, although sparsely populated upland areas and some islands were excluded.[40] As most of the landed-estates maps were computed at the larger scale (1:2,500) but then reduced to six inches to the mile, the role played by the courts in the promotion of large-scale rural mapping is not widely appreciated.

The later revisions of the six-inch series were based on the fieldwork undertaken for the 1:2,500 'resurvey'. In terms of cartographic records, 'area books' for Dublin, each titled *Book of reference to the plan of the parish of* ———, County of Dublin were printed 1868–72, running to seventy-nine volumes. Dublin is the only county fortunate enough to have published area books. These can be consulted in the NLI. They note the reference number of each parcel of land (usually a field), its state of cultivation (distinguishing between arable and pasture), and its acreage. These 'books of reference' or 'area books' are prefaced by remarks directed at the lay user, explaining that the scale approximates one square inch to one acre, 'the length of each sheet is little more than a mile and half, the width not quite one mile, and each full sheet contains 964 acres'.

The sheet numbering system is explained, and some of the symbols, abbreviations and font styles employed by the OS, such as braces (ʃ), used 'to indicate that the spaces so braced are included under the same reference number' and the symbol [used to show the point at which there is a change in the boundary. It is the insertion of acreages that distinguishes the 1:2,500 'parish plans' from the earlier six-inch maps; the explanation through a worked example of how to covert decimal fractions of an acre (the style employed) to the more commonly understood acres, roods and perches, is therefore especially valuable to the local historian; this

38 Andrews, *A paper landscape*, pp 258–9, 287–8. **39** Ibid., pp 257, 263, 8–9. **40** See Andrews, *History in the Ordnance map*, p. 43.

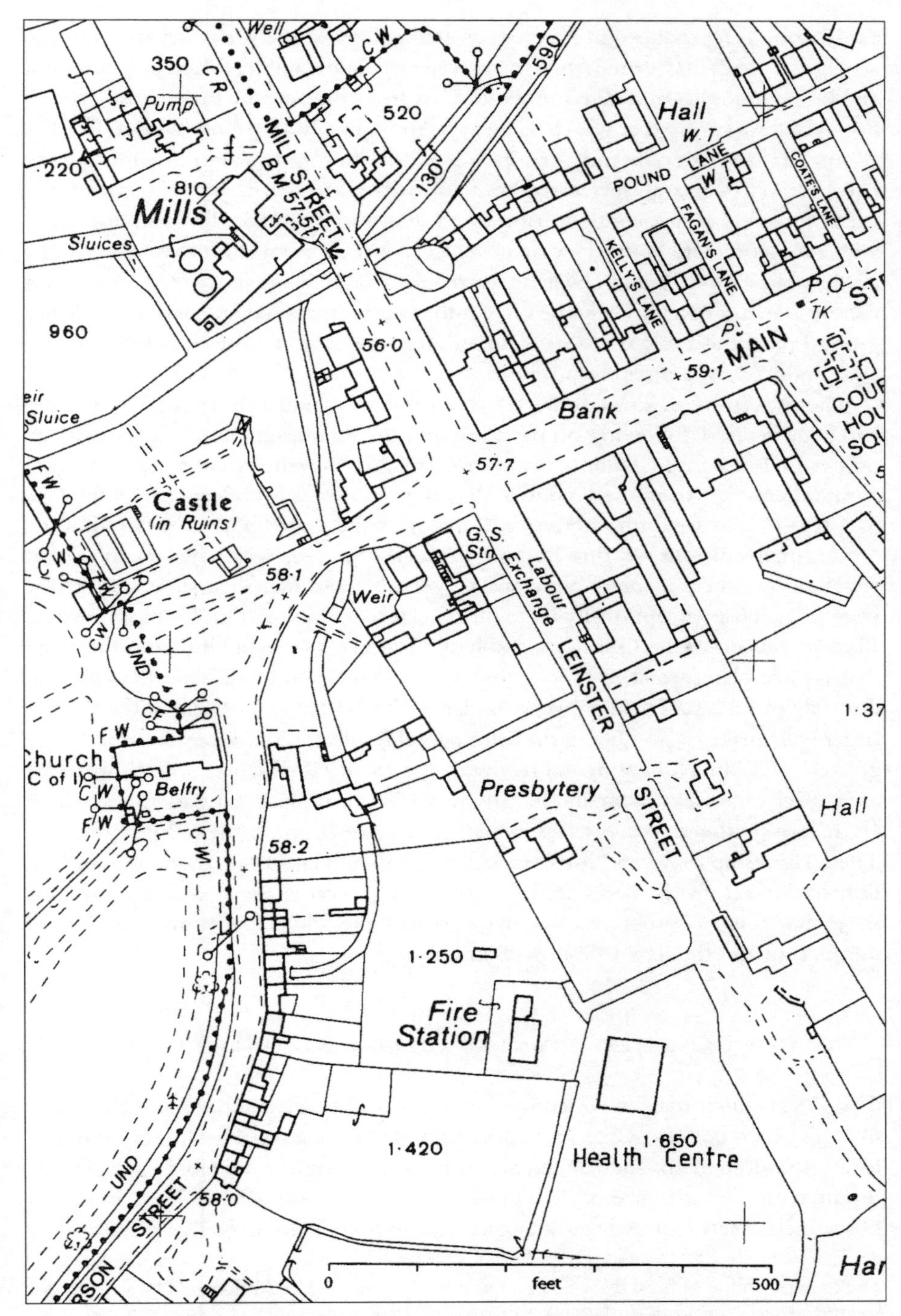

36 OS 1:2,500, Maynooth, County Kildare, sheet 5.16 (1977 revision, published 1981)
Mapping by Ordnance Survey Ireland, Permit no. APL 0000104 © Goverment of Ireland

explanation is reproduced in chapter 2.[41] It also allowed the OS to separate out the acreage of roads and water from the acreage of 'land' (arable, tillage and pasture), and to exclude the land taken up by ditches from the acreage proper, the type of fine-tuned surveying that was demanded by those involved in land transfer. The OS administrative materials and correspondence relating to this monumental survey have not yet been transferred to the National Archives of Ireland. However, the scale of these maps has ensured that they are a favourite resource for local historians: with near-complete country-wide coverage, covering rural and urban centres in a predictable sequence, and covering a sizeable area on a single sheet. And it is not merely a matter of presenting the same information (from the six-inch) on a clearer ground plan; the 1:2,500 is a new survey in its own right, with new information and produced for a different purpose.

The pattern of revisions of the 1:2,500 sheets was erratic. By 1924 nine counties had been revised; following on partition and the establishment of distinct survey offices in Belfast and Dublin, revision of the six Northern counties was more comprehensive than in the Republic. A major revision was undertaken by the OS*i* in 1939–40 and for some urban areas again in the 1970s. The OSNI published 'provisional editions' at this popular scale in 1972–9, with the warning that vegetation cover was incomplete and parcel areas had not been measured. These were printed on exceptionally large sheets (2400 x 1600 metres). From June 1981, all new editions of the OSNI 1:2,500 Irish Grid plans were published in the more manageable plan format of 1200 x 800 metres. OSNI also published urban plans at the scale of 1:1,250. These are very similar to the NI 1:2,500 maps, but the slightly larger scale makes it possible to show topographical features in more detail and with greater precision. Dispensing entirely with the county boundaries, OSNI opted for an entirely metric system in the 1970s, with the replacement of the six-inch (1:10,560) by the 1:10,000 series, which is very much more streamlined (fig. 37). The relationship between Northern Ireland's six-inch and 25 in. sheets is preserved only in the fact that it takes sixteen 1:2,500 sheets to cover the area of one 1:10,000 sheet, but there is otherwise no link between the 'old' OS county-based sheet numbering and the new OSNI grid numbering.

OTHER OS SERIES: COUNTY INDEXES AND ONE-INCH MAPS

The OS six-inch maps were completed on a county-by-county basis; the index maps for each county, while most obviously designed to enable the researcher to locate the desired six-inch sheet, are in their own right a valuable cartographic resource. Because the size of each county in Ireland varies drastically, the number of six-inch sheets required to cover each county in full also differs greatly. To allow

41 For an example of how this information can be used, see J.H. Andrews, 'The struggle for Ireland's rural commons' in Patrick O'Flanagan, Paul Ferguson and Kevin Whelan (eds), *Rural Ireland, 1600–1900: modernisation and change* (Cork, 1987), pp 11–12.

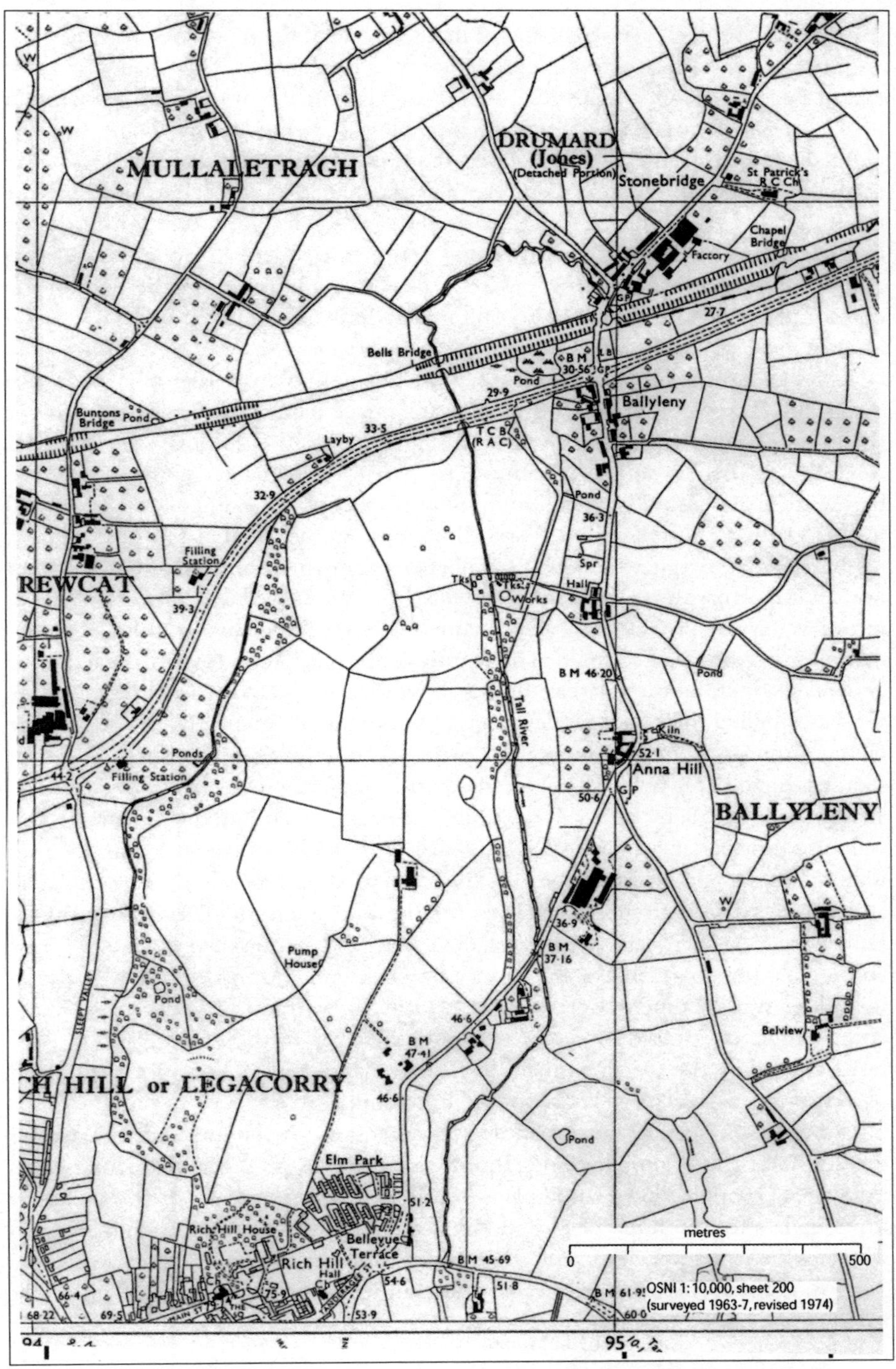

37 Drumard, Ballyleny, OSNI 1:10,000, sheet 200, grid squares H and J (surveyed/revised 1963–7; edition of 1974)

for this variation, the six-inch index maps are at different scales, allowing each county, regardless of size, to be represented on a single sheet. The smallest county, Louth, takes 25 six-inch maps, and its county index map is at the scale of one inch to 1½ miles (1:95,040); the same scale is employed for the Dublin index (twenty-eight sheets). The largest county, Cork, requires 153 six-inch maps, and its index map is at a scale of one inch to three miles (1:190,080). This scale is also used for the index maps of Donegal, Kerry, Mayo and Galway, and for County Antrim, the first county to be completed. Down and Tyrone are indexed on the scale of two and a half miles to one inch (1:158,400). The scale employed for the bulk of the county index maps is half inch to one mile, which at 1:126,720 is a most useful intermediate scale for many local history purposes.

The index maps are most useful as stand-alone county maps (fig. 38) as they give a very full if crowded picture of roadways, drainage patterns and relief, settlement (including landlord demesnes, which are lightly shaded) administrative divisions (including parishes) and placenames. The parish of Kilmalkedar (fig. 38) in the barony of Corkaguiny (Corca Dhuibnne) for example is distinguished by several historic ruins (church, castle, Chancellor's House), penitential stations, oratory, Saint's Road, coastguard station, RC chapel, two named houses and a cluster of houses at Ballynagall overlooking Smerwick Harbour, as well as a network of roads, minor waterways, rocky cliffs and a named lake, an impressive amount of information in a very small compass. The county index maps were revised as part of the general six-inch revision; later editions show the new railways. In latter years the OS has supplied the county index maps as photographic copies, but these lack the clear definition found in the copper plate (in some cases up to 1912) and subsequent zincograph versions. Where the parish boundaries are edged in colour (as was done in fig. 38, by the OS), the units are easily picked out; the distinctive font style used for parish names (*KINARD, GARFINNY*) also assists. Alongside each index map is a table naming the area (in acres, roods and perches) for each parish and barony, with a single grand total for the entire county. Thus the parish of Dingle has 9,097 acres and 27 perches, the barony of Corkaguiny of which it is part comes to 138,990 acres, 1 rood and 17 perches, and the total county of Kerry, including water bodies, comes to 1185917 acres, 2 roods and 6 perches. These county index maps also carry the first published key to the writing and characteristics used by the OS in its main (six-inch) map; a separate characteristic sheet (for extracts see figs. 56–58) was not published until 1889.[42]

Other supplementary information to be found on the county index maps (for the south of the island) includes thumbnail sketches of the parliamentary and municipal boroughs, and ground plans of principal towns, at scales of two inches or three inches to one mile. The 1846 index of Kerry, for example, includes a map showing the parliamentary borough of Tralee (three inches to one mile), while the 1845 index map of Cork carries an inset showing the parliamentary borough, parishes and ward boundaries in the city.

42 Andrews, *History in the Ordnance map*, p. 22.

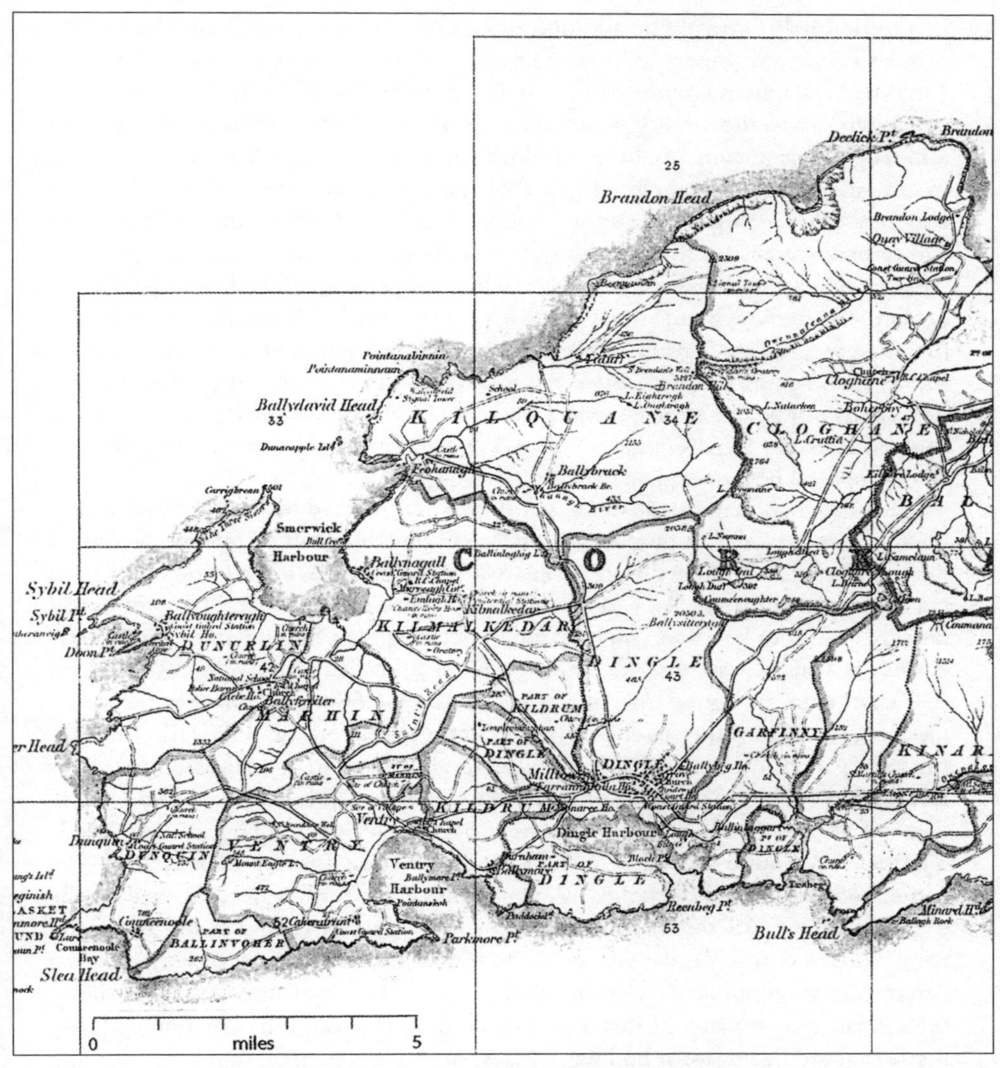

38 Index to OS six-inch series, Kerry county. Scale of original index map is one inch to three miles.

The index maps to the six-inch or townland survey, despite the variety in scale and in sheet size, and the crowded nature of the detail, could be brought together to form an all-Ireland atlas, as in the *Atlas of the counties of Ireland, published by order of the Honourable Board of Ordnance, coloured to indicate the extent of bogland, for the Irish Amelioration Society* (London, 1847).[43] In this case the barony boundaries are tinted, all areas of water are coloured blue, and the bogland ('bog or uncultivated ground')

43 See CUL, Atlas 1.84.4.

is coloured light brown. The addition of colour selects out information from the mass that is already present in these maps; similarly, other elements such as demesnes could be highlighted county-by county enabling intense all-Ireland study.

Drawbacks to the county index maps are apparent when attempts are made to join neighbouring counties. In the six-inch survey each county had its own 'county origin', and all points were related to it; there cannot therefore be an exact fit between counties. The Geological Survey of Ireland (GSI) was foremost in demanding an authoritative, seamless OS map (disregarding county boundaries), at a medium scale, which might act as a base map. An inquiry into the progress of the Ordnance Survey held in 1846 opens with the reminder that a 'general' one-inch map of Ireland, similar to that in progress in England, had been ordered from the start of the survey. Work on the six-inch maps took precedence, but it had never been intended to supersede the general map, or be the 'ultimate object' of the survey.[44] The one-inch project therefore was considered by some Ordnance Survey's critics as its proper aim, and great pressure was brought to bear to explain delays.

The one-inch map (made up of 205 sheets) was printed by the OS on behalf of the GSI, but the OS first produced its own one-inch black and white topographical map, in both an 'outline edition' from 1854 to 1862, and a 'hill edition', published between 1855 and 1895. An early economy measure, the omission of contours, had quickly proven to be a short-sighted move, obliging the OS to complete its outline edition in parallel with a series showing hill-shading. An inquiry of 1853 outlines the arguments in favour of contours.[45] Andrews' *History in the Ordnance map* (pp 34–41) explains the sheet numbering, survey sources, and publication history (including revisions) of this complex series; his essay in *Sheetlines* (1991) gives further detail on the alterations introduced to the hill map, based on a bound record copy of all 205 sheets held by the Ordnance Survey as its Phoenix Park headquarters.[46] A bound set is also available in the National Archives of Ireland. The names and boundaries of civil parishes, baronies and counties were entered on the one-inch maps. All townland boundaries were excluded, and many townland names omitted; townlands were named in 'floating' mode wherever there was a danger the map would look empty. Figure 39 is taken from sheet 171 and shows the information available at this scale. The communications infrastructure is clearly marked as are institutional buildings. Mills, smithies, defences, signal stations and coastguard stations are named, while antiquities are especially prominent. The Upper Colony and Lower Colony at Dingle refer to the Protestant missions established here in the 1840s. The one-inch map incorporated new information; it was not merely a reduction of the six-inch.

A valuable use of the one-inch series was the construction of RC parish maps by the ESB in Ireland for the Rural Electrification Scheme, set up in March 1944

44 *Report from the select committee on Ordnance Survey*, 1846 (664) xv. **45** *Report of the select committee on the map of Ireland, 1852–53* (921) xxxix. **46** J.H. Andrews, 'A record copy of the one-inch Irish hill map' in *Sheetlines*, 30 (1991), pp 4–5.

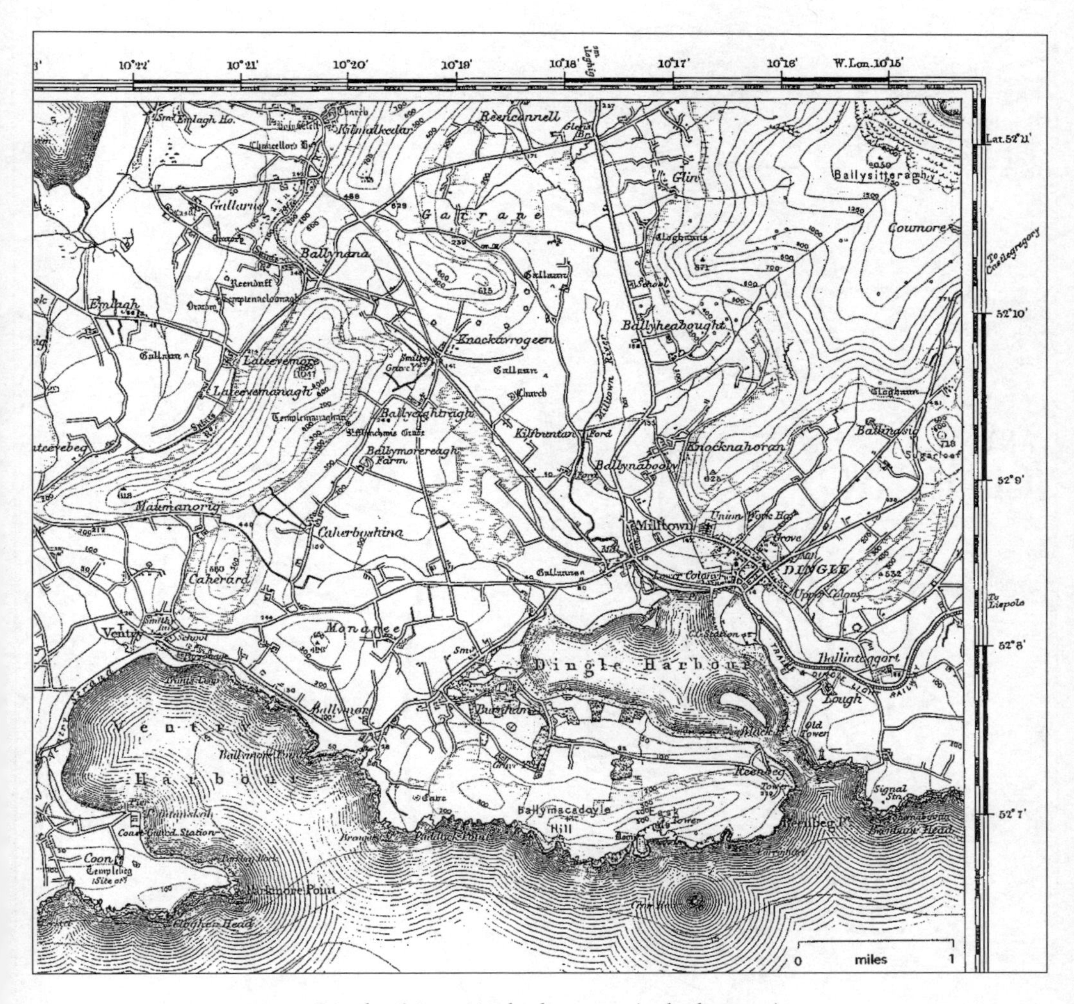

39 Dingle, OS one inch, sheet 171 (3rd edn, 1910)

and operating most extensively 1945–5 (fig. 40). The state electricity network was extended on a parish-by-parish basis throughout rural Ireland, in an extremely decentralised process which required the closest co-operation of local committees and voluntary groups (the Irish Countrywomen's Association was one of the most significant) with the ESB engineers and workers. Exhortations to co-operate, so that all parishioners might have the option of being included in the heavily subsidised system, were made from church pulpits.[47] In terms of communication, the Roman

47 Maurice Manning and Moore McDowell, *Electricity supply in Ireland: the history of the ESB* (Dublin, 1984), pp 129–36.

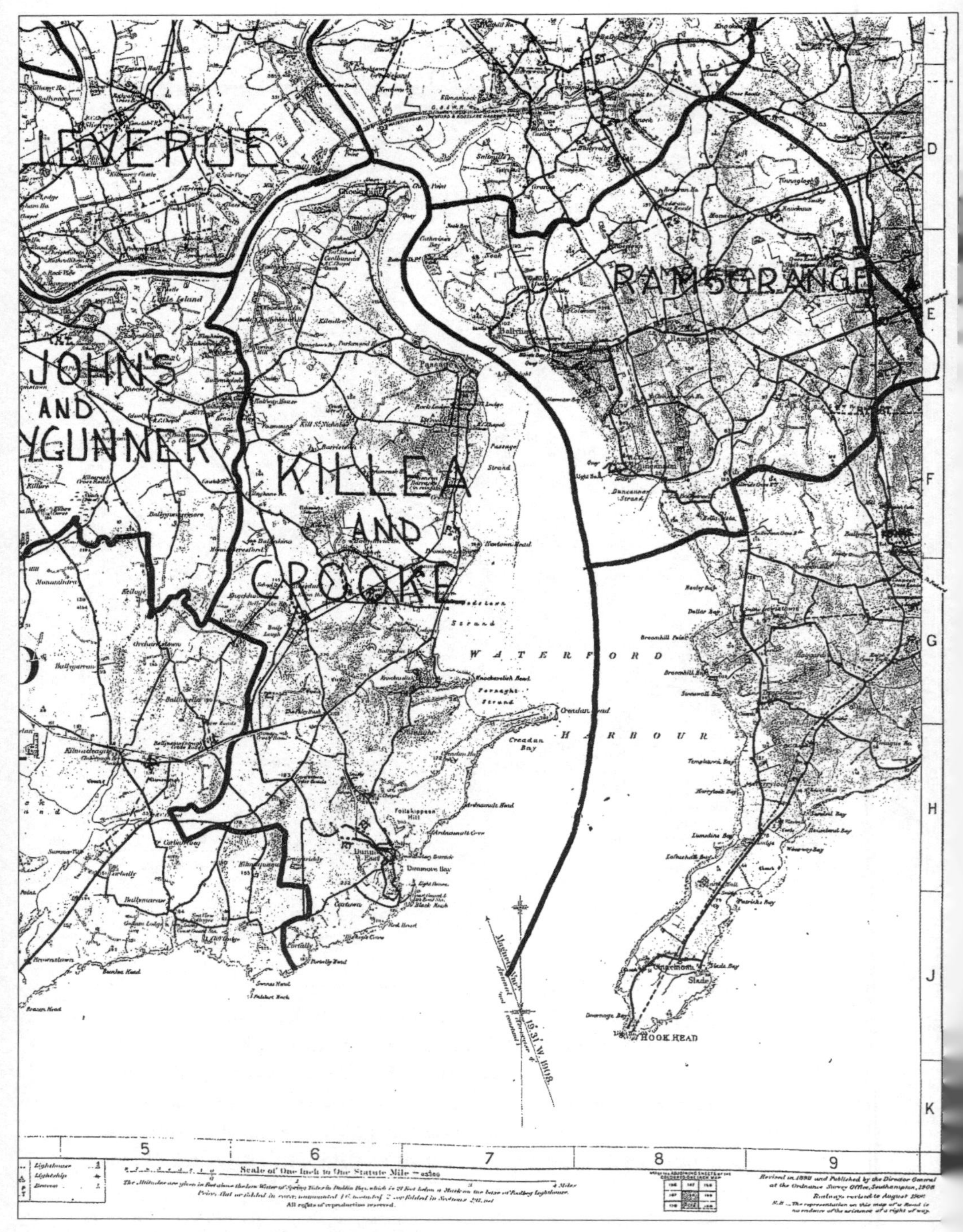

40 RC parishes south-east Ireland, ESB series, *c.*1945, sheet 38; base map: OS one-inch sheets 168, 179 (revision of 1906)

Catholic parish structure was important to the ESB – around 900 parishes were involved – but cartographically none of the small-scale administrative maps produced by the OS (quarter inch, half inch or ten mile series) were suitable as they did not include RC parish boundaries. The ESB therefore joined a number of one-inch OS sheets to make a series of base maps which were independent of county boundaries, reduced these photographically, inserted RC parish boundaries by hand, and renumbered the maps according to its own independent system. Thus figure 40, based on OS one-inch sheets 168 and 179, is sheet 38 in the ESB system. The originals of these ESB parish maps, complete with hand-drawn index map, are in the Russell Library, Maynooth; coverage is extensive but not complete for the 26 counties. They are strictly rural maps covering a rural scheme; electricity was already available in all the larger towns, in many cases the result of schemes undertaken by the local municipal authority under a special act of parliament. Electricity transmission is shown on the 1938–39 revision of the 1:2,500 OS sheets. Maps and plans for various urban 'electric lighting' schemes (pre-independence) are in the House of Lords Record Office, Westminster and in the records of the Clerks of the Crown and Peace (in the NAI), as explained in chapter 3.

An enterprising and probably illegal use of OS medium-scale mapping was made by the county surveyor of Westmeath, Arthur E. Joyce, in response to the reform of local administration ushered in by the 1898 act. He was author of a two-volume atlas simply titled *County council maps* (Dublin, n.d.).[48] This atlas consists of 32 maps on a scale of one inch to four miles, noting county electoral divisions, district electoral divisions, rural districts and union boundaries. It is printed in seven colours, and includes principal roads and settlements. Facing each map is a template on which the purchaser inscribed names of the newly-elected councillors (for each county electoral division), the names of those co-opted, and grand jury nominees; also the names of chairmen of rural districts. No acknowledgement is made in the atlas but Joyce relied on OS mapping for information, probably redrawing the county index maps to a common scale (one inch to four miles). It is a cartographic record of the massive re-organisation which the 1898 act brought, and a good example of mapping expressly for electioneering.

ANTIQUITIES AND ARCHAEOLOGICAL MAPPING

Antiquarianism, archaeology and mapping have long been intertwined. In terms of cartographic history, most of the major classes of maps here discussed include 'ancient monuments' or other historical data on their maps. Circular enclosures, earthen mounds, castles, towers; ecclesiastical enclosures, associated buildings and monuments; relict field patterns, old homesteads and mills; town walls and gates – a full list would cover the entire range of study that is the discipline of field archaeology. Features of this type may be named or unnamed, symbolised by

48 See CUL Atlas 3.89.22; the printers are Cherry and Smallridge Ltd, Dublin.

realistic-looking drawings (which may be field sketches or entirely imaginative) or by other specialist symbols. They may have been purposely included at the time of the map's creation – as with the Ordnance Survey's topographical maps at the six-inch scale (figs. 1, 20, 21) – or they may appear as an aside, a welcome but fortuitous addition in a cartographic enterprise that served other more utilitarian interests, as in plantation maps, estate maps, or railway maps.[49] Petty instructed his surveyors to 'record the situation of townes, castles, mills, raths, notable houses &c.' (figs. 2, 5, 6) but unfortunately only some of his men took this directive to heart.[50] Taylor and Skinner's *Maps of the roads of Ireland* (1778) have symbols for 'churches in ruins' and 'castles in ruins' in the legend key, and in addition mark in prehistoric forts on the maps themselves (fig. 10), all matters of interest to the educated gentleman class that was their target market. In the quarter-inch map produced for the railway commissioners in 1838 (fig. 43), 'a few remarkable antiquities have been inserted'. Those 'of pagan origin' are indicated by the use of Gothic script, military sites are denoted by the letter M, while 'where round towers occur the letters RT are added', all based on information furnished by the renowned scholar George Petrie. There is, however, a sense in which all topographical maps serve the purposes of archaeology. While their function is not to display historical data, each is produced in history, and acts as a record of how humans left their mark on the landscape over time; settlements built or planned, mountains traversed, water-power harnessed, places made sacred by churches, burial grounds, crosses, holy wells. In that sense, current Irish maps follow the tradition established in this country in the sixteenth century, and will (in digital form) provide source material for future archaeologists. But while maps are a major source for archaeology, they are also indispensable for the analysis and presentation of research results. The spatial distributions of 'findings', whether of ringforts or megalithic tombs island-wide, or of medieval house plots found below the level of an eighteenth-century streetscape, can be brought to the public's notice only through mapping. In combination with sketches and field notes, mapping at both large and small scales has long been one of the methodologies adopted by those devoted to making sense of material remains, becoming quite scientific (along with geological mapping) in the early nineteenth century.

Sarah Gearty, cartographic editor of the *Irish Historic Towns Atlas* project, has drawn attention to the ways in which pre-OS maps can yield more information on antiquities than might first be realised,[51] while Patrick O'Flanagan has demonstrated how much can be learned of early settlement by the close examination of non-OS maps.[52] The role of the cartographer as antiquarian in pre-OS mapping is also explored by John Andrews.[53] When maps from the period preceding the OS

49 Terence Reeves-Smith, 'Landscapes in paper: cartographic sources for Irish archaeology' in Terence Reeves-Smith and F. Hammond (eds), *Landscape archaeology in Ireland* (B.A.R. British Ser., 116, London, 1983), pp 119–77. **50** Thomas A. Larcom, *The history of the Down Survey* (Dublin, 1926), p. 123. **51** Sarah Gearty, 'Irish historical maps as a source for the archaeologist' in *Trowel*, viii (1997), pp 1–6. **52** Patrick O'Flanagan, 'Surveys, maps and the study of rural development' in Donnchadh O'Corráin (ed.), *Irish antiquity* (Dublin, 1981), pp 320–7. **53** J.H. Andrews, 'Mapping the past in the past: the cartographer as antiquarian

can be placed alongside the OS sheets, as in the *Irish Historic Towns Atlas*, the intersection of cartography and archaeology is manifest in a most practical way. In the *Kells* fascicle, by Anngret Simms with Katherine Simms, the ninth-century monastic enclosure is traced in the street patterns on the 1:2,500 OS (edition of 1970); in the *Kilkenny* fascicle by John Bradley, the line of the medieval town wall, the position of the gatehouses, towers, churches and castles have all been reconstructed from close analysis of Rocque's 1750s map. In the *Maynooth* fascicle by Arnold Horner, the extent of the massive Fitzgerald castle, part of which still stands at the college gates, is reconstructed from an early sketch map, and written accounts (fig. 59).

Antiquities and the OS

The Ordnance Survey record dominates archaeological mapping in Ireland. The initial instructions in 1825 by Thomas Colby to the OS officers noted 'antiquities' as one of the headings under which they were to collect additional 'local information'. Thomas Larcom's 1834 'heads of inquiry' produced for the *Memoir* scheme (Appendix 4) elaborated on this: 'topography: ancient, the history of the parish, as shown by objects of antiquity (pagan, ecclesiastical, military), and ancient buildings which remain'. The result was the first systematic survey and cartographic record of archaeological remains, including every townland and parish on the island and immediately offshore. John O'Donovan dedicated his seven-volume edition of the *Annals of the Four Masters* (Dublin, 1848–50) to his 'friend, Captain Larcom R.E.', describing him as 'the active promoter of Irish literature, antiquities, and statistics', who has 'exerted himself most laudably to illustrate and preserve the monuments of ancient Irish history and topography'. Archaeologists have commented on the importance of the six-inch maps for 'scientific archaeology' even while warning students that coverage is incomplete.[54] The fieldnotes, letters, and other materials gathered for the production of the ambitious OS *Memoir* series, and in connection with the standardisation of placenames, have already been discussed, while the 'regular' materials produced in the course of the Ordnance Survey, the Boundary Survey, and the General Valuation (Griffith's) can also yield information on what was included, and the manner in which it was represented. And as soon as they became available, extracts and tracings from OS six-inch maps were used by accomplished archaeologists in their more intense inquiries, as seen in the illustrations they produced for the publications of local historical and archaeological journals, a practice which continues today.

The way in which antiquities were represented, and the degree of coverage on any given OS map, are issues for continuing debate. In the case of ringforts, for example, Matthew Stout compares the OS record, from the 1:2,500 series, with his own extensive field research, to determine how useful the OS plans are for determining their 'true' characteristics.[55] The boundary surveyors, who by virtue of

in pre-Ordnance Survey Ireland' in Colin Thomas (ed.), *Rural landscape and communities: essays presented to Desmond McCourt* (Dublin, 1986), pp 31–63. **54** Ibid.; Seán Ó'Ríordáin, *Antiquities of the Irish countryside* (London, 1966), preface to the first edition, p. xii; Michael Herity and George Eogan, *Ireland in prehistory* (London, 1977). **55** Matthew Stout, 'Plans

their task were always in advance of the Ordnance Survey officers, were given a rather fussy selection of symbols with which to embellish their work (fig. 27); the system adopted on the final engraved sheets by the Ordnance Survey is more restrained. Features are named where possible and not just mapped, church dedications are often supplied, the names of owners of castles, and 'in ruins' or 'site of' added where necessary (fig. 1). The use of an 'antique' font highlights features such as 'Castle' or 'Tower' that might otherwise be overlooked by the local historian. The incompleteness of their task, especially in mountainous areas, has also been criticised. The officers recorded only what was conspicuous and over-ground, and some complained of the unwillingness of their men to concern themselves with what they regarded as superstitious nonsense.[56] In consequence certain features were quite simply ignored or misunderstood, as later more professional fieldworkers have found, while aerial photography has revealed sites that could not have been visible to even the most zealous of the Survey's officers.

Despite its limitations, the six-inch survey remains the most important cartographic source for the identification of 'noted antiquities', and the basis upon which subsequent discoveries continue to be mapped (fig. 41). Although the 'topographical department' was disbanded in 1842, the OS has continued archaeological research well beyond the initial six-inch survey, most notably in its work on Ireland's megalithic monuments.[57] The revision of the OS six-inch series undertaken in 1931 provided an opportunity to update the archaeological content. A system of 'field memoranda' was initiated by which all antiquities met with by the revisers in the field were noted and recorded; these are housed in the National Monuments section of the Department of the Environment, Heritage and Local Government, and though of uneven quality (the surveyors were not trained archaeologists) they are still of interest to local historians.[58] The antiquities marked on the 1:50,000 Ordnance Survey *Discovery* sheets (fig. 47), have been taken from its six-inch survey, and as confirmed by later research, largely the Sites and Monuments Record (SMR), as explained below.

From Office of Public Works to the Department of the Environment, Heritage, and Local Government

The connection between the Commissioners of Public Works (the OPW or 'Board of Works'), archaeological mapping, and statutory responsibility for the protection of 'national monuments' from the 1870s through to the present day needs to be expounded. Rena Lohan's *Guide to the archives of the Office of Public Works* (Dublin, 1994, chapter 3), is essential reading in this regard. In brief, the highly complex rearranging of the Church of Ireland's temporal affairs on Disestablishment (effective 1 January 1870) involved the Board of Works in the protection of any

from plans: an analysis of the 1:2,500 OS series as a source for ringfort morphology' in *Proceedings of the Royal Irish Academy*, xci, C (1992) pp 37–53. **56** Andrews, *A paper landscape*, p. 127. **57** Ruaidhrí de Valera and Seán Ó Nualláin, *Survey of the megalithic tombs of Ireland* vols i–iv (Dublin, 1981–2); Seán Ó Nualláin, *Survey of the megalithic tombs of Ireland* vol. v (Dublin, 1989). **58** I am indebted to Paul Walsh for this information.

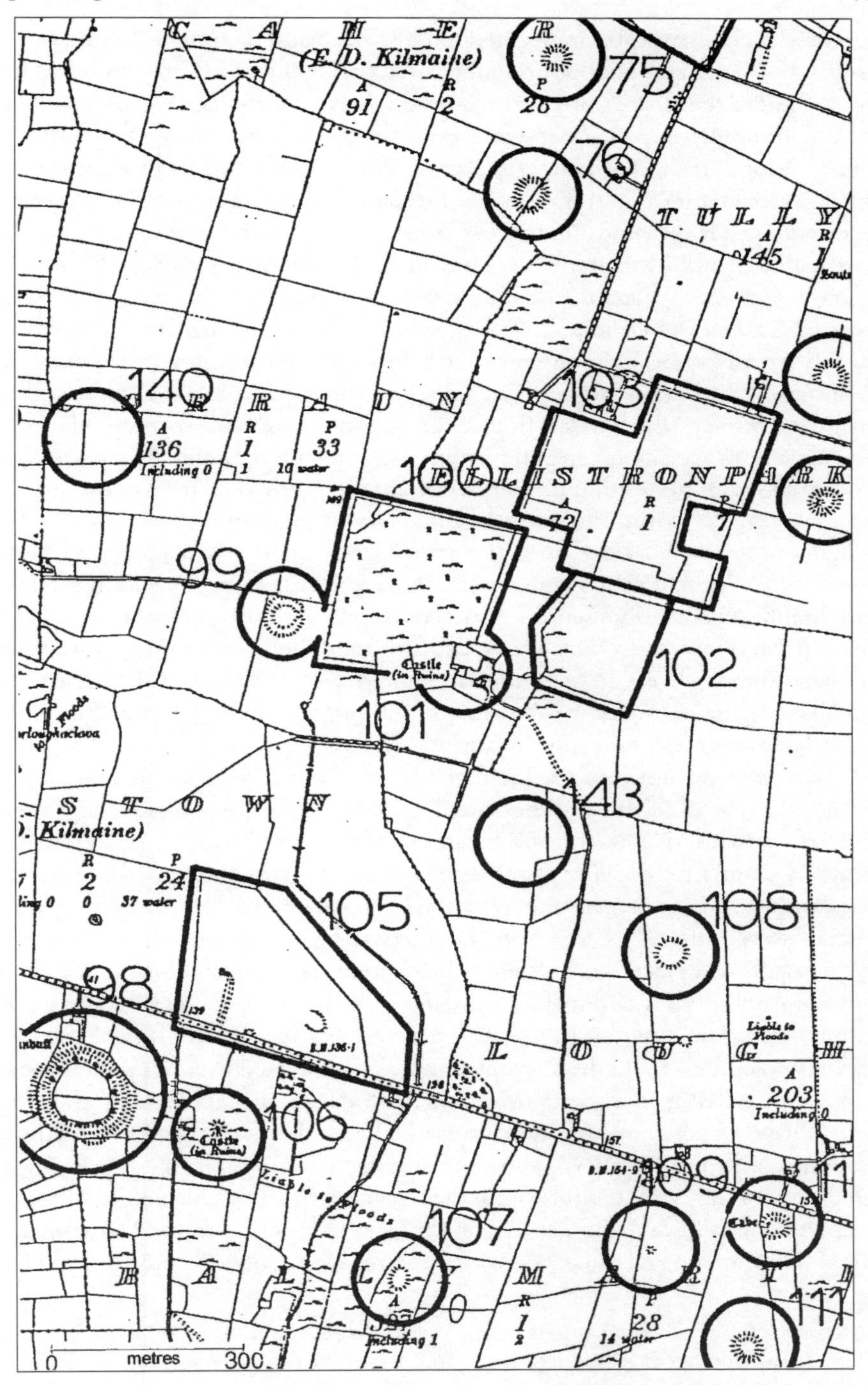

41 OPW archaeological survey of Ireland, Mayo sheet 118, date/edition of map 1929, date of SMR detail 1991

'church or ecclesiastical ruin deemed to be of historical or antiquarian interest and in need of conservation, but no longer used for public worship' which had been placed under the Board's protection by the Church Temporalities Commissioners. It was limited to works that were essential for the purposes of security and preservation, such as paying a caretaker, erecting a boundary fence or repairs that would halt further deterioration. By December 1877, over 100 monuments, in every county, had been vested in the Board, including the major sites of Cashel, Glendalough and Monasterboice. The range of monuments for which the Board could be responsible, and the definition of what was to be protected, was expanded beyond Church of Ireland ecclesiastical buildings by further legislation, culminating in the 1930 National Monuments Act.[59] From the local history perspective, the 'registers of inspection', surveys (including drawings, plans, sections, photos, sketch maps and tracings and extracts from OS maps) and associated correspondence and accounts from the late nineteenth century onwards are all useful. As the OPW was concerned with preventing further deterioration, it was keen to map the layout of the monuments within individual cemeteries or enclosures, and as a result has lengthy notes on what repairs were deemed essential. As noted already, from 1874 (when the ecclesiastical remains on the Rock of Cashel were vested in the state), any building habitually used for ecclesiastical purposes cannot have the status of 'national monument'.[60] This partly explains why there are numerous important heritage sites, such as Christ Church in Dublin, with which the OPW has had no involvement (and hence has no maps), although it could – and did – place 'guardianship orders' on major sites which were not vested in its care.[61]

Between 1984 and 1992 the Commissioners of Public Works issued the *Sites and Monuments Record* (SMR) for the state. The SMR list was accompanied by a set of annotated maps, that is, the latest edition of the OS six-inch maps (1:10,560), reduced slightly for copying purposes to a scale of c.1:12,600. With the passing of the National Monuments Amendment Act (1994) the Commissioners of Public Works were obliged by the new statute (section 12) to 'maintain a record of monuments and places where they believe there are monuments'; the record was to comprise both list and map(s) 'showing each monument and such place', on a county basis. The resulting *Record of monuments and places* (RMP) is the basic listing of archaeological sites in the Republic, and was completed over the period 1995 to 1997. The RMP is in effect a subset of the SMR, and includes only those sites which have a known or suspected location. The work of compiling the RMPs was undertaken by staff at OPW who were transferred in 1996 to the newly-formed Department of Arts, Culture and the Gaeltacht.[62] On the abolition of this Department in 2002 its heritage functions were transferred to the Department of the Environment and Local Government in 2003 (subsequently renamed the

59 Lohan, *Guide to the archives of the Office of Public Works*, pp 85–9. **60** David Sweetman, 'The man-made heritage: the legislative and institutional framework' in Neil Buttimer, Colin Rynne, Helen Guerin (eds), *The heritage of Ireland* (Cork, 2000) pp 527–33. **61** Ibid., p. 527. **62** Since 1997 the Department of Arts, Heritage, Gaeltacht and the Islands; Mary

Department of Environment, Heritage and Local Government). The archive of the SMR/RMP is housed at 6 Upper Ely Place, Dublin 2.

On the maps of the SMR and RMP, the position of each known archaeological site is circled or boxed (see fig. 41) on the appropriate OS six-inch sheet (slightly reduced). The accompanying list gives the monument number, OS sheet number, National Grid co-ordinates (up to ten-figure co-ordinates), townland name, and classification. The sheet/plan/trace system (explained below) which is employed on the SMR and RMP was in fact pioneered by the Archaeological Survey of Northern Ireland, where the 1932 six-inch survey sheets are the basis upon which historic monuments are marked and numbered on a county basis. On Antrim sheet one, for example, which includes Rathlin Island, antiquities are numbered from 1 to 83, and given the simple reference Antrim 1:5, Antrim 1:6, Antrim 1:7 etc. The full Irish Grid reference, to eight digits where possible, is also noted. The Northern Ireland 'sites and monuments' record can be consulted in hard copy at the Environment and Heritage Service office at 5–33 Hill Street, Belfast, BT1 2LA [http://www.ehsni.gov.uk]. It will also be found online at the Archaeology Data Service site, managed by the Department of Archaeology, University of York [http://www.ads.ahds.ac.uk/index.html].

Figure 41 illustrates the system in place in the Republic. The circled or boxed areas are merely the 'approximate location' of both antiquities and places, and 'not to be taken as indicating their precise extent or boundary'. The sheet/plan/trace system is a simple way of reckoning the general position of a monument on the OS six-inch sheet. Mentally divide each full sheet into 16 equal parts or plans, and each of these plans into six equal parts or traces.[63] In the case of County Mayo (fig. 41), the base map is the edition of 1929, but the archaeological detail is from August 1991. The *Record of monuments and places* list to accompany it bears a publication date of 1996. The site numbered 98, for example, is coded MA118–098 (Mayo six-inch sheet no. 118), with the grid reference 12342/26037, townland of Ballymartin, and classification of 'hillfort'. On the sheet/plan/trace system its location is referenced 118–/15/6. It is thus near the bottom of sheet 118, to the right of centre. Other archaeological features include a castle (MA118–101 and 106); an enclosure (MA118–96, 107, 108 and 111); a *fulacht fia* or early cooking site (MA118–140 and 143). Where there are several associated sites in close proximity, the *Record* gives one overall number and then identifies individual features by sub-numbering. Thus the feature numbered 102 (fig. 41) on Mayo sheet 118, townland of Ellistronparks includes an enclosure (MA118–10201), a hut site (10202), and a field wall (10203). Similar groupings are covered by no. 103 (a field system, a cairn, a road and a possible enclosure) and by 105 (an enclosure and a field system). Other

B. Deevey, *Irish heritage and environment directory* (Bray, 1999), pp 2–3. **63** Each six-inch sheet is derived from sixteen 25-inch plans, which in turn are based on six pieces of tracing paper, the 'field traces'. With the sheet/plan/trace reference it is possible to access the OS 25-inch archive and find the original trace on which the monument was plotted in the first place.

classifications or categories from the *Record* for County Mayo include ecclesiastical remains, church, graveyard, cashel, mound, souterrain, holy well, earthwork, fishpond, nunnery, children's burial ground, and megalithic tomb.

The *Record of monuments and places* and its forerunner, the *Sites and monuments record*, is complemented by the Urban Archaeological Survey, comprising reports on 240 towns, and issued by the OPW between 1985 and 1995. Unpublished urban archaeology surveys are held by the Department of the Environment Heritage and Local Government for all counties in the republic; these may be consulted at 6 Ely Place Upper Dublin 2. Anyone hoping to undertake redevelopment work in town or country needs to consult these records before making a planning application, and before commencing any work at or in relation to any listed monument or place. Current legislation specifies that two months' written notice must be given to the Minister for the Environment, Heritage and Local Government of any proposed work. Failure to do so is an offence, carrying the possibility of heavy fines and imprisonment.[64] But the statutory *Record of monuments and places* is essentially a list. Much more comprehensive are the county-based archaeological inventories, which have been published to date for 14 counties: Cork (4 vols), Galway (2 vols, west and north), Cavan, Carlow, Laois, Leitrim, Louth, Meath, Monaghan, Offaly, Tipperary (vol. 1, North Tipperary), Waterford, Wexford and Wicklow. The inventories have much fuller descriptions of each feature, explanatory essays on the groupings employed, a townland name index, and a local and historical name index. They also feature outline road maps, at the back of each volume, locating the sites, at least in a very general way. The inventories have their own numbering system but these are cross-referenced to the earlier *Record of monuments and places* so that it is still possible to use what were formally called the 'recorded monument register maps' (fig. 41) in conjunction with the newly-published inventories. The local authority planning office, the local Teagasc office, and the county library hold copies of the relevant volume of the *Record of monuments and places* or, where published, the archaeological inventory, alongside the annotated register maps, which are strictly for reference only. These records can also be consulted by prior appointment at 6 Ely Place Upper, Dublin 2 and it is intended that much of this information will be made available online. By simply clicking on a county the researcher will be able to access the archaeological information for his or her local area. The National Inventory of Architectural Heritage (interim county surveys) uses this type of technology to the fullest, publishing its searchable database in CD format, along with location maps (read only), also on CD. The rural maps are based on the six-inch (OS 1:10560) and urban maps on the twenty-five inch (OS 1:2,500) series. The counties of Carlow, Fingal, Kerry, Kildare, Laois, Meath and South Dublin have been published to date.

The cartographic sources for Irish archaeology are substantial, as for example listed by Terence Reeves-Smith (1983)[65] and Cherry Lavell (1997).[66] However, this

64 National Monuments (Amendment) Act (1994), section 12 (3). **65** Terence Reeves-Smyth, 'Landscapes in paper, cartographic sources for Irish archaeology' in *British Archaeological Records*, cxvi (1983), pp 119–77. **66** Cherry Lavell, *Handbook for British and Irish archaeology,*

is very much an expanding field, as GIS and satellite technology enable previously unidentified sites to be recognised, and known sites to be pinpointed more exactly, opening up new riches to the local historian and archaeologist.[67] Innovative ways of depicting landscape and archaeology continue to be explored, most impressively in the celebrated Connemara (1990), Aran Islands (1980) and Burren (1977) maps of Tim Robinson. And increased sophistication in map technology allows 'layers' of material to be presented, separately and in combination. This shift is exemplified in H.B. Clarke's map, *Dublin c.840 to c.1540*, which was first produced on behalf of the Friends of Medieval Dublin in collaboration with the OS (Dublin, 1978), and has recently been re-issued by the Royal Irish Academy and the *Irish Historic Towns Atlas* (2nd edn, Dublin, 2002). Clarke superimposed the medieval street pattern, wall towers and gateways, churches, chapels and religious houses, water courses, and other key features, in colour, on the 1:2,500 OS base map. Number 11 in the *Irish Historic Towns Atlas* series is *Dublin, part I, to 1610* by H.B. Clarke (Dublin, 2002), in which digital mapping has allowed several 'layers' of information to be shown, with and without the plot detail. Early maps of Dublin were among the principal sources for these topographical reconstructions, along with the results of recent archaeological excavations. Through the process of mapping, the citizen and visitor can today try to discover 'the medieval town in the modern city', as we walk at least some of the same streets today, albeit several metres above that early ground level.

GEOLOGICAL MAPPING

Although there had been some attempts at geological mapping pre-1800, and bogs and rocky terrain have long been among the features noted by cartographers, serious geological mapping is very much a nineteenth-century phenomenon.[68] Tied in with the search for mineral resources, notably coal and iron, to fuel the industrial revolution which contemporaries hoped would occur in Ireland, road, canal and later rail engineering all gave an important impetus to mapping the underlying rocks of this island. Geology was a consideration from the very foundation of the Ordnance Survey for practical reasons: the conversion of astronomical readings to terrestrial distances required knowledge of the specific gravity of the local rock mass and the role nearby landmasses played in refracting light.[69] Geology was one of the headings under which the first OS officers were ordered to collect additional 'local information' while in the field, and the hope was held out that the Ordnance would be able 'ultimately to accompany their map of Ireland with the most minute and accurate geological survey ever published.[70]

sources and resources (Edinburgh, 1997), pp 179–87. **67** Paul Gibson, 'Radar detection and ranging: the implication of spaceborne imaging radar for archaeological investigations' in *Archaeology Ireland*, xiii, no. 3 (1999), pp 8–11. **68** Andrews, 'Paper landscapes, mapping Ireland's physical geography' in Foster (ed.), *Nature in Ireland*, p. 209. **69** Clarke and James, *Ordnance trigonometrical survey of Great Britain and Ireland*, pp xvi–xvii. **70** Thomas Colby

The Ordnance Survey's greatest geological achievement was the *Report on the geology of the county of Londonderry and of parts of Tyrone and Fermanagh* (by J.E. Portlock, 1843); this was tied into the Survey's over-ambitious memoir scheme.[71] Though a monumental piece of scholarship, like the *Londonderry* memoir, the cost of this supposedly peripheral Survey work ensured that it too would be a stand-alone volume. From 1845 the geological survey was placed under the supervision of the Office of Woods and Forests (rather than the military), with a new director, Henry James.[72] Between 1856 and 1890, all 205 sheets (at first hand-coloured) of the geological survey of Ireland were published at the scale of one inch to one mile.[73] The OS connection was maintained in one respect only: all these maps were printed in the Phoenix Park, and hand watercoloured at Southampton (to ensure the consistency of colouring across the UK). Figure 42 illustrates the Hook peninsula, County Wexford, a favoured geological study-area as it is a distinct, manageable unit with outstanding fossil-bearing limestone. The asterisks * indicate 'fossil localities', in addition to the general note, 'fossils most abundant and beautiful along the shores of Hook Head except where the Black Dolomite prevails'. The boundaries between geological units are indicated by light dotted lines and letters: Lower Limestone (d^2); Magnesian Limestone (md); Bala beds (b^3); Old Red Sandstone (c^2), while the direction of dip of the beds is indicated by a small arrow, 'the figures denoting the angle below the horizon'. The definitive study of the geological mapping of Ireland is provided by G.H. Davies in *Sheets of many colours, the mapping of Ireland's rocks, 1750–1890* (Dublin, 1983), and *North from the Hook, 150 years of the Geological Survey of Ireland* (Dublin, 1995). The latter disentangles the early history and activities of the GSI from its parent body the Ordnance Survey, and recounts its subsequent development. Along with maps and sections, the Geological Survey of Ireland holds field sketches of geological formations by expert artists including du Noyer (1836–69), the GSI correspondence books 1845–*c.*1900, papers from the Portlock survey of the 1820s–40s and associated pre-1900 geological literature. Researchers may view the six-inch field sheets, upon which the geology was entered, as well as the one-inch published sheets. These have been scanned to a high resolution and can be consulted in the public reading room; print-outs and copies on CD can be procured at a modest cost.

From the perspective of the local historian, geology maps can be identified by pursuing those institutions and individuals with a specialist interest in mapping the physical endowment, or alternately, by first focusing on some striking local feature. Lough Neagh, the Giant's Causeway, the Castlecomer coalfield, the discordant river patterns of south Munster, the most impenetrable of Ireland's bogs, and any place

1826, quoted in Andrews, *A paper landscape*, p. 145. **71** Andrews, *A paper landscape*, pp 163–5, 168–70. **72** Herries Davies, *Sheets of many colours*, p. 127. **73** For an account of the progress of geological map publishing, see 'Abstract return of skeleton map of Ireland for Irish Geological Survey, statement of progress in publication of maps', H.C. *Command papers – accounts and papers*, 1854–5 (278) xlvii. 469.

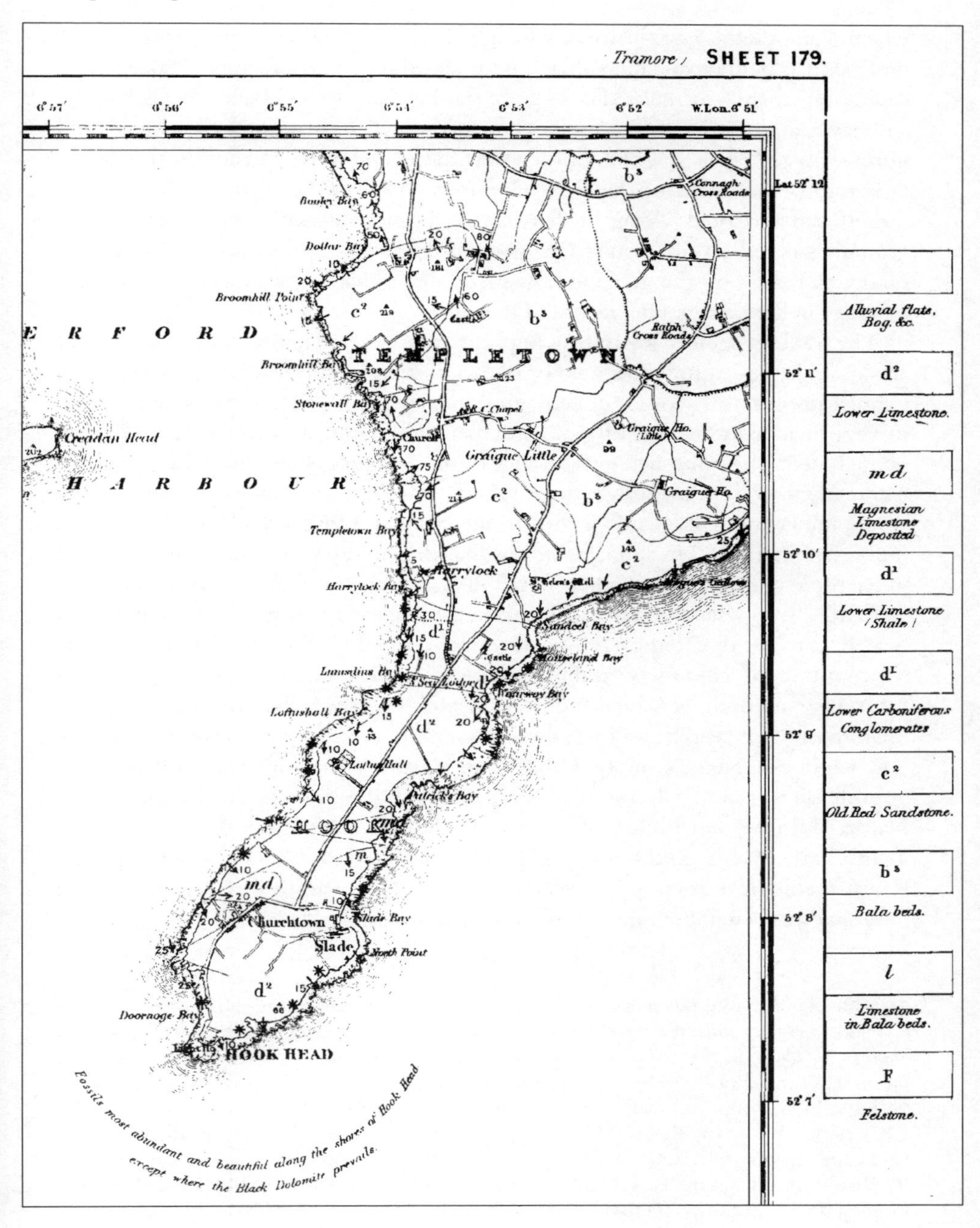

42 Hook peninsula, Geological Survey of Ireland one inch (Tramore) sheet 179 (1857)

where there was the least chance of finding gold all provided early geologists (amateur, professional, 'gentleman' and others), with absorbing areas of study. Drainage and navigation schemes formulated in 1847 for the 'Ballinamore and Ballyconnell drainage and navigation district', for example, generated excellent maps of the coal fields and iron works of Arigna as well as of Newtownforbes and other parts of north County Longford.[74] The maps of the Bogs Commission (1810–19) with which Richard Griffith was associated (see fig. 11) have already been discussed.[75] 'Bog or uncultivated ground' is noted on the regular OS six-inch sheets and the county indexes to this survey, as well as on the geological maps, ensuring that bogland receives extensive coverage in the cartographic record of Ireland.

The 'Dublin society for improving husbandry, manufactures, and other useful arts and sciences', through its county 'statistical surveys' commissioned in 1800, was the first institutional sponsor of geological mapping. Four of the completed 'county surveys' included some sort of geological map, although that of Kilkenny is regarded as the best.[76] The published proceedings of the Dublin Society from 1807 chart its increasing interest in matters geological, and its decision in 1809 to commission the young mineralogist, Richard Griffith, to undertake a study of the Leinster coal field (published in 1814). His sponsors hoped that this survey would set a high standard in scientific research which others could then emulate. Griffith's Leinster researches were followed by fieldwork in coalfields in each of the other provinces, and further field research as directed by the Society.[77] The Royal Dublin Society, through its 'Committee of chemistry and mineralogy', continued to sponsor geological surveying, and both its *Journal* and later *Scientific Proceedings* acted as major vehicles for publishing research findings, such as Griffith's 1821 study of the Kilmaleady 'moving bog' in King's County (Offaly),[78] or Kinahan's 'map of Croghan Kinshelagh gold district' (1882).[79] Geological surveying was also encouraged by the Belfast Natural History and Philosophical Society, by the Royal Irish Academy, by the Trinity-based Geological Society of Dublin (founded in 1831 and from 1864, the Royal Geological Society of Ireland), and by the Institute of Civil Engineers; published maps will be found in all of their journals.[80]

74 John Mac Mahon, *Ballinamore and Ballyconnell drainage and navigation district, report to the Commissioners appointed under the provisions of the 5 & 6 Vict., c.89, on the drainage of the flooded lands in the above district, and on the line of the junction navigation for connecting Lough Erne and the river Shannon and thereby the whole of the inland navigations of Ireland* (Dublin, 1845). **75** For a full account of Griffith's wide-ranging career see Gordon Herries Davies and R. Charles Mollan (eds), *Richard Griffith, 1784–1878* (Dublin, 1980). **76** Andrews, 'Paper landscapes, mapping Ireland's physical geography', p. 209. **77** See Gordon Herries Davies, 'Richard Griffith and the Royal Dublin Society' in Herries Davies and Mollan (eds), *Richard Griffith*, pp 123–42. **78** Richard Griffith, 'Sketch of the bog of Kilmaleady in the King's county showing the moving bog' in *Journal of the Royal Dublin Society,* i (1821), p. 144. **79** G.A. Kinahan, 'Map of Croghan Kinshelagh gold district' in *Scientific Proceedings of the Royal Dublin Society,* new series, iii (1882), pl. 21. **80** For example, F.T. Hardman, 'Age and mode of formation of Lough Neagh' in *Journal of the Royal Geological Society of Ireland*, iv (1875), pl. xi.

The *Instructions to the valuators and surveyors* ... issued by Richard Griffith (edition of 1853), gives a lengthy exposition on the economic justification for geological mapping, and encloses a small-scale loose-leaf geological map of Ireland. In the complex task of assessing the 'intrinsic worth' of soils, a knowledge of rock types and the composition of drift material is urged, 'and a reference to the geological map will frequently assist his judgment in this respect' (art. 59).[81] While ostensibly about the business of fixing the limits of holdings and assigning a fair and transparent value to each unit, valuators and draughtsmen in the office's employment also made geological observations, that contributed to Griffith's long-running and much cherished ambition of creating the first accurate all-Ireland geological map, as noted below under 'railway mapping'. Copies of his early small-scale geological map were used by other authors to argue the case for investment in Ireland, as in William Bullock Webster's polemical text, *Ireland considered as a field for investment and residence* (Dublin, 1852).

Great hopes continued to be placed on the mineral resources of Ireland into the twentieth century; maps were as always central to that discourse. Volume II of the *Memoir on the coalfields of Ireland*, launched in July 1921, consists of eight maps, showing 'the coalfields of Ireland, correlated with the coalfields of Scotland and Wales' (25 miles to one inch), and larger-scale maps of the individual major coalfields of Ireland (Leinster, Connacht, Slieve Ardagh, Munster, Ballycastle, Lough Neagh), concluding with a reproduction of Griffith's 1818 cross section of the Leinster coalfield, with additions from 1920. The minutes of evidence 1919–21 were also published, greatly enhancing the value of the maps to historians of these areas, which cover Northern Ireland as well as the Republic. Maps were also produced to illustrate the *Report on water power* (January 1922), showing water power stations, and in the *Report on peat* (December 1921), showing the peat bogs of Ireland. These reports are held by the NLI as well as the RIA.

Recent publications by the Geological Survey of Ireland of interest to local historians are the bedrock geology 1:100, 000 series (large fold-out map and booklet), which are available directly from the GSI (Beggars Bush, Dublin 4), while the metallogenic map of Britain and Ireland shows the range of metalliferous and related mineral deposits at a scale of 1:500,000. Contemporary geological mapping can act as an entry point to a range of study topics, including the history of the numerous small-scale mining enterprises throughout the country.

RAILWAY MAPPING

The first line of modern railway opened in Ireland was the short Dublin to Kingstown (Dún Laoghaire) line in 1834.[82] By 1836 this was still the sole line.[83] The four commissioners appointed in October 1836 'to consider and recommend

81 There is a copy of this map in RIA, Fr.c/sect.4. **82** Extensions to Dalkey proposed in 1845 and again in 1846; George Smyth, 'Lines proposed by private enterprise in 1836, a general system of railways' for Ireland approached the subject from the perspective of a government willing to subsidise railway development in Ireland (though not in Britain), on the basis that 'the backward state of the country and its not unfavourable position, presents a stronger obligation, as well as a wider scope for improvement'.[84] The chairman of the commissioners was Thomas Drummond, under-secretary of Ireland who had worked on the primary triangulation of Ireland (see above). He and his engineering colleagues saw rail as 'an invaluable source of general wealth and prosperity', as they contemplated the improvements that had come in its wake in the US and in Britain.[85] The advantages which a railway network (linked to steam-packet ports) would bring to Ireland were perceived as boundless, boosting commerce, agriculture, and tourism, hastening the dispatch of letters, and even influencing 'the security of the public peace' as troops could be deployed so much more quickly and efficiently by rail.[86] All of these interests are represented in the map-making that was part of railway planning.

Alongside the general encouragement of railway development, it was clear that some sort of strategic long-term plan was required if the lines were to be commercially successful. Geological mapping was regarded by the commissioners as essential to their work, to serve as 'a valuable guide in determining the best lines for improved internal communication, whether by ordinary roads, by railways, or canals'.[87] The commissioners appointed a team of surveyors to produce large-scale surveys of individual routes.[88] They also produced a road traffic map to determine the flow of goods and people between the principal towns of Ireland, a remarkably sophisticated example of thematic mapping, and part of the atlas of maps appended to the second report of the commissioners (1838) discussed below. Most importantly, they arranged for the reproduction of Griffith's geological map of 1835 on a superior base map, specially compiled by Thomas Larcom at the Ordnance Survey from original field survey materials, including its own triangulation and the six-inch maps already completed, namely those of the northern counties, along with private survey materials.[89] First published at a scale of about 1:633,600 (one inch to ten miles), it was appended to the *Second report of the Irish railway commission* (HC 1837–8, xxxv, p. 472), under the title *A general map of Ireland to accompany the report of the Railway Commissioners shewing the principal physical features and geological structure of the country.* The map itself was not completed until 28 March 1839 (according to Griffith's signature in the margin), a few months after the report was submitted.

Kingstown and Dalkey railway', H.C. 1844 (265) xliii.191. **83** *Second report of the Irish Railway Commission*, H.C. 1837–8 (145) xxxv, part III, p. 88. **84** Ibid. **85** These were John Burgoyne (of the Irish Board of Works, appointed Inspector General of Fortifications in 1845), Richard Griffith (of the valuation survey), and Peter Barlow, from the Royal Military Academy (Woolwich) who was an expert in the mechanics of railway locomotion. See Andrews, *Shapes of Ireland*, p. 287. **86** *Second report of the Irish Railway Commission*, p. 93. **87** Ibid., p. 34. **88** These are listed in *Railways (Ireland) return … of the expenditure of the Commission in Ireland*, H.C. 1839 (88) xlvi; letter book of Charles Vignoles (NAI, OPW 1/10/3); report of surveys by John McNeill (NAI, OPW 1/10/4). **89** Andrews, 'Paper landscapes, mapping Ireland's physical geography' in Foster (ed.), *Nature in Ireland*, p. 210.

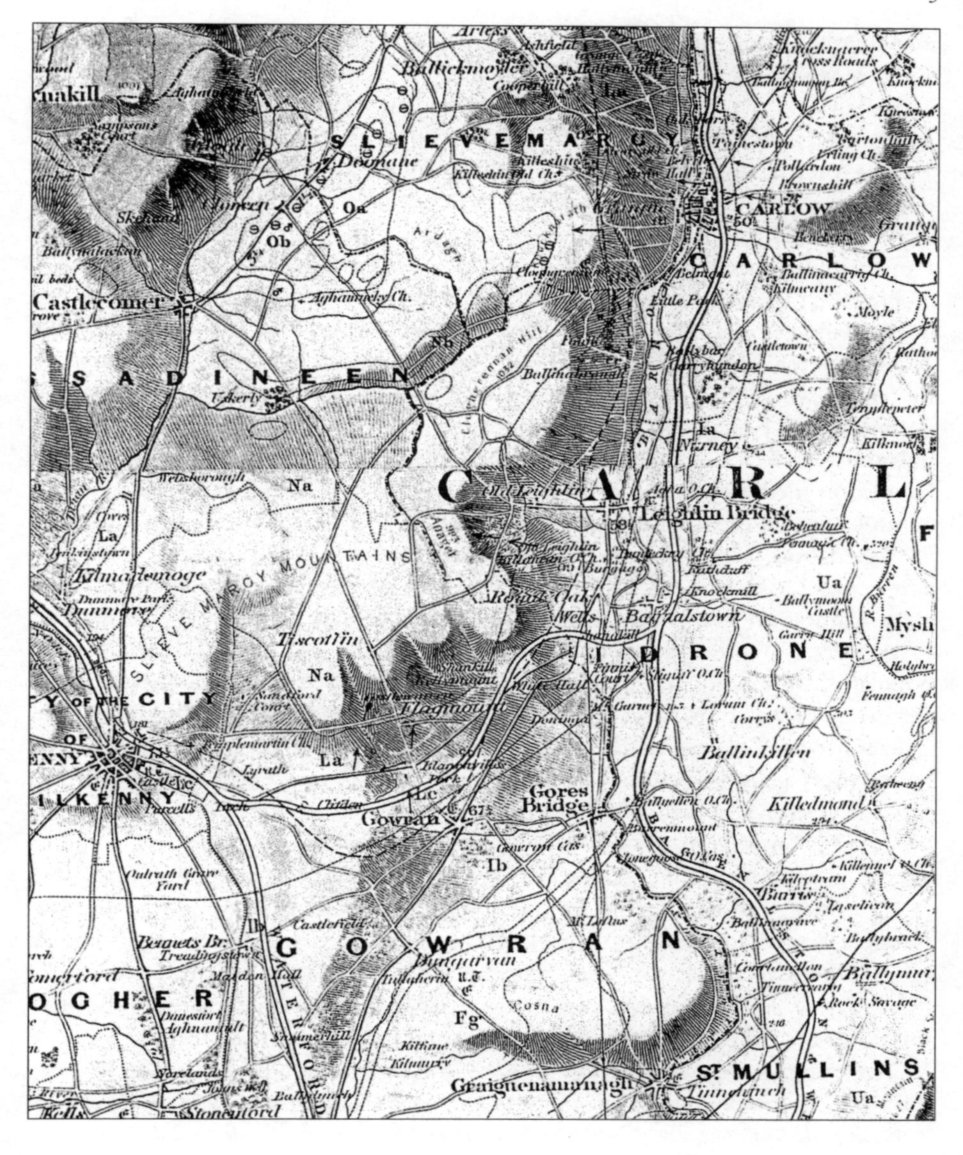

43 Extract from the railway commissioners map, 1838

The following year, the full-size monumental version was launched to great public acclaim, at a scale of four miles to one inch, or more popularly called the 'quarter-inch map' (1:253,440). Taking six sheets to cover Ireland, it was the most detailed and accurate small-scale map of the entire country yet produced,[90] and valuable well beyond the realm of railway planning. This map makes an impressive wall display, as in the TCD map library, but is more commonly viewed as separate sheets, as in the NLI, or in fold-out book form, as in the RIA Library. Cambridge University Library holds a set dated 1849–60, and a revised version of 1891.[91]

Figure 43 is an extract from the Larcom and Griffith map of 1839 covering the border between counties Kilkenny and Carlow. The geological boundaries ('laid down by Richard Griffith') are indicated by light dotted lines, the dip in the strata by arrows, while relief, drainage, bogs and general topography are also represented, on what was – briefly – the most up to date map yet available. The secondary triangulation was not yet completed; progress may be gauged from the inclusion of spot heights on the railway map given in feet, as in (fig. 43) Cloghgrennan Hill (1032), Mount Leinster (2610), Blackstairs (2411) and White Mountain (1627). Churches are noted, both 'old' and more recent (Ballyellen O. Ch., Templeshambo Ch.) and a few chapels (Ratharoge Chap.). The rank of the settlement is indicated by the font style, from county towns (KILKENNY), other principal towns (*NEW ROSS*), market towns (Gores Bridge), and 'villages, gentlemen's seats etc.' (*Tinnehinch*). Strict population thresholds did not apply, and according to Andrews alongside a few instances of market towns having larger populations than a 'principal town', there are miniature central places passed off rather freely as villages as the compilers struggled to ensure every placename was tied to a particular spot on the map.[92] The 'gentlemen's seats' may be picked out from the ornamental trees (*Castlegrace*). The barony names are rather conspicuous (IDRONE, ST. MULLINS); these boundaries pre-date reforms started at the end of 1839.[93] There is some sense of hierarchy among those roads shown, with trunk roads from Dublin indicated by thickening one line (for example, from Castlecomer to Kilkenny), while distances are given from Dublin in statute miles (Carlow 50, Leighlinbridge 58). The proposed railway lines are indicated by double lines, the few lines already approved by parliament (but not surveyed) by dashed lines (as between Kilkenny and Gowran and Carlow town).

The railway map, without the proposed railway lines, was reduced and reproduced as *A general map of Ireland for the use of the Commissioners of Public Works, exhibiting the boundaries of counties, baronies, Poor Law unions, and electoral divisions with the principal physical features of the country, 1847*.[94] It was issued in six sheets, and published in 1852. On this less cluttered map, mileage from Dublin is noted (in figures over the towns) and from town to town (over the road), with a circle for the

90 Gordon L. Herries Davies, 'Richard Griffith, his life and character' in Herries Davies and Mollan (eds), *Richard Griffith*, pp 16–18. **91** One inch to four miles, by T.A. Larcom, CUL Maps 171.84.2 and 171.89.1. **92** Andrews, *Shapes of Ireland*, pp 297–8. **93** Ibid., p. 297. **94** CUL Maps 171.84.7.

location of workhouses. Distances mattered to the poor law planners, as no single place should be more than a day's journey from the nearest workhouse. This administrative version of the railway map is without the arrows showing geological 'direction', and there is a changed font style to distinguish the poor law unions from the baronies (now in italics). The map demonstrates how central government, in the shape of the Poor Law commissioners, literally created an urban hierarchy even where there was no urban centre of consequence, as in Glenties in County Donegal.

State-sponsored high-quality mapping, as in the work of the railway commissioners, is but one aspect of the explosion in cartography that accompanied the advancement of the railways. The private sector was the chief sponsor of the new railways, albeit with state support and some funding. Proposals to create railway companies, to construct and run lines, to add branch lines or extensions, to deviate from an original planned route, and for other minor railway works all required – in Ireland as in Britain – that these sponsors deposited large-scale plans with their applications to parliament.[95] The procedures date back to 1794 when all promoters of canal or water bills involving new works were obliged to submit plans of any land that might be needed for compulsory purchase under the bill, along with a book of reference to the owners and occupiers (and usually also lessees) of such land, and list of owners, occupiers and lessees consenting, a statement of any sums subscribed for the purpose, and an estimate of the expense of the project.[96] In brief, the 'standing orders' of the House required that any person or body seeking to have legislation passed granting them compulsory purchase powers in respect of their canal or water project, needed to submit very full details of the project – as maps, tables and lists – to parliament before their request could be considered. 'Private legislation' of this type confers special powers or benefits on an individual or group (such as the directors of a canal company), in excess of or conflicting with the general law,[97] and thus is treated in a very serious manner. This approach was soon expanded beyond canal and water projects, with similar deposits required in the case of docks and harbours (from 1800), railways and tramways (from 1802), town improvements, streets and paving (from 1811), and turnpike roads and bridges (from 1814). With reference to railways, from 1838 deposits of small-scale maps of the entire project were required, along with enlarged plans of built-up areas, and sections of the whole works, together with a copy of the subscription contract. The abbreviations PSR (plan, section and book of reference); L, O&O (lessees, owners and occupiers), D&E (declaration and estimate), CSC (copy subscription contract), SC (subscription contract), EE (estimates of expense) are entered after the scheme name in the manuscript 'Index to House of Lords Plans' from 1794 onwards.

Figure 44 is an extract from the House of Lords Record Office (HLRO) deposited plans showing the 'Dublin to Drogheda railway with Clontarf deviation' dated 17 May 1836, at a scale of 'four inches to one British mile. The line of the railway is given in red on the original, and the parish boundaries in green. The

95 David Smith, *Maps and plans for the local historian and collector* (London, 1988), pp 127–33.
96 Maurice F. Bond, *Guide to the records of parliament* (London, 1971), p. 71. **97** Ibid., p. 70.

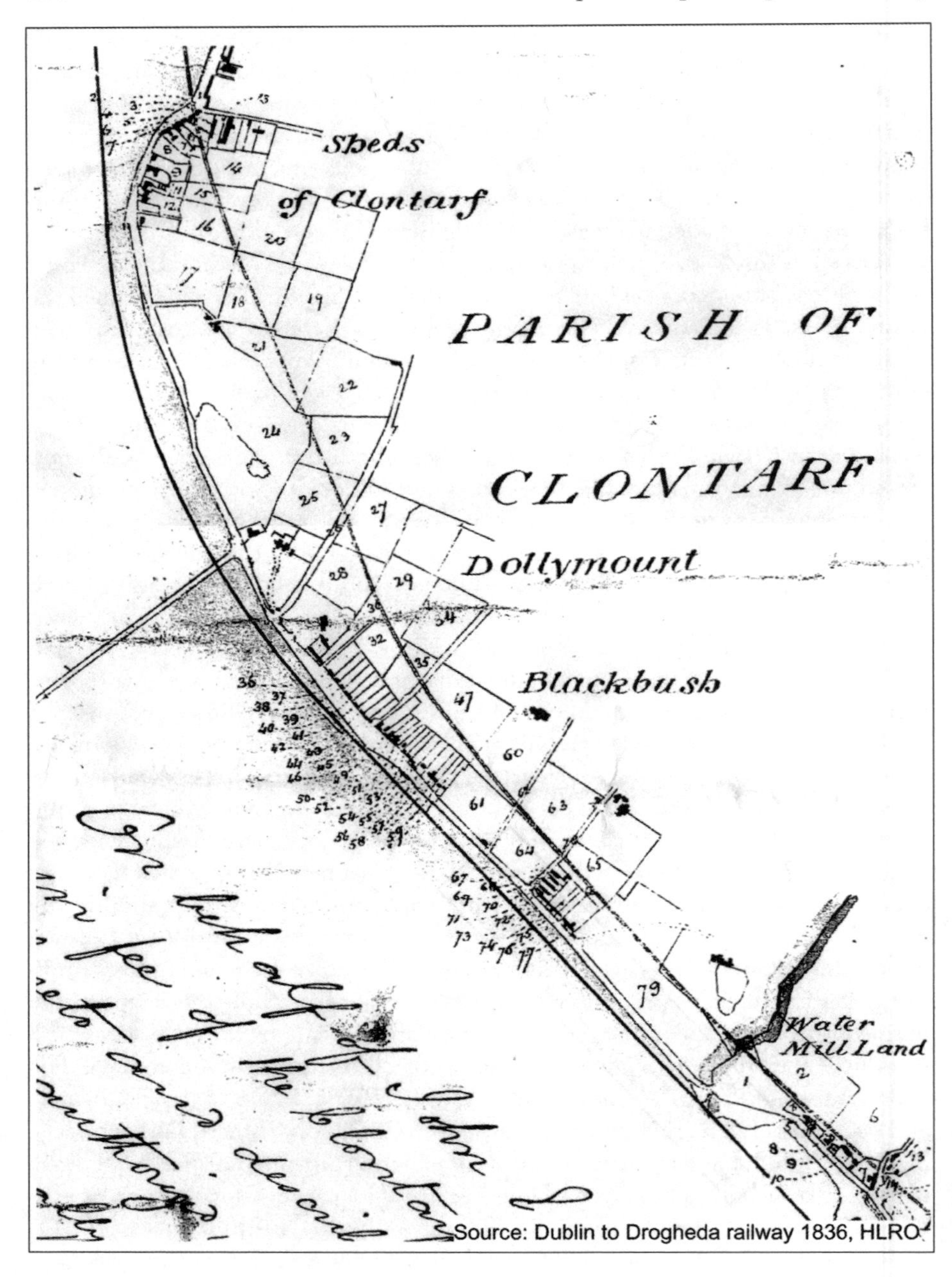

44 Dublin to Drogheda, with Clontarf Deviation, 17 May 1836, HLRO

numbered units refer to the manuscript 'book of reference', which gives map number, name of owner or reputed owner, lessees, occupiers, townland, parish, barony and description of property ('small house and garden', 'stripe of ground', 'a field', 'waste ground'). A separate 'list of owners', keyed to the map in the same way, records assent/dissent/neuter with a column headed 'observations', left blank except for a few cases in which doubtful ownership is confirmed or corrected. There is also a list of occupiers, again keyed to the plan, but as one name is used to cover several houses this must be a record of the 'head tenant' rather than of the families who were resident in each house. Units numbered 1–17 in the parish of Clontarf (fig. 44) were the property of John E.V. Vernon, who 'dissents'. The map records the agreement made with him through his agent, James Sullivan, and confirmed in London by Charles Nashe, chief justice, in the presence of three witnesses, by which Vernon forced the planners to deviate from their preferred line (through the 'sheds of Clontarf') avoiding his own demesne (now St Anne's Park), before taking up the planned line again after Kilbarrack. There is a dynamism about the deposited railway plans as they record alternatives, compromises and aborted routes as well as the final 'decided' routeway. Other changes to the Dublin-Drogheda line which are recorded on these plans are the intention to continue the line right though to Sackville (O'Connell) street, with a grand terminus almost opposite the GPO, rather than in Amiens Street, where the station was eventually built. Large-scale numbered plans and lists show how far this plan for a central site had advanced, before the decision was noted, 'this part of the line has been abandoned'. The plans for the Drogheda – Dublin line are accompanied by sections prepared by the engineer William Cubitt, numbered at intervals along the line. The total estimated cost is given at £600,000, and the subscribers are named, with the sum each invested (and hence number of shares), their 'place of abode' and 'quality or calling'. Lord Talbot de Malahide put forward £5,000 entitling him to 50 shares; the hotelier J.M. Gresham of Sackville Street invested £3,000, S.K. Mulholland, flax spinner, Belfast, risked £1,000, while among the fifty-three enterprising residents of Drogheda who invested in the project were merchants, farmers, victuallers, a linguist and a printer. Some of the major contributors came from Manchester, Liverpool and Salford (in that order); tea dealer, merchant, corn factor, gentleman and surveyor are among these occupations. It is this type of detail that makes the railway plans so valuable to the local historian.

The principal drawback to the House of Lords Record Office railway plans is that they are concerned with a narrow strip of land on either side of the proposed line. They are also rather cumbersome to work with; *Plan of a proposed railway from Dublin to Drogheda prepared January 1836*, at a scale of four inches to one mile (fig. 44 is an extract from the revised version, May 1836), is over nine feet in length and 2½ feet wide. The accompanying overview map, 'shewing the adjacent countryside', is at the scale of one inch to one mile and so is more manageable. Before the OS six-inch survey sheets were available for the district in question, the creation of these maps was a costly and time-consuming process. Many of the later railway maps are 'cut and paste' productions, with the relevant sections of the OS map numbered and coloured by hand. However, they do provide an insight into the process by which railways were developed at a very local level, how they were funded, who were the local entrepreneurs, and indeed who opposed the line. The property listings predate the general valuation (Griffith's) in some cases, and so provide a plot-by-plot account of land ownership and occupation which can be compared with Griffith's. The early maps are good examples of map-making by local surveyors and engineers, who had to rely on existing maps, at least in part, and even where the OS sheets were available to them, the amendments and notes can be valuable additions. By 1910 much of the country was in reach of at least a branch line so that the coverage is quite good.

There was a long, tortuous route from initial planning to the day when the company's directors and select visitors embarked on the first journey along the newly-laid track. Maps were the core document along the way: in the formal submission by the railway promoters to parliament (explained above), in the prospectuses, published reports, timetables and guidebooks of the individual companies, and in the official reports and special inquiries set up by the House of Commons and published as parliamentary papers. The NLI pamphlet collection, for example, holds many samples of promotional or advertising material.[98] There are also the records of the railway companies themselves; in the Republic these were subsumed into CIÉ, the state transport company in 1950. Some have since been transferred to the National Archives of Ireland. In Northern Ireland the Ulster Folk and Transport Museum sponsors research and publication in this area. In Ireland, the production of a geological survey of the entire country was hastened by the pressures of railway construction, as discussed above. Alongside the Railway Commissioners archive, featuring surveys (including maps) from 1837, as well as associated minute books and correspondence[99] there were numerous experts ready to comment critically, and publicly, on their activities.[1] The extent and pace of rail development beyond the planning stage may be ascertained by examination of guidebooks, some of which were marketed under company names, such as the *Dublin, Wicklow and Wexford Railway, arrangement of trains* (1 August 1870, NLI P 383). Others were published under the names of their authors, such as Wyer, Fisher, Walsh and Berry, as in *Berry's Irish Railways* (1850, NLI JP 3834–5). Detailing connections to Holyhead and thus the link-up with the advanced British rail network was a recurring theme, a reminder of how local rail developments – the building of a new station, or laying of a branch line – were played out on an international stage. An early update on the extent of rail development, including information on junctions and shared lines, may be gained from the maps accompanying the *Irish railway guide* (Dublin, 1847, NLI P 924). Comparison between the recommendations of the railway commissioners, so meticulously mapped and carefully argued, and the

98 *Proposed railway line to Kilkenny*, 1836 (NLI, P.2139); *Prospectus of the Dundalk Western Railway*, 1837 (NLI, P.926). **99** See railway commissioners, secretarial branch (NAI, OPW 1/10) and accountants' branch (NAI, OPW 2/10). **1** George Smyth, *Observations upon the report of the Irish Railway Commissioners*, 1839 (NLI, P.472).

network which ensued, is instructive, especially at the local level. A comprehensive summary of the railway network by 1906, showing lines completed, those under construction, coach routes, and joint lines, is found in the *Report of the viceregal commission on Irish railways, 1907*.[2]

Although the fold-out maps which were part of railway handbooks may appear to be rather modest in terms of cartography, they merit the close attention of the historian for the unexpected local detail which they record, the interests they served, and the light they shed on the hopes engendered by the advent of the railway. When examined against the more detailed backdrop provided by the HLRO files, they become even more intriguing. The Dublin and Drogheda railway handbook of 1844, for example, with its well-designed and compact route map, prepared from the information already to hand in the large-scale plans, aims:

> to accompany the traveller on his journey from Dublin to Drogheda – to point out to him every object worthy of engaging his attention – to describe the various localities through which he has to pass – and supply such information generally, respecting the country, its trade, commerce, manufactures, statistics, historical associations, traditions and antiquities, as cannot fail to prove at once instructive and interesting.[3]

In the guidebook map the stages are numbered and stations highlighted to enable the traveller to follow the route exactly, moving from Amiens Street station northwards. The colourful and partisan commentary fills out the sparse map work. The passenger is treated to impressive – and reassuring – facts on the construction of the railway as the journey progresses, and historical notes on the most spectacular views, notable residences and villages to be seen from the carriage. Members of the Hamilton family had forced the engineers to replace the straight-line distance from Lusk to Balbriggan (the 'intended line') with a wide, and expensive, loop to the east, which was very much to their advantage. When the Hamilton properties come into view, the commentary runs quite out of control, extolling the 'handsomely situated' fishing port of Skerries, the newly 'fashionable resort' of Balbriggan whose prosperity was attributed to the 'spirited enterprise, fostering encouragement and munificent liberality of the Hamilton family', and glimpsed by the passenger between these two towns 'the elegant and spacious mansion of Hampton, delightfully situated in an extensive and richly-wooded demesne of five hundred acres', which is 'the beautiful residence of the excellent chairman of the Company, George Alexander Hamilton, Esq. MP'.[4]

Although most railway schemes were initiated by private bodies, the role of the Office of Public Works (Board of Works) in the creation of railways in Ireland should be given special acknowledgement. Its maps and associated records represent

2 *Vice-regal Commission on Irish Railways*, appendix to first report, 1907, v, H.C. 1909 (Cd.3633) xlix. **3** *The hand book to the Dublin and Drogheda Railway, containing a description of the scenery, towns, villages, and remarkable places lying along and contiguous to the line from Dublin to Drogheda, with all necessary statistical, historical, traditional, and antiquarian information, and also a guide to Drogheda and its environs*, 2nd edn (Dublin, 1844), p. 8. **4** Ibid., pp 49–54.

one of the most important archives in the study of railway development from the local to the national level. The minutes of the railway commissioners are deposited in the OPW archive (NAI, OPW 1/10/1, 1/10/2/1 and 2), and the annual reports of the OPW from the outset include reference to railway schemes alongside summaries of harbour, piers, road, drainage and other projects. While the OPW had no powers of development, control or regulation in the case of ordinary railways, under section 8 of the *Drainage (Ireland) Act*, 1842 (5&6 Vict.c.89) the board was required to investigate the adequacy of any 'bridges, culverts, tunnels and other works required for the passage of water under or across the line of railway'.[5] From 1871 the OPW advanced railway loans in Ireland; previous to this railway loans were made by the English Public Works Loans commissioners. However, from its establishment in 1831 the OPW possessed powers to make loans for the advancement of public works in general; under this heading it had, for example, granted £100,000 towards the development of the Dublin and Kingstown railway, 1834.[6] From 1851 the OPW arbitrated between landowners and the railway companies over the valuation of land; from the 1860s it inquired into the financial arrangements made by the promoters of tramway companies who were seeking government loans; and from 1889 it was the body through which state assistance to tramways and light railways was regulated. It had additional powers of investigation and supervision in the case of the Midland Great Western Railway (1849), linking Mullingar to Galway, where very large sums of public money were at stake. In these and other complex matters, the OPW required massive documentation on each applicant company, its financial security, surveyors' reports, maps, plans and estimates of the proposed construction, including longitudinal sections, transverse sections, plans for bridges and culverts.[7] Its papers are now held by the National Archives of Ireland, and catalogued online (under electronic catalogues).

TOURIST MAPPING

The railway guide-books could be considered simply as tourist literature; indeed local promoters were very anxious to ensure that existing or potential tourist sites, with their coastal, mountain or lakeland scenery, would be reached by the new 'iron road'. But even in advance of the railways, every encouragement was given to potential visitors, with the publication of travellers' accounts and road maps (as already discussed) from at least the mid-1770s. Promotional tourist literature made good use of maps, including both 'scenic route' maps and local town plans. Premier visitor destinations such as Killarney and Glengarriff harbour were amply covered.[8] Figure 45 is a typical extract noting the routes by train and by road from Belfast to

5 Lohan, *Guide to the archives of the Office of Public Works*, p. 269. **6** Ibid., p. 2. **7** Ibid. pp 269–84. **8** For example, *Notes on Glengarriff and Killarney* (London, 1878), includes route map to Glengarriff, Killarney and SW Ireland (eight miles to one inch), and inset map of Glengarriff harbour and vicinity (one inch to one mile).

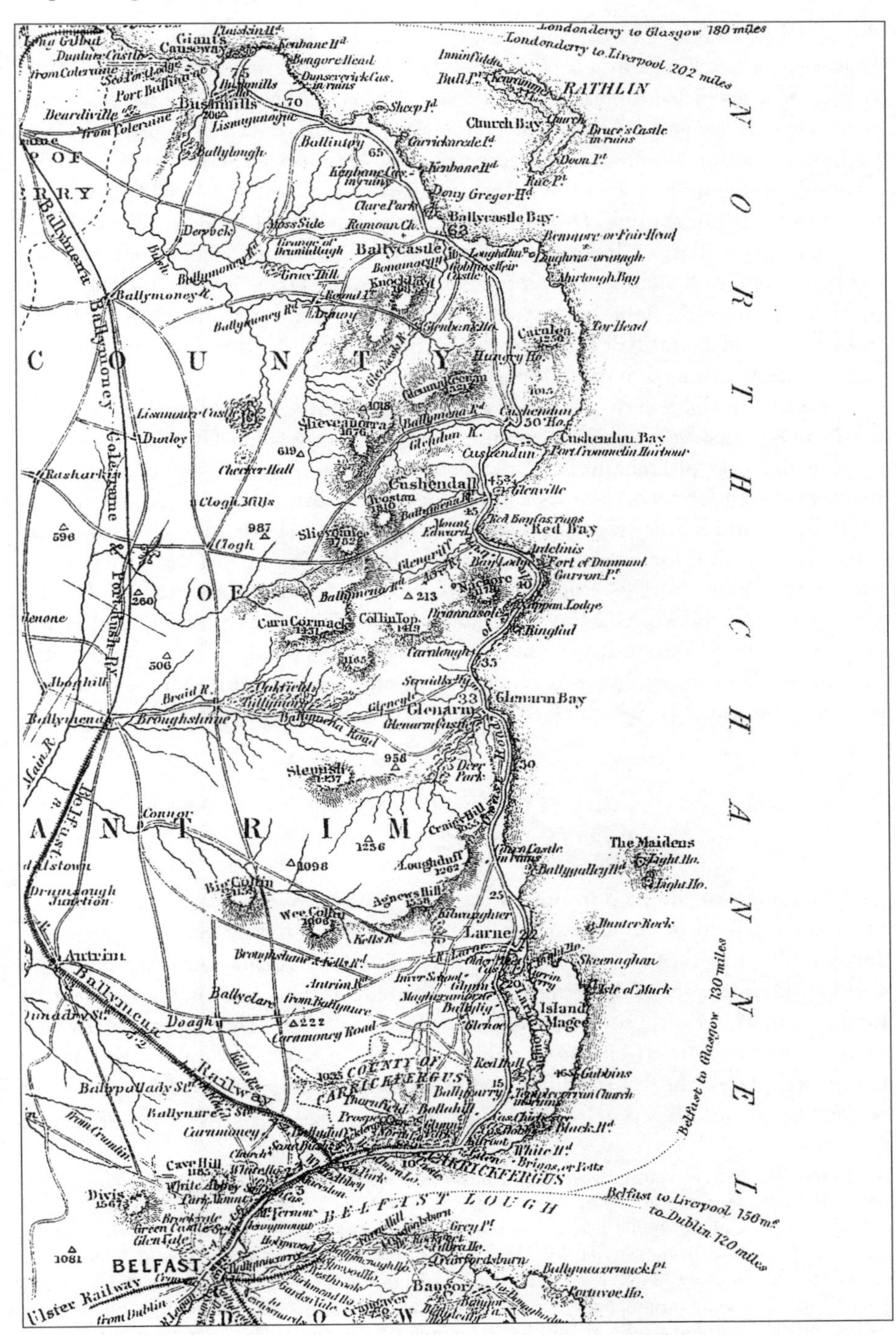

45 Belfast to the Giant's Causeway, County Antrim, from James Fraser, *Handbook for travellers in Ireland* (Dublin, 1854)

the acclaimed Giant's Causeway, seventy-five miles from Befast, in James Fraser, *Handbook for travellers in Ireland, descriptive of its scenery, towns, antiquities, etc. with all the railways now open, and various statistical tables, Also, an outline of its mineral structure, a brief view of its botany, and information for anglers* (Dublin, 1854). The intrepid Halls, professional visitor-commentators, adorned their *Pleasure tours in Ireland* (1827) with a ¾ inch to one mile map of the Lakes of Killarney, credited to Mr S. Hall. An 1892 reprint of Arthur Young's *Tours in Ireland, 1776–79*, edited by A.W. Hutton, was enhanced by a map of Ireland, noting the route Young had taken. Howth was exceptionally well mapped: to appreciate the route taken by the mail road; to promote its development as an asylum harbour for Dublin; to develop a rail link from a local quarry at Kill Rock to the harbour; and to further its popularity as a day-trip destination, all required map-making.[9]

A lot of late eighteenth- and early nineteenth-century tourist literature is superficial and predictable; the political commentator and cartographer Thomas Newenham subjects that class of literature to a merciless pounding in 1809.[10] The quality of the maps produced for such volumes is therefore highly variable. However, the visits of professional and scholarly groups were in a different category; such meetings were often the occasion for specially-commissioned maps. The August 1910 meeting of the Junior Institution of Engineers was held in Ireland. Visits to the Dublin Corporation waterworks, Messrs Harland and Wolff's works (Belfast), and to the Giant's Causeway, Portrush were all enhanced by maps produced to introduce the visitors to these impressive sites.[11] Map-making here was part of the promotion package, a lesson with which all tourist boards are long familiar.

MAPS TO ACCOMPANY PUBLISHED REPORTS AND PARLIAMENTARY INQUIRIES

The usefulness of the map in presenting the results of research or to describe the geographical field of endeavour has long been appreciated by private individuals and volunteer groups as well as by the State. The later plantation maps could be considered as *resumés* of the progress in settlement to date, and encouragement to further settlers. Mission societies were very keen on maps, which represented part of the overall effort to engender support for their work in faraway and as yet benighted places; the NLI holds a good selection of their published reports and tracts. The twentieth report of the Island and Coast Society (1855), whose object

9 Thomas Rogers, *Remarks on a road or safe anchorage between Ireland's Eye and Howth*, 1800 (NLI, P. 1610); *Report from the Select Committee … on Howth harbour*, H.C. 1810 (203) iv; *In considerations … of an asylum port in the bay of Dublin*, 1811, (NLI, P. 767); *Map shewing the situation of Howth Harbour, also the Mail road leading from it to the city of Dublin* and *Ninth report … on Holyhead road*, H.C. 1831–2 (584) xxiii. **10** Thomas Newenham, *A view of the natural, political and commercial circumstances of Ireland* (London, 1809), pp x–xii; Newenham's 'new map of Ireland' is appended to this pamphlet. **11** *Junior Institution of Engineers, journal and record of transactions*, 29th session, xx, part 11 (1909–10), p. 473.

was 'the glory of God and the instruction to life eternal of the poor ignorant people' is prefaced by a hand-coloured map which identifies the 37 stations where their agents labour 'in a field the least attractive and most difficult of any in the country'. The report laments how many 'very efficient teachers' had resigned 'whose wives could not be prevailed on to expose themselves to all the dangers and discomforts of those isolated and sea-girt abodes', as picked out in red on the map.[12] But many of the maps to accompany reports go well beyond merely locating and naming places. The early nineteenth-century 'statistical revolution' emphasised quantifiable amounts – tonnage, land valuation, deaths per 1,000 persons, percentage population in fourth-class accommodation, number that could read and write or read only – and the 'exact knowing' that this new paradigm held out as the way to advancement was complemented by the 'exact placing' that the map offers.

Some state inquiries and key legislative reforms were associated with mapping. The Poor Law (Ireland) Act 1838 required large-scale mapping: to define the limits of each union, the bounds within which the poor rates would be collected, and to authoritatively value individual holdings so that the tax burden within each union might be equitably spread, resulting in the tenement valuation, which has already been discussed. The new administrative divisions were to be 'compact and convenient units', precisely defined, an amalgamation of townlands (where such were defined) or of parishes, but freely dispensing with parish boundaries,[13] a sweeping territorial re-organisation that would have major consequences for public health administration and local politics into the next century. Maps of poor law unions, boundary revisions and later amalgamations will be found with the reports (parliamentary papers) of the poor law commissioners, and those of successive 'select committees' appointed to investigate the system.[14] The workhouse layout was also the subject of maps, plans, cross-sections and architectural drawings, most disturbingly in the thirteenth annual report of the poor law union commissioners for 1847, when famine was at its height. George Wilkinson, the architect of the system (for both England and Ireland), shows by means of coloured plans how a workhouse designed for 1,600 could be extended to take 2,500 persons, the lines of drains and location of cesspits and privies, the design of fever wards, temporary dormitories and a more 'economical

12 *Missionary progress of the Island and Coast Society, twentieth report and yearly statement* (Dublin, 1855), p. 11. **13** George Nicholls, 'First report, Poor Laws (Ireland) 1836' in *Three reports by George Nicholls Esq. to HM Principal Secretary of State for the Home Department* (London, 1838), p. 34. **14** For example, 'Map shewing the 130 new workhouses' in *Report of the Commission for inquiring into the execution of contracts for certain union workhouses in Ireland*, H.C. 1844 (C.562) xxx.495; 'Map of Ireland showing the present unions and the new unions which it is proposed to form, coloured green' in *First report of the commissioners for inquiring into the number and boundaries of poor law unions and electoral divisions in Ireland*, H.C. 1849 (1015) (1015–11) xxiii.369, 393; *Select committee on poor relief and medical charities (Ireland)*, H.C. 1846 (694) xi has maps of the following unions: Londonderry, Kilkenny, Sligo, Rathdown, Cootehill, Newtownards, Ballinrobe, Navan, Tralee; 'Return of unions and electoral divisions in Ireland which extend into two or more counties' *Command papers – accounts and papers*, 1872 (399) li.827, *Poor Law Reform (Ireland) Commission*, vol I: report, H.C. 1906 (Cd.3202) li, shows 'counties, poor law unions, and sites of the public institutions situated therein'.

bedstead'. The maps and plans impose some kind of surreal order on the human tragedy that was played out at the local level throughout the country.

Probably the most valuable collection of maps for Irish local history purposes to be found among parliamentary papers is the atlas of town plans produced for the purposes of municipal reform. On 12 February 1836 the undersecretary, Thomas Drummond, appointed a special Ordnance Survey team, under the command of a Captain Jones in the field, backed by the expertise of Thomas Larcom and his staff in the Phoenix Park, to the task of producing town plans at a scale of four inches to one statute mile.[15] They were to propose new boundaries for 67 named towns,[16] and to divide the cities and large towns into wards. Considering the pressure of time, the maps are exceptionally detailed, and further enhanced by individual commentaries. While the bill to regulate municipal, corporate and borough towns in Ireland could not proceed until the town boundaries had been definitively mapped, the commissioners did not expect the surveyors to do any extraordinary reworking of existing boundaries. Wherever the 'ancient boundaries' could fairly be said to 'embrace all that may properly be considered as the town' along with a sufficient space to allow for any probable increase, they could be adopted without further ado. The prospects of the town were judged on the basis of local economic activity, population and infrastructure. Wherever the town had expanded beyond these 'ancient limits' then a new boundary was to be proposed. Whether or not to include a nearby suburb was to be taken on a case-by-case basis, but the economic connections of the suburb with the town, and 'the extent and occupation of the ground separating such suburb from the town', were to be taken into question. Thus Rosbercon, in County Kilkenny, was to be embraced by the town of New Ross in County Wexford.

The surveyors did their best, but against serious odds in places. Wherever the OS six-inch survey had been completed their task was straightforward. Ardee, Armagh, Cavan, Newtown Limavady and Drogheda were all to hand, while Castlebar was also available to them, though not yet published. In other places (and mostly outside Ulster) they relied on 'local surveys', as in Athlone, Athy and Galway. All the maps are oriented north, following OS standard practice. The 'ancient boundaries' were established 'as accurately as personal investigation and tradition would admit'. There were a few instances where the ancient boundaries could not be determined at all, as in Charleville, County Cork. In some extreme examples the 'ancient boundaries'

15 *Instructions given … with reference to the boundaries and division into wards of the several cities, boroughs and towns corporate in Ireland, with letters and reports and plans received in response to such instructions*, H.C. 1837 (301) xxix.3, report no. 119. CUL Hib 3.8374. **16** Antrim, Ardee, Armagh, Athlone, Athy, Ballyshannon, Banagher, Bandon, Bangor, Belfast, Belturbet, Boyle, Callan, Carlow, Carrickfergus, Cashel, Castlebar, Cavan, Charleville, Cloghnakilty (*sic*), Clonmel, Coleraine, Cork (3 plans), Dingle, Donwpatrick, Drogheda (2 plans), Dublin (3 plans), Dundalk, Dungannon, Dungarvan, Ennis, Enniscorthy, Enniskillen, Feathard (*sic*), Galway (2 plans), Gorey, Granard, Kells, Kilkenny, Kinsale, Limerick (3 plans), Lisburn, Londonderry, Longford, Mallow, Maryborough, Middleton, Monaghan, Mullingar, Naas, Navan, New Ross, Newry, Newtown Ards, Newtown Limavady, Portarlington, Roscommon, Sligo, Strabane, Thomastown, Tralee, Trim, Tuam, Waterford (2 plans), Wexford, Wicklow, Youghal.

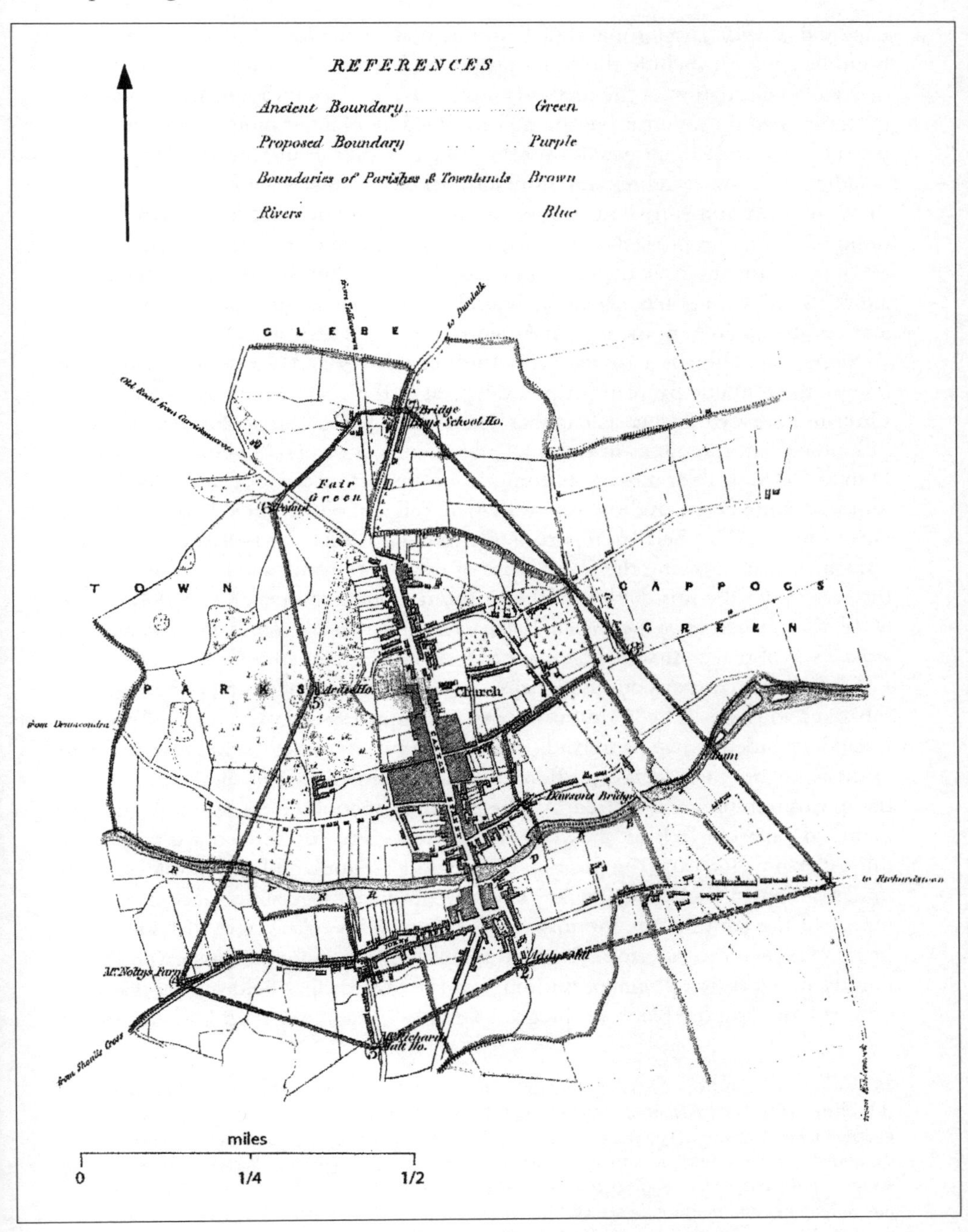

46 Ardee 1837, plan of proposed new municipal boundaries produced by Thomas Larcom, OS on behalf of the Under-secretary for Ireland, Thomas Drummond, in connection with the reform of municipal corporations

enclosed a very large rural area (as in Kinsale, Wicklow, Clonmel); the new boundary was to include the town proper and no more. Ardee (fig. 46) may be taken as an illustration of the method employed. The 'ancient boundaries' could not be ascertained on any map, but the officers relied on oral testimony, noting how the town had expanded southwards over the river, beyond the ancient boundary, before deciding on a much tighter and more rational bounded area which included these 'new' suburbs and sufficient ground to allow for modest expansion. To end all dispute, boundaries marked on the maps were to be complemented by a descriptive account so unequivocal 'that the same may be traced on the ground without the aid of a plan', with points where the boundary line changes direction clearly named and numbered (1–8, in the case of Ardee).

Signed by Thomas Larcom, with the boundaries and wards overprinted in colour, these municipal plans maps are typical of the 'special requests' made of the Ordnance Survey by other state bodies. The 1841 'digest' of the 'act for the regulation of municipal corporations in Ireland, Dublin', includes a pre-OS boundary map of Dublin.[17] The tolls and markets commission, reporting in 1853, included a map 'distinguishing between those [towns] where toll is abolished or still levied at their fairs or markets'.[18] The 1859 report on the municipal affairs of Belfast city included a six-inch map showing the boundaries of the 'lighted and watched' districts and the area under the jurisdiction of the Harbour Commissioners.[19] The 1881 *Report of the royal commission to inquire into boundaries and municipal areas of cities and towns in Ireland* was also accompanied by maps of individual urban areas.[20]

Several state commissions used maps to enhance their progress reports in a rather self-conscious way. The 'Commissioners for inquiring into the state of poor in Ireland' included a map of Ireland (one inch to twenty miles) to indicate the state of their progress in 1830: the 'portion of Ireland perambulated is coloured … (blue), that portion in progress … (red)'.[21] The National Schools commissioners similarly included a map of Ireland 'showing the districts visited by the ten assistant commissioners in 1868', and the position of the model schools and agricultural schools under the board (1868 and 1869).[22] The map produced specially to accompany the report of the land tenure commissioners (1845) shows the places in which they heard evidence or undertook 'inspection' visits; the heavy emphasis given to the railway network is a poignant reminder of the great hopes held out by this new infrastructure, on the brink of the great famine.[23] The Congested Districts Board

17 William B. Gannon, *Digest of the act for the regulation of municipal corporations in Ireland* (Dublin, 1841). **18** *Report of the Royal Commission to inquire into the state of fairs and markets in Ireland*, H.C. 1852–3 (1674) xxxxi, map by J.A. Humphreys. **19** *Report of the commissioners appointed to inquire into the state of municipal affairs of the borough of Belfast, part II, minutes of evidence and plans*, H.C. 1859 (2526.Sess.2) x.57. **20** *Report of the Royal Commission to inquire into boundaries and municipal areas of cities and towns in Ireland*, part 1: H.C. 1880 (2725) xxx; part 2: H.C. 1881 (2827) l. **21** *Report on the state of the poor in Ireland*, 3rd report, minutes, H.C. 1830 (665) vii. **22** *Royal Commission on primary education (Ireland)*, H.C. 1870 (c.6) xxviii, part ii. **23** *Map to accompany the report of the land tenure commissioners*, H.C. 1845 (C.605) xix; copies will be found in NAI, OS 111 no. 74, and in PRONI, OS/18/12 A–B.

for Ireland and the Royal Commission on Congestion in Ireland also supplemented their reports with maps, as in 1908 including a map to show 'the principal tours of inspection, and the country sittings of the Commission' (one inch to ten miles).[24] Maps were important in the transition to independence and in the first decades of the new state; the report and maps of the Boundary Commission was perhaps the most contentious of all and was not published until 1969. Maps showing the natural resources of Ireland (1921) have already been introduced under geological mapping. Maps were used in a sophisticated manner in the *Comisiún na Gaeltacht: Report* (Dublin, 1925). The number and percentage of Irish speakers in each district electoral division (DED) as returned in the census of 1911 was mapped by the Commission (map no. 1, over four sheets, one inch to four statute miles). The areas with high returns (counties Donegal, Mayo, Galway, Kerry, Cork and Waterford, and parts of counties Sligo, Roscommon, Clare, Limerick and Tipperary), were the focus of the 'special enumeration' of 1925. The number and percentage of Irish speakers in these areas was mapped (over several sheets); in the Dingle peninsula, for example, the DED of Dunquin returned a population of 460, nearby Ventry had a population of 848, and Kilmalkedar has a population of 1,002, all 100% Irish speaking. Based on its extensive researches, the Commission defined the 'Gaeltacht', as 'Irish-speaking districts' and 'partly Irish-speaking districts', mapped at the scale of ten miles to one inch (map no. 3). The printed report explains the criteria employed; children under seven years were defined as 'Irish speakers or non-Irish speakers' depending on whether older siblings were fluent in Irish, or where there were no older siblings, whether their parents were fluent speakers. The maps accompanying the *Comisiún na Gaeltachta* reports are large scale, but in most cases the maps appended to reports are more modest productions. The information in this type of publication is often given in list form within the report so that the map itself cannot always be regarded as an important source for the local historian. Nevertheless, the same information when viewed spatially can stimulate new lines of inquiry.

Maps were part of the planning of new infrastructure: improved roads, new bridges, well-constructed piers and harbours, drained bogland, canal link-ups, fisheries, and (later) railways. The reports of the Office of Public Works give a good idea of the range of works in which the board was involved, but are best viewed as the 'tip of the iceberg' in terms of sources. Fortunately, some at least of the NAI's enormous collection has been computer catalogued with a sophisticated on-line search facility, accompanied by Lohan's *Guide to the archives of the Office of Public Works*. This publication opens with essential historical background on the OPW's formation, activities and administrative structures, and an overview of the complex referencing systems created by the OPW (Board of Works) to manage its ever-expanding records. The responsibilities of the OPW with reference to railway development, inland navigation and archaeology have already been outlined; these however were but three of its many functions, most of which generated a rich crop of cartographic materials. The ubiquitous Richard Griffith features here also (from

24 *Royal Commission on congestion in Ireland*, final report, H.C. 1908 (4097) xlii, p. 746.

1846), most significantly, along with Thomas Larcom of the Ordnance Survey, in charge of administering famine relief works.[25]

In terms of cartographic history, the types of maps that have ended up with the OPW vary immensely. Many are simply OS maps or geological maps used in connection with drainage or inland navigation applications (see 'Engineering and architectural drawings – roads, bridges and canals' index, for example file OPW 5HC/6). Extracts from the six-inch sheets proved indispensable in planning new piers and harbours.[26] Other maps predate the OS, and could be considered under the headings of regional, county or road maps.[27] But even 'regular' OS maps in the OPW archive may prove invaluable, as from the 1870s certain categories of sensitive military information was excluded from the OS maps in the public domain, but included on those sheets prepared by the OS for 'secure' uses. OPW files on the refurbishment of military or constabulary barracks, for the RIC and later for the 'civic guard', can fill in these blanks. For example, in the case of Gort, County Galway, a series of OS maps (of the 25in. sheet 122.v), have numerous manuscript additions and corrections, dated from 1883 to 1933, outlining the area to be retained for police purposes, the refurbishment required, the new buildings to be erected and drains to be laid, marking the transformation from a military barracks to a Garda Síochána station.[28]

The context within which each map in the OPW archive was created and the purpose for which it was intended is made clear from the material with which it is lodged. Memorials or applications to the OPW for a loan to erect a harbour, pier and quays in Ballywalter County Down are accompanied by a map of the district, but also by correspondence, public notices, draft and final declarations, details of loan charges, and notes on the district to be assessed.[29] The annual *Public Works (Ireland)* reports (discussed above) bring map and text together, though obviously without the correspondence, gossip and trivia that lighten historical research. Additional responsibilities, such as the famine relief schemes, generated new cartographic materials. The *Public Works (Ireland)* report for 1847–8 includes a map 'exhibiting the numbers employed … during the relief operations of 1846–7' while the report for 1880–81, when another catastrophic famine was widely feared, also includes a distribution map of the relief works in hand.[30] That the OS co-operated with the OPW at times of national calamity is evident in the willingness of Captain Larcom to supply 'office proofs' of OS maps of Kerry in January 1846 'to enable the county surveyor to lay down public works projects for the county', though he insisted that these as yet unpublished OS maps be returned to headquarters when the county surveyor had completed his own maps.[31]

25 Ronald C. Cox, 'The engineering career of Richard Griffith' in Herries Davies and Mollan (eds), *Richard Griffith*, pp 41–9. **26** For example, *Twentieth annual report from the Board of Public Works in Ireland*, maps, H.C. 1852–3 (1569) xli. **27** For example, NAI, OPW5HC/6 no. 0306 also NAI, OPW5HC/6 no. 1339; *Report on the southern district*, H.C. 1824 (352) xxi. **28** NAI, OPW5HC/4 no. 192, 1911–1933. **29** From 'pier and harbour structures' index 1846–54, NAI, OPW8/30. **30** *Public Works (Ireland), sixteenth report, 1847–8; Public Works (Ireland), forty-ninth report*, 1880–1. **31** Relief Commission papers,

Important inquiries which required or produced maps were not limited to government-appointed bodies, though the publication of their formal reports as parliamentary papers ensures they are widely accessible. The reports of individual researchers typically reached public notice in pamphlet form, or as papers in the journals of scholarly associations.[32] It is common for the same report to appear both as a stand-alone pamphlet and within the Houses of Commons or local authority series. But regardless of how they were presented to the public, the availability of high quality base maps gave a real boost to researchers to use maps in their analysis and presentation. The OS was the major source of both small-scale and large-scale maps, producing maps to accompany official inquiries, such as the Dublin main drainage inquiry of 1879. However, private cartographers and county surveyors also produced their own maps (as seen in the OPW files), and in many cases out-of-date maps, and maps produced to accompany commercial directories, were also employed. Some topics were particularly suited to mapping, such as public health issues. The landmark report on the sanitary state of Belfast by Dr Malcolm (1852) is accompanied by five maps.[33] The problem areas are mapped, lest there be any uncertainty about the locations, extent and causes of 'the great epidemic of 1847'. The 'monster grievance' of the Blackstaff river threatens 'a large proximate and increasing population, a great property, and indirectly, all Belfast'; as is made clear from the map series.[34] So too the solutions are represented cartographically, as in the Great Victoria Street scheme, where 'a splendid array of marts and emporia of trade and commerce' have replaced the former slums 'where nought but vice, death and poverty held their fearful orgies'.[35]

In some cases the maps that accompanied published reports were the product of private surveyors who reworked existing maps to fit the purpose in hand. In other cases the Ordnance Survey itself produced the maps to order. However, some freelance map-makers simply made their own tracings from OS plans, overlaid these with new information, and made no reference to their master or base OS maps at all. This is very obviously the case with the maps accompanying the 1913 local government board inquiry into the state of Dublin housing, where slum housing is identified premises by premises,[36] a survey made possible in the first place only by the existence of large-scale authoritative mapping. The North Dublin Survey of 1918, in this case a local authority (corporation) venture, produced a massive land utilisation survey on the 1:2,500 sheets.[37] The significance of maps in local administration and planning should not be underestimated.

distress reports January 1846, Z series (NAI, RLFC2/Z762, nos. Z762, Z974, Z1514); Kerry first edition OS six-inch was published in November 1846; Andrews, *A paper landscape,* p. 333. **32** For example, railway pamphlets in the Haliday collection, 1845, vol. 1950, RIA, include several maps. **33** A.G. Malcolm, *The sanitary state of Belfast, with suggestions for its improvement, a paper read before the statistical section of the British Association*, 1852; this is reprinted as document 25 in the PRONI collection, *Belfast, 1780–1870, Problems of a growing city* (Belfast, 1973), pp 156–166. It is also held by many libraries, including NLI. **34** Ibid., p. 163. **35** Ibid., p. 157. **36** *Report of the Departmental Committee appointed by the Local Government Board for Ireland to inquire into the housing conditions of the working classes in the city of Dublin* H.C. 1914 (Cd. 7273) xix, pp 61–106. **37** 'Report of the housing committee,

THE 1:50,000 'DISCOVERY/DISCOVERER' SERIES

The full-colour *Discovery/Discoverer* 1:50,000 series (2cm to 1km), though designed for the tourist market, is also of value to the local historian. It is widely available through local booksellers, as well as directly from Ordnance Survey map sales in both Belfast (Colby House), and Dublin (Phoenix Park). The dual version (Irish and English) of major placenames on the *Discovery* sheets provides the official translation in an easily-accessible form, while the inclusion of townland names (though not of boundaries) will assist in locating specific places on earlier larger-scale maps. Antiquities are subgrouped as 'named antiquities', 'enclosure' and 'battlefield'; these are taken from the townland six-inch survey (published in 1833–46), and as confirmed by later research undertaken by the archaeology section of the Ordnance Survey in collaboration with the staff of the SMR (Sites and Monuments Record) office, in what was then the OPW (and is now part of Dúchas, the Heritage Service). The SMR maps and manuals, upon which the archaeological content of the 1:50,000 series largely depends, were published between 1984 and 1992; the nature of the digital database, however, allows additions and minor amendments (subject to the limitations of scale) to be made on an ongoing basis. Changes in relief are readily discerned by the use of contour lines at ten metre intervals, with every fifty metre contour in a heavier line, and isometric shading changing every 100m. Spot heights are given in metres, and the familiar triangle sign for trigonometrical stations is a welcome reminder of the difficulties overcome in the making of the network of points, distances and angles upon which the first all-Ireland 'interior survey' relied. On the Dublin city and district sheet 50, these include the Ben of Howth, and Three Rock Mountain, but areas with more rugged relief can boast more stations. Replacing the old one inch to one mile 'district' or black and white outline series (1:63,360, for example fig. 39), the 1:50,000 is a collaborative venture between the Ordnance Survey of Ireland (*Discovery* series) and the Ordnance Survey of Northern Ireland (*Discoverer* series). There are 89 sheets in all, 71 produced by OS*i* and 18 by OSNI. The Northern Ireland sheets are completed to the same specifications as those in the Republic, although there are some stylistic differences. An index to the full series is printed on every map, along with the principal authorities and sources used in its compilation (which historians will welcome), and dates of various revisions.

The Ordnance Survey undertook a dynamic process of consultation with user-groups, including hill-walkers and tourist bodies, at the early design stages, which is reflected in the addition of otherwise unexpected features: the mile posts along the railway lines (10MP, 11MP), walking tracks, electricity transmission lines and canal lock gates. In high moorland areas, where human habitation is scarce, it is

being a survey of the north side of the city of Dublin ...' in *Reports and printed documents of the Corporation of Dublin*, I, no. 13 (1918). Held by the Dublin City Library and Archive, Pearse Street, Dublin 2.

even possible to show prominent fences and walls.[38] Figure 47, an extract from sheet 50, demonstrates the massive amount of information which can be shown on a small-scale single sheet, due to the new digital technology combined with sophisticated printing techniques. The use of colour, carefully graded, allows overprinting on quite complex backgrounds, while the very fine contour lines ensure that the relief information does not intrude. The walking tracks in the vicinity of of the reservoir (*Gleann Smól*) and the Hell Fire Club are marked by dashed lines, individual antiquities (standing stone, barrow, holy well) are noted in red, and the use of at least some Irish – *Cluain Dolcáin, Cill na Manach, Sliabh na mBanóg* – can provide today's researcher with an entry point into local history, as with the early memoir-writers and placename specialists of the Ordnance Survey.

This new map series is a landmark in Irish cartographic history, the first country-wide survey based on aerial photography at the remarkably detailed scale of 1:30,000, entirely computerised, and in colour. As explained on each sheet, the information was collected digitally, and compiled into a structured vector database, that is, a 'link and node' structure, in which every point and line segment can be defined by a unique co-ordinate address. This information is entered in 'layers', allowing countless different combinations; the information displayed in the 1:50,000 sheets is not at all an exhaustive or full compilation. The entire computerised database survey is in essence a seamless whole, allowing any particular place to be 'centred' on a sheet, which is valuable where the area of interest crosses the margins between sheets. This has made possible the generation of maps from a digital topographical database at different scales, what OSNI term their 'ACE' series (Address Centred Extract), and OS*i* market as its 'PLACE map series' (Planning, Legal, Agricultural, Construction, Engineering). These customised extracts are typically for farm management, conveyancing, and planning application purposes; an example for Tallaght village, at the scale of 1:10,000, is reprinted as figure 53. The customer supplies an address or Irish Grid/National Grid reference, specifies coverage, scale and paper size (from A4 to A0), and the map is generated to order, centred exactly on the premises in question. While the price for this service may at first appear expensive, the costs behind such a high-quality product, though perhaps not apparent to the customer at the desk, are enormous. The Architecture Library, UCD, Richview, Dublin 4, is one of a small number of agents licensed by the OS*i* to provide the PLACE service on their behalf, at commercial rates, while the service is also offered directly to the public at the Phoenix Park headquarters; a full list of OS agents will be found at the http://www.osi.ie. Digital topographical database has already been used by the OS*i* to generate maps at a larger scale, for example the 1:25,000 hill-walkers' maps produced for Oileáin Árann and Killarney National Park, and the *Wayfarer* series, which include text commentary (by Michael Fewer), also at the 1:25,000 scale (the Wicklow Way and the Western Way South). Fluctuating demand for the different products is no longer an issue, as maps are

38 Introduction to *Discoverer* series, http://www.osni.gov.uk/.

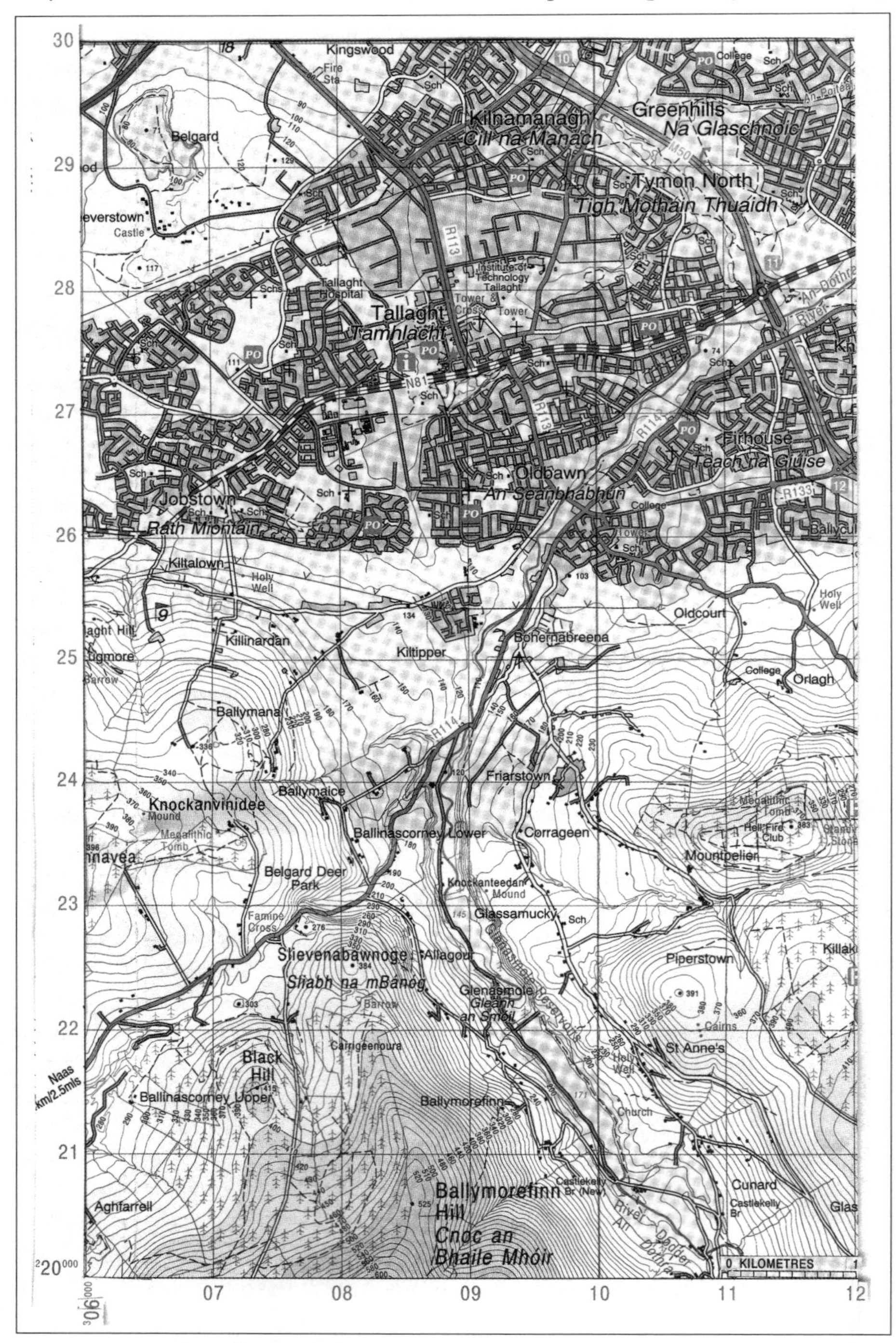

47 OS*i* *Discovery* series, 1:50,000, extract from sheet 50 showing Tallaght, County Dublin (2nd edn 2001)
Mapping by Ordnance Survey Ireland, Permit no. APL0000204 © Government of Ireland

produced in direct response to market demand, and there is no longer the need to hold immense paper stores. Even when maps have not been digitised, 'master copies' are now stored in raster form, as pixels or picture elements, from which copies are printed in response to demand. The digital mapping revolution has made possible the rapid revision of the 1:50,000 and other maps, by simply amending the database, and the easy production of 'special interest' or 'special event' maps, as in the competition venues map (at a scale of 1:50,000), produced by the OS*i* for the 2003 Special Olympics world summer games in Ireland. This also has implications for map archives, as maps are revised much more quickly, and each 'edition' must be saved electronically; there may not be a paper record at all. The digital databases held by the Ordnance Survey (OSNI and OS*i*) have immense commercial value, and licensing the use of its data for a wide range of computer-based applications, such as CAD (computer aided design) and GIS (geographic information systems) moves the OS into quite literally different realms.

This is a good point at which to conclude this introduction to Irish cartographic history. The digital age is transforming map-making in ways that could not have been conceived even a decade ago, and the demand for maps has never been so great. The local historian and geographer need to keep abreast of these trends, appreciative of the ways in which maps are now based upon electronic databases, speedily updated, produced to different criteria (with features included or excluded, changing areal coverage, various scales and page sizes etc.), and printed to impressively high technical specifications. But the critical stance is what the local historian brings to the world of map-making, always questioning its aims, audience, author and larger context (see Appendix 1), its silences and loud noises, the language in which it advertises itself, and the way in which it is marketed. Current state-sponsored and commercial maps promote their own world view as effectively as did any early modern map with its walled towns and flag-flying forts; a sensitivity to this will greatly enrich local studies.

CHAPTER 2

The map as a coded world: essential map-reading skills

The map is a selective and conventionalised representation of reality, but one which rarely represents a single moment in time. Most maps depend on fieldwork which stretches over days, months or even years, while many maps quite overtly refer to long periods of time, such as Clarke's *Dublin* c.*840*–c.*1540*. And the type or 'product' may look very different over time. The 'bird's eye view' characteristic of early modern maps may be compared with the more familiar two-dimensional ground plan (for example, figs. 13 and 14). Yet the underlying concept is identical: 'the representation of features (places, people, phenomena, real or imagined), in their relative or actual spatial location'.[1] Each is written in its own code, and cartography as both science and art has developed a sophisticated language, to communicate, highlight, make connections, separate out, and group phenomena. Data collection and manipulation, information display and image processing, are all part of the cartographic process; all involve making choices. What will one include? Why? Where can the map-maker get this data? How should it be represented? At what scale? How legible will it be? The conventional symbols used on maps, as noted in the 'legend', 'key' or 'characteristic sheet', and (more rarely) explained in an accompanying manual,[2] are the most obvious entry point to map reading. But, as with the acquisition of any new language, there are further subtleties that are grasped only from close and sustained engagement over time with different examples from various historical periods and geographical regions. Map reading can be likened to the analysis of literary texts,[3] where the 'language' of the map is explored for meaning, intent, creative context, stylistic predecessors, genre and so on. As with the 'deconstruction' of a poem, novel or picture, layers of meaning may be found behind the written text, through a process of close analysis of each component, drawing comparisons with forerunners in the field (earlier maps), questioning the readership for which it was produced, and making reference to the intellectual, political and artistic milieu in which it was created (see Appendix 1). As emphasised in the opening chapter of *English maps: a history*, 'the economic, cultural, political and demographic features of each period are as much a part of map history as are

1 Rebecca Stetoff, *The British Library companion to maps and mapmaking* (London, 1995), p. 1. **2** For example, see Daniel Augustus Beaufort, *Memoir of a map of Ireland; illustrating the topography of that kingdom, and containing a short account of its present state, civil and ecclesiastical; with a complete index to the map* (Dublin, 1792). **3** Harley, 'Maps, knowledge and power', pp 277–312.

the more commonly-stressed scientific, technical aspects of map-making'.[4] But the map must also be approached from a technical point of view. How can complex reality be represented in two dimensions? What is its mathematical framework? How are absolute location and relative distance conveyed? To what reference points is the map tied? What are the possible ways of conveying changes in height and depth? What area is to be covered, at what scale, and at what level of generalisation? What units of measurement to employ? On what basis will the map sheets be organised? The inclusion of boundaries and the naming of places are technical as well as philosophical and indeed political questions.

A direct way of entering into an appreciation of the map as a coded world is to compare a recent large-scale map firstly with an aerial photograph of the same area (fig. 48), then compare with other landscape views, and finally one should walk the land, before repeating the exercise. Through this process, the map-reader becomes more conscious of how 'removed' the map is from the real world. It reduces very different quality housing to similar flat shapes, lofty public buildings to site outline, ignores 'temporary' structures and presents roads as unobstructed routeways. The aerial photo presents all the visible details of the land surface, without discrimination (fig. 48a); the map shows point, line and polygon data, as in site of antiquity, road and property boundary, church outline (fig. 48b). In addition it shows information not physically present on the ground, such as parish boundaries and placenames. The instruction of Major Thomas Colby that 'everything attached to the ground is to be inserted on the [six inch] plans', with the (initial) exception of field boundaries (*Instructions*, 1825, no. 65), was laudable, but humanly impossible. The map extracts, aerial photos and characteristic sheet or legend key in the OS*i* educational map pack for the Leaving Certificate geography curriculum provide ideal materials to embark upon this type of serious map study. Absence from the paper cannot be equated with absence from the real world, while the inclusion of certain features on maps constructed in the twenty-first century has as much to do with the long history of cartography – what we have come to 'expect' on a 'good' map – as it has to do with what is perceived as being of contemporary significance. Barony boundaries were a prominent feature on eighteenth- and nineteenth-century atlas maps even when they had little relevance to human geography. Similarly civil parish boundaries on maps have long outlasted their relevance as an administrative unit; township boundaries and electoral subdivisions did in time appear, but fortunately for the historian the erasure of redundant units from the map takes time. The map cannot give a sense of the worshipping congregation but merely locates a church building, which may be derelict, refurbished or a replacement building. The distinctions selected by the map maker – such as between coniferous plantation, natural woodland and mixed woodland on the new OS 1:50,000 *Discovery* sheets (fig. 47) – may or may not matter to an individual map-user. In most cases the map-reader does not have access to the principles or instructions governing the map-making process; he or she is faced solely with the finished

4 Catherine Delano-Smith and Roger J.P. Kain, *English maps: a history* (London, 1999), p. 1.

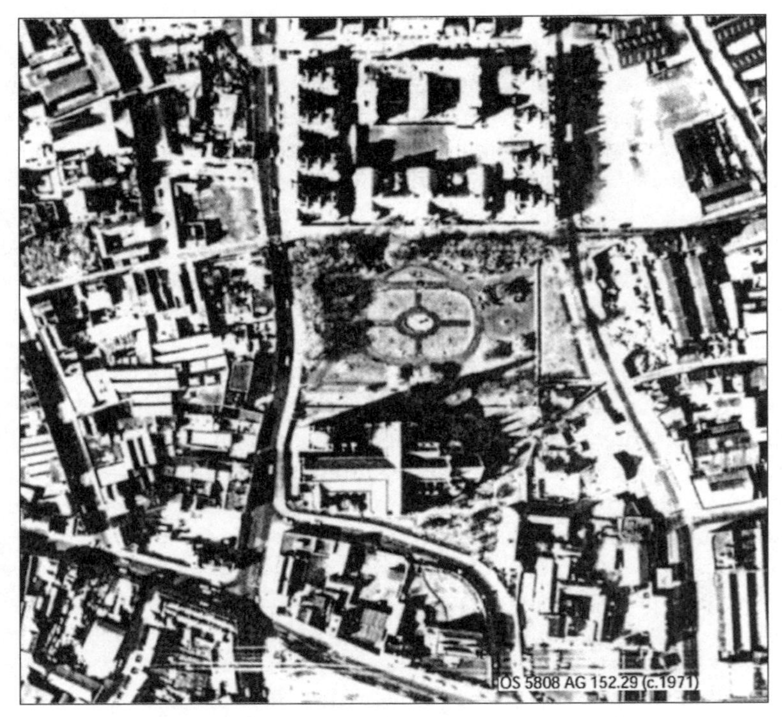

48a Aerial photograph of St Patrick's Cathedral, Dublin (1971)

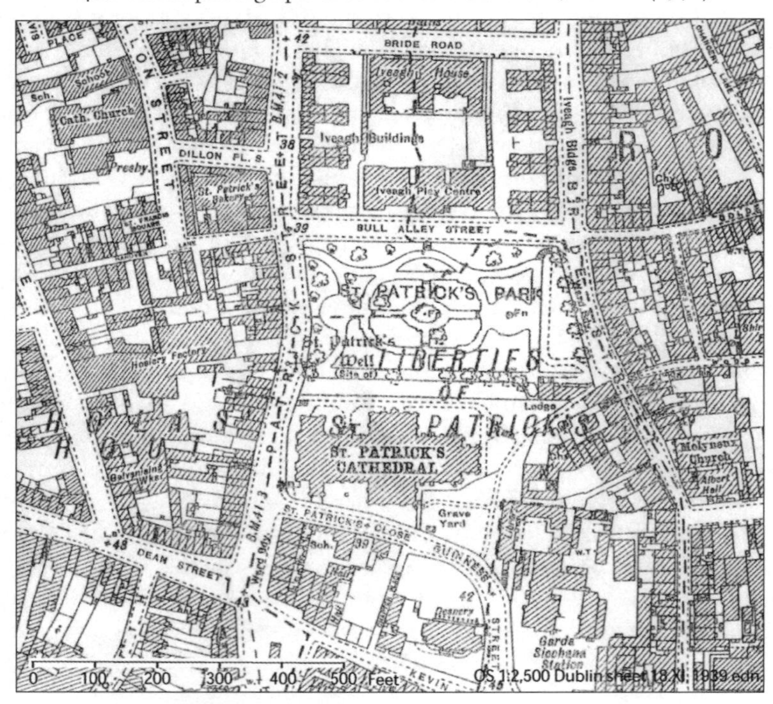

48b Planimetric view of St Patrick's Cathedral, Dublin (1939)

product, left to unravel a complex document without any context or back-up. There is also the question of empty space. As J.H. Andrews notes in relation to Tudor cartography, most cartographers 'regard blank paper as a discreditable admission of ignorance',[5] which allied to the natural inclination to use all the space available to the artist, allows some features to 'overspill' or dominate, while otherwise blank areas are filled with marginal drawings and commentaries. The restraint of Petty's surveyors is noteworthy, but not all map-makers even in the nineteenth century could resist the temptation to fill the page. Where space is left empty, it prompts multiple questions as to what is not seen, denied, purposely excluded, overlooked in error, or incomprehensible? What is beyond the cultural constraints of this surveyor to even imagine?

The map is therefore far from being merely a 'snapshot', however simplified, of reality. It is a much more sophisticated construction, requiring some skill in deciphering the ways in which it communicates. An appreciation of mathematics (geometry, algebra, trigonometry), draughtsmanship and graphic design, a readiness to question each component and its place in the whole, and an ability to enter imaginatively into the early stages of the map-making process and the administrative scaffolding that made its creation possible will all assist in decoding this world. The means of reproduction also need to be considered, most strikingly the moves from entirely hand-copied maps to copper plate printing and on to cheaper lithography which made low-cost mass production possible. This opened up new and different markets for map products, but also affected the science of map-making itself, as engraving made maps 'more professional and more international', the symbols becoming smaller, neater and more standardised, and the cartographer's subject-matter itself 'shrinking' to a 'low common denominator comprising coasts, rivers, hills, woods, settlements, and territorial divisions', with main roads (from the end of the seventeenth century).[6] By acquiring a familiarity with the range of highly technical field notebooks, plots of small areas, draft maps, boundary remark books, engraving journals and other manuscript materials in the Ordnance Survey archive (National Archives of Ireland), along with the fine collection of surveying instruments held by the Ordnance Survey itself and at the National Science Museum, Maynooth, the researcher can appreciate the new world of 'scientific accuracy and unity of principle' that was the ambition of the Ordnance Survey.[7]

The nineteenth-century six-inch townland survey of Ireland is by far the largest and best documented cartographic undertaking in Irish history. However, the production of even the most humble estate map required skills in the mathematics of surveying, and the arts of converting numbers (heights, distance, direction, location, area) into coherent, and representative patterns on a page. The following is an introduction to some of the key concepts in modern cartography. The aim is to assist the local historian to appreciate the map as an artistic, mathematical and

5 Andrews, 'Paper landscapes, mapping Ireland's physical geography' in Foster, *Nature in Ireland, a scientific and cultural history,* p. 201. **6** Ibid., p. 203. **7** 1824 Spring Rice report, quoted in Andrews, *A paper landscape*, p. 305.

technological achievement, but also to highlight the skills required in his or her own cartographic endeavours. And before trying to correlate points and lines on an early map (such as the Down Survey maps), with an up-to-date OS map of that district (which is aligned to the national grid), the researcher needs to consider some of the technical points which are basic to map-making.

The production of a geometrical frame within which detail may be plotted is the first task of the cartographer, therefore the mathematical basis of any map must be considered at the outset. Terms with which the reader should be familiar include scale, projection, graticule, grid, neat lines, and sheet numbering system.[8]

MAP SCALES

The scale of a map is the ratio of distance measured upon it to the actual distances which they represent on the ground. In the six-inch to one-mile series, a straight line distance of six inches, measured with a ruler on the map, represents one mile in the 'real' world. Taking a metric scale, if a straight line distance on the ground (the 'real world') measures 2.5km, and this length is shown on the map by 2.5cm, the scale can be expressed as 1:100,000, based on the following calculation:

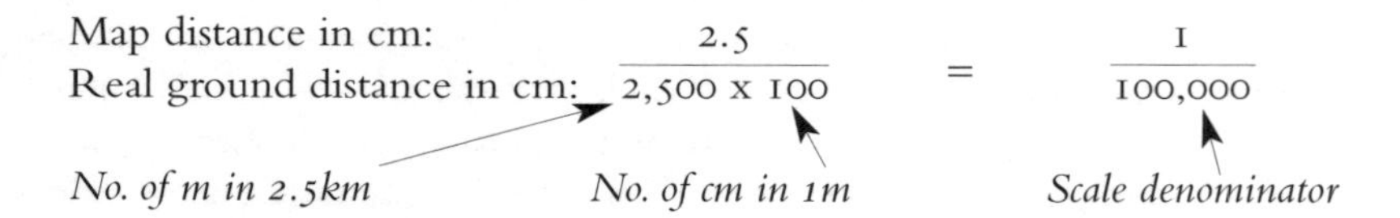

The local historian needs to be able to work in an assured manner with maps of different scales. The type and nature of generalisation, that is, what can and cannot be represented, and the way in which this is done, is greatly dependent on the scale at which the original survey was plotted. Obviously, on a large-scale map such as a town plan at 1:500 or 1:1,000, more detail can be shown than on a smaller scale map such as a school atlas map which shows all Ireland at a scale of 1:2,000,000. The smaller the scale the greater the amount of detail which must be generalised or omitted, but the larger the area covered on a single sheet. Large-scale maps cover smaller areas but in more detail. In figure 49a for example, the extract from the County Galway index sheet, at a scale of 1:190,080 (one inch to three miles) locates the settlement of Mount Bellew, local road network, other settlements and townland boundaries at a scale that allows the village to be studied in relation to its hinterland but does not provide any great detail on the village itself. At the scale of 1:10,560 (six inches to one mile, figure 49b) individual buildings and field boundaries can be distinguished, but a smaller area is shown, while at the scale of

8 These notes are based on the International Cartographic Association's *Multilingual dictionary of cartographic terms* (Wiesbaden, 1973) and on R.W. Anson and F.J. Ormeling (eds), *Basic cartography for students and technicians*, i, 2nd edn (London and New York, 1993), pp 19–63.

1:2,500 (25.344 inches to one mile, figure 49d) a close-up view of the village is obtained, allowing field acreages and pathways to be shown, but this larger scale is at the cost of areal coverage. A single sheet at the scale 1:2,500 (fig. 49d) shows a much smaller geographical area, but in greater detail, than will be seen on the same size sheet of paper focused on the same settlement at a scale of 1:190,080 (fig. 49a). Appendix 6 is a listing of scales which are commonly encountered in local history. Comparisons can only be made when all the measurements are expressed in the same units, hence the need to convert all measures (miles, furlongs, chains, yards, feet) to simple numerical ratios, in the form 1:XXXX. This has been done for figure 49, so that a progression from 1:190,080 (one inch: three miles) to 1:10,560 (six inches: one mile) to 1:2,500 (25.344 inches: one mile) can be noted. There are 63,360 inches in one statute mile; if this alone is memorised, further calculations are straightforward.

Map scales are represented in three ways: as a representative fraction (RF) or ratio (1/2500 or 1:2,500); in words (25 inches to one mile), and graphically as a graduated bar, subdivided into units of ground distance. The ratio format (1:1,000; 1:9,000; 1:63,360), which is often termed the 'natural scale', allows the reader to compare maps internationally, without having to concern him or herself with the 'local' units of measurement, such as Irish miles, statute miles, French paces, Indian versts and sajenyams, and so on. The *International Historic Towns Atlas* project, for example, requires that each participating national atlas committee (including Ireland), produce three core maps at the same scales: a reconstruction of the town *c.*1840 at 1:2,500; a map showing the town and its environs at a scale of 1:50,000; and a modern town plan at 1:5,000.[9] Thus it is possible to compare urban centres from very different parts of Europe, including Ireland, and from North America, each of which could be using their own very different 'local' units of measurement. Many libraries (including the Map Library in the British Library) catalogue maps primarily under the representative fraction system, a further incentive to the researcher to ensure that they can readily move from imperial units into ratios. When searching for the 'modern' equivalent of an earlier map it is necessary to operate through ratios; it can be quickly observed that the 1:50,000 is fairly near to the 'one inch to one mile' (1:63,360) scale, or that there is very little difference in scale between the current 1:1,000 and the old 'five feet to one mile' (1:1,056) town plans. The difference in scale between the various OS county index maps (introduced in chapter I), is easily understood if one first changes the scale given in words into a representative fraction, thus the largest scale, 'one and a half miles to one inch' becomes 1:95,040 (for the smallest counties of Louth and Dublin), half inch to one mile becomes 1:126,720 (used for most of the counties), the intermediate scale 'two and a half miles to one inch' used for Tyrone and Down becomes 1:158,400, and the smallest scale, for the largest counties 'three miles to one inch' becomes 1:190,080. Writing map scales in words alone therefore has

9 Terry Slater, 'The European historic towns atlas project' in *Journal of Urban History*, xxii, no. 6 (1996), pp 739–49.

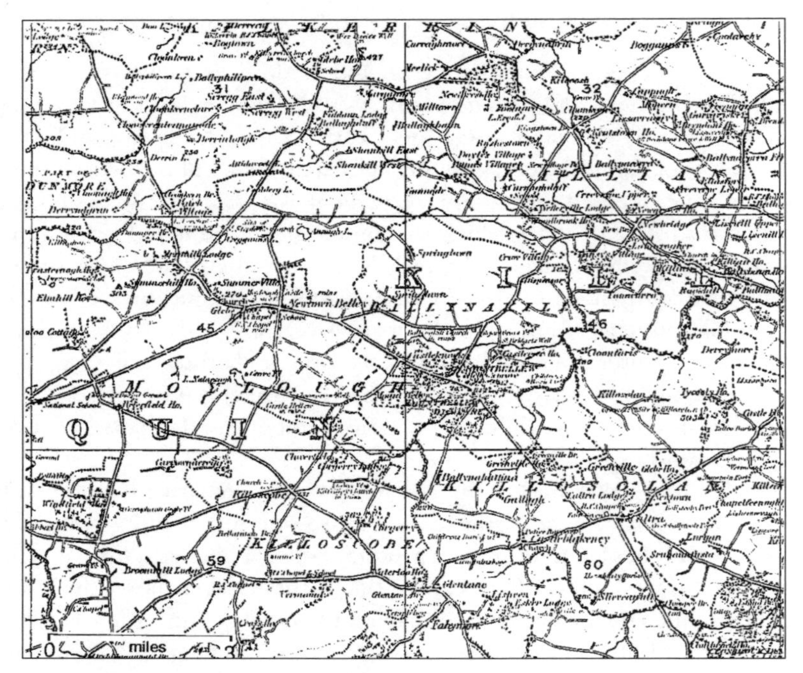

49a Extract from OS index to the townland survey of the county of Galway, 1841

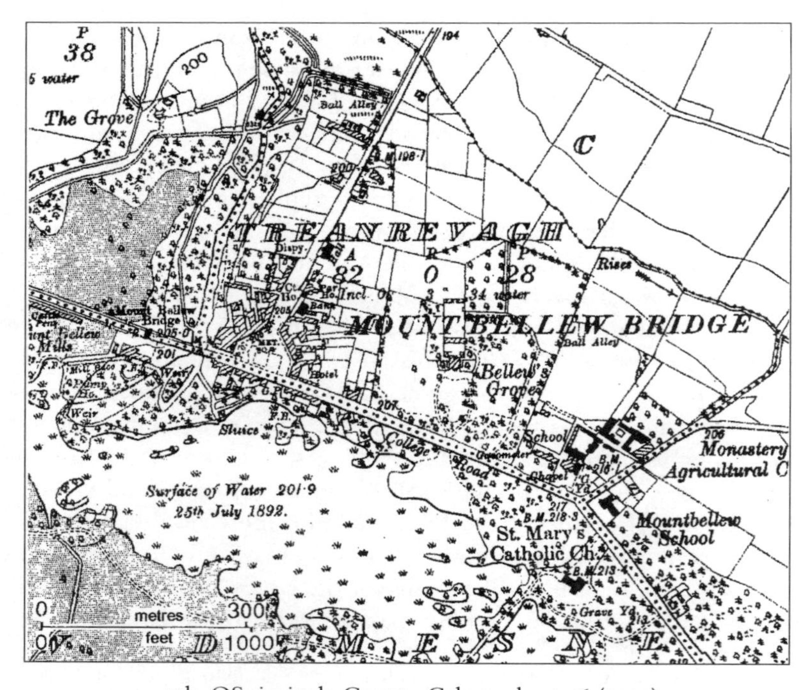

49b OS six-inch, County Galway, sheet 46 (1932)

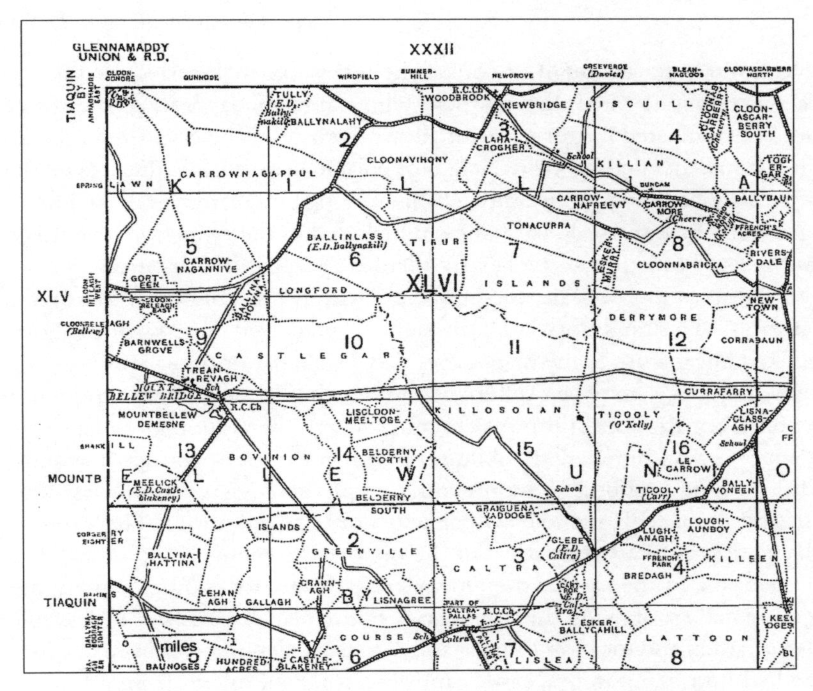

49c Extract from OS index to 1:2,500 (25 inch) series, County Galway

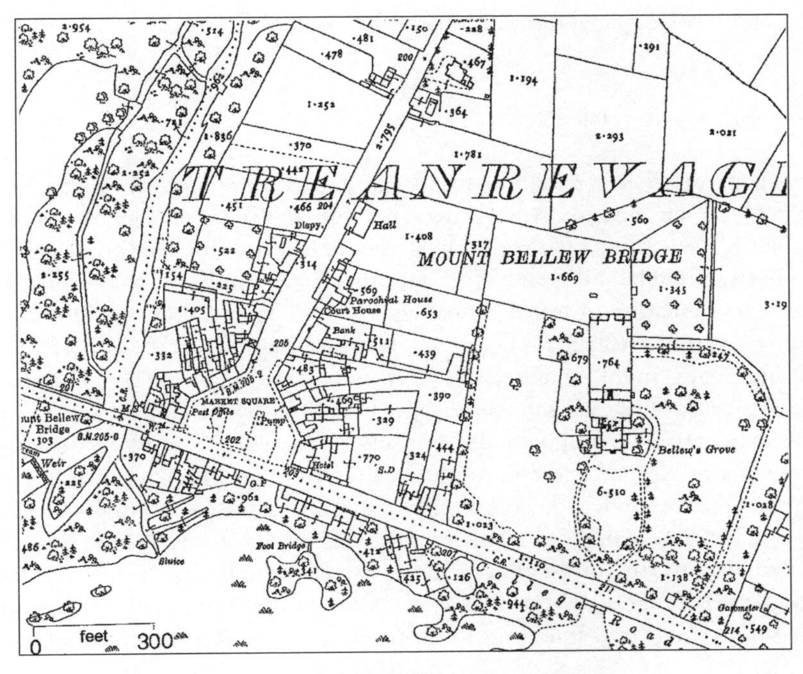

49d OS 1:2,500, Galway (Killian) sheet XLVI.9 (1930)

serious limitations, although of course the terms 'six-inch series', or the '25-inch maps' will continue to be used by those who can readily appreciate these measurements. The graduated line or scale bar allows distances to be calculated even in cases where the map has undergone reduction or enlargement; the line distance on the map represents a given distance (in yards, chains, furlongs, miles, metres or kilometres) on the ground. Once the scale bar has been included in the reduction or enlargement it continues to function as the key to calculating real world distances.

On the question of scale, the above discussion presumes that the map under investigation is a straightforward 'plan view', as perceived directly 'from above', and has a fixed linear scale. Some maps never had a scale bar or other indication of what any given distance on paper, vellum, wood or other surface might represent in the real world, but they are still most certainly maps. There are many maps in which plan view is combined with oblique or bird's eye perspectives, a wonderfully versatile way of combining architectural or pictorial views of houses, churches, forests and so on with a guide to their actual or relative location. The 'ground plan' might be drawn to a fixed scale, but the pictorial elements could be more artistic than scientific. In the case of the Down survey maps, for example, the underlying plan is created by drawing the boundaries of townlands, parishes and baronies, the course of rivers and location of bogs and forests 'as seen directly from above', while individual houses, churches, castles and forests are all drawn from a bird's eye or oblique view. While it is fruitless to try making a strict division between a landscape view or 'picture' and a map, what distinguishes the latter is that it reveals ground layout and relative spatial location, at least in essentials.[10]

GRATICULES AND PROJECTIONS

Many maps, upon closer examination, can be seen to be crossed by a network of lines (for example, figs. 37, 38, 47). This is tied up with the challenge of representing this spherical earth, or a part of it, on a flat surface, and the need to identify each place both uniquely, and relative to each other place on the earth's surface. In the primary triangulation of Ireland, great energy was expended by the OS on taking definitive astronomical readings of a select number of places, so that the final published maps might be as accurate in this as in all other aspects. The invisible mesh of parallels of latitude and meridians of longitude (fig. 22) known as the graticule of the map, will be familiar to all from school atlases. Longitude is measured in degrees east or west of a certain meridian; lines through the Azores, Tenerife, Paris and through Greenwich outside London are only some of the lines which have been employed as 'prime meridians' by map-makers. Each meridian of longitude is oriented 'true north' on the globe, that is, the distance between the meridians of longitude gets progressively smaller as we move further north, before meeting at a single point, the North Pole (fig. 22). Latitude (measured in degrees, minutes and

10 For a fuller discussion of this see Delano-Smith and Kain, *English maps: a history*, p. 1.

seconds) is the distance north and south of the equator. If the earth was a perfect sphere the parallels of latitude would remain at the same distance from each other; in practice the earth is a spheroid and the distance in miles of a degree of latitude is a little longer as one travels closer to the poles. Meridians of longitude and parallels of latitude intersect at right angles to each other. Moving from the spherical reality to the flat page for the purposes of map-making requires some 'stretching' or distortion of the mesh of meridians and parallels; the type of projection chosen will determine how exactly this network is 'stretched' or 'torn', and hence the appearance taken on by each country and continent.

For the purposes of local history research, the question of projection applies largely to all-Ireland maps, and in particular where the researcher wishes to 'overlay' early maps for the purposes of correlating exact location. What may appear on first reading to be errors may in fact have something to do with the projection employed. Andrews explains this point very clearly in chapter I of *Shapes of Ireland*, including illustrations of different projections. It is worth reiterating that it is mathematically impossible to represent a curved surface upon a plane without some kind of deformation. In the case of Ireland, if a single parallel of latitude is drawn as a straight horizontal line south of the island, another parallel of latitude is drawn as a straight horizontal line north of the island, and two meridians of longitude are drawn to intersect at right angles to these parallels (to the east and to the west of the island), the resulting quadrilateral (termed an equirectangular cylindrical projection) may look very exact on paper, but is highly problematic. The east-west distance along the northern boundary has been 'stretched' or distorted. On a map at a scale of 10 miles to one inch, the error would be approximately one inch, a substantial figure. Remedies for this error (that is, alternative projections) are explained by Andrews, while an appendix by Matthew Stout (pp 327–9) introduces the procedures which computer technology has made possible for 'transforming' or adjusting maps drawn to different projections at different scales, and oriented differently, by correlating a large number of identifiable points, as evenly distributed as possible. This procedure (using the TRANSFORM function of the GIS system Arc/view) allows a fairer assessment of accuracy to be made. Where early maps (as in the Down Survey barony maps, or grand jury county maps) can be transformed onto the National Grid (the projection now used by the OS*i* and OSNI, and explained below), the local historian is very well placed to exploit the map evidence to the fullest. However, on a large scale map such as a town plan or when dealing with a small geographical area such as a parish, the extent of variation is so small that the question of projection may be ignored.

ORIENTATION

True North has already been introduced as the direction of any meridian of longitude as it 'arrives' at the North Pole. Magnetic North is the direction indicated by a magnetic compass (assuming that it has not been disturbed by any local

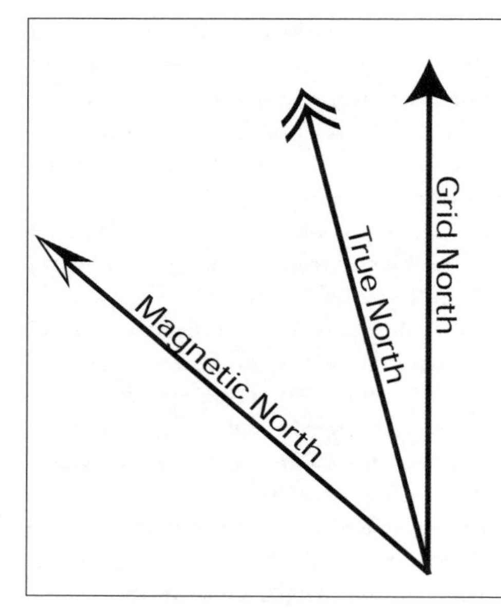

Diagrammatic only, from OS*i* 1:50,000 *Discovery* sheet 50 (includes Dublin district). At the mid-point of the eastern edge of this sheet True North is 01° 35' west of Grid North. At the mid-point of the western edge of this sheet True North is 01° 07' west of Grid North. At the centre of this map Magnetic North is 09° 06' west of Grid North (1990). Annual decrease 09.0'.

50 North points

attraction). This is the north point most usually encountered in pre-OS mapping, such as on the bogs commission maps. Magnetic north moves slowly with a variable rate, therefore the year in which the value was compiled, and the annual rate of decrease (or increase) must also be stated. The horizontal angular distance between True North and Magnetic North is called magnetic variation, and is generally indicated diagrammatically on the edge of OS sheets.[11] For orientating yourself in the field, using compass and map, an understanding of magnetic variation, and the differences between True North, Magnetic North and Grid North (explained below) is essential. Full utilisation of an early map depends on being able to superimpose it on a modern base, therefore knowing its orientation is vital. If its north was compass north such superimposition could be greatly facilitated by knowing the contemporary magnetic variation. On most Irish estate maps, for example figures 15–18, the indicators are compass-only (Magnetic North), and in the late eighteenth century, for example, the variation was more than 20 degrees. The declination therefore cannot be ignored by the local historian.[12]

The practice of orientating large-scale maps with north at the top does not become standard in Ireland until the published sheets of the Ordnance Survey first appear, however, the pre-eminence of the compass in navigation ensured that nautical charts and maps in early modern atlases were more than likely to have North at the

11 The vertical angular distance between True North and Magnetic North is called declination or 'dip', and can be ignored for most local history and leisure map-using purposes. **12** I am grateful to J.H. Andrews for explaining this matter and its importance in local history, and also for clarifying points on the use of projections.

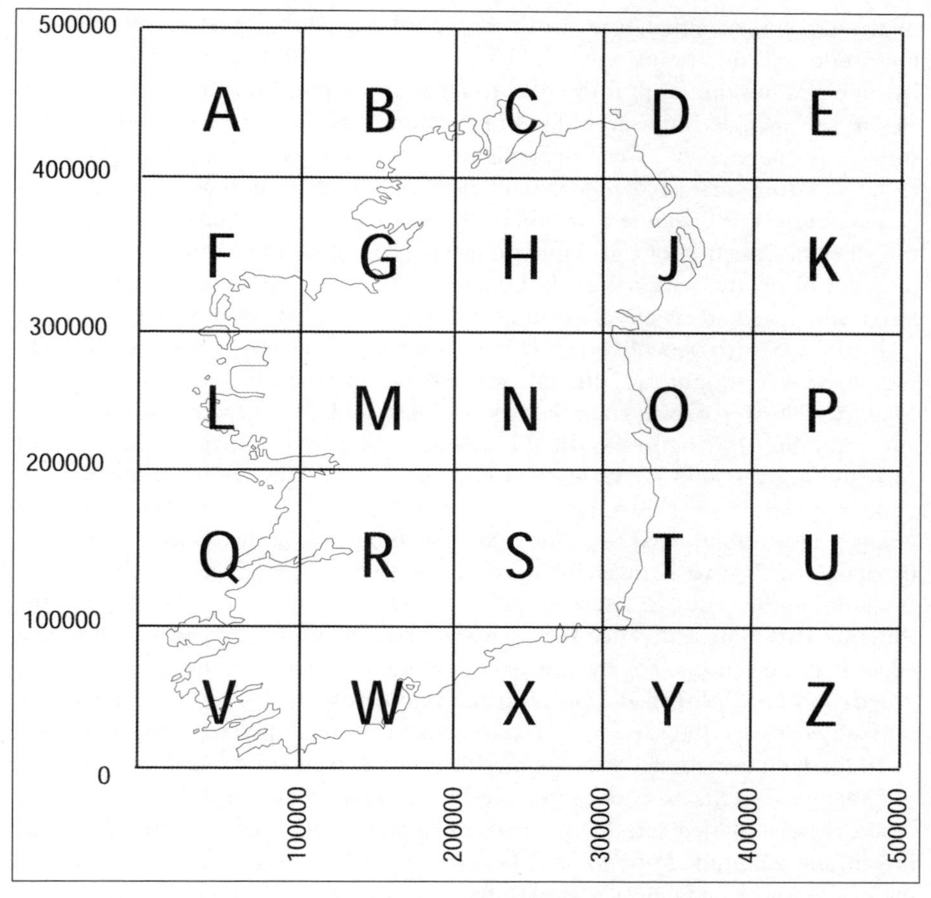

51 National or Irish Grid

top, particularly where the shape of the country or region being mapped fitted neatly into a standard one-page format, as did Ireland. The researcher therefore might profitably consider the orientation of the map under scrutiny, why this particular page layout, and what the implications are for its content and language.

THE NATIONAL OR IRISH GRID

Latitude and longitude are properties of points on the earth's surface, and their usefulness in global mapping and navigation are self-evident. However, for mapping at the national, regional and local scales this graticule has many drawbacks. Map-makers have developed a complementary system of defining exact location by constructing separate national grids. This type of 'mesh' consists of straight lines intersecting each other at right angles (fig. 51). As this involves plane geometry, the mathematics are simpler than with spherical geometry. One line only (at the centre of the grid) is oriented true north in accordance with a certain meridian of longitude; and one point only (the base of this central line, defined in terms of latitude and longitude) pins the grid to a particular point on the earth's surface. Beyond this single point and line where (alone) the grid is 'tied' to the earth's surface, it operates its own independent system of criss-crossing lines, creating rectangles (or squares) of equal size, each of which is numbered or labelled consecutively. The grid references therefore are properties of points on a map. Grid North is the direction of a grid line parallel to the central meridian (the axis of the projection) on the national grid; the angular difference between a meridian of longitude (oriented True North) and Grid North is called convergence.

In the OS 1:50,000 *Discovery/Discoverer* series, for example sheet 50 (Dublin region, extract as figure 47), the difference between True North, Grid North and Magnetic North is explained in the marginal data under the heading 'North Point', and reproduced as figure 50. In this case, the Magnetic North to Grid North variation is given as 9° 06' W in 1990 and decreasing at the rate of 09 minutes per year. By the year 2004 therefore, the magnetic variation is 2° 06' (that is, 09' x 14 years).[13] This is subtracted from the figure for 1990 to give the magnetic variation for 2004 as 7° W, that is, magnetic north in the year 2004 is 7° west of Grid North at a point in the centre of *Discovery* sheet 50, namely, the north-eastern sector of the Phoenix Park. The difference between True North and Grid North at the sheet edges is also given (fig. 50); by interpolation therefore the difference between True North and Grid North, at a point in the middle of any 1:50,000 sheet must be mid-way between the two values given for the eastern and western sheet edges.

In the Irish or national grid, the meridian which runs close to Athlone (8° West of Greenwich) is the 'base' or central meridian, and the true origin (the single point where the grid is 'tied' into the graticule of latitude/longitude) is at latitude 53° 30' North, and longitude 8° 00' West of Greenwich. Its false origin (the point at which the grid numbering begins) is 200 km west and 250 metres south of the true origin (fig. 51). The Irish Grid, the framework upon which new maps of Ireland have been 'hung' by both the OSNI and the OSi since the mid-1970s, is a mesh of 5 x 5 equal divisions creating x25 boxes, each designated by a letter from A–Z (excluding the letter I). The use of a unique letter for each 100km box reduces the number of digits required for a grid reference. Distances are calculated along a line running east of 0, that is, along the X axis, called the Easting. The resulting (x,y) co-ordinate or grid reference locates a particular place exactly.[14] It is always given in an even number of digits (four, six, eight).

Despite the care invested by the Ordnance Survey in ascertaining the latitude and longitude of selected places, the sheet lines of the first edition six-inch townland maps had no direct relationship to the Athlone meridian or indeed to any

13 Students should note that longitude is written in the form 00° 00' 00", that is, degrees, minutes and seconds, East or West of Greenwich; there are 60 minutes in one degree of longitude and 60 seconds in one minute. **14** For a detailed account of the creation of a national grid, and the mathematical calculations required, see *Basic cartography* i, 2nd edn, pp 32–3.

other system of meridians and parallels. In a sense, each county had its own 'grid', that is, the sheet lines of the six-inch maps as they appeared on the county index map (as in figs. 38 and 49a). It could be said that the first OS grid was in fact 32 grids, as each county had its own point of origin.[15] In the 1850s, the county maps were combined into a single general (one-inch or 1:63,360) map of Ireland divided for latitude and longitude (see fig. 39). This was truly an all-Ireland grid, disregarding county boundaries. For scales of 1:63,360 or less the differences between counties had been eliminated. Over this map (made up of 205 one-inch sheets) it is possible to draw the graticule of latitude and longitude, or indeed to use as a grid the sheet edges of the 205 one-inch sheets as these would appear on a small-scale index. There were few further developments in OS grid matters relating to Ireland until the 1940s.

The relationship between the sheet lines of the six-inch series and the later Irish or national grid is illustrated in figure 52, an extract from the OSNI index sheet 1 for County Antrim (1969). The larger numbers and heavier lines indicate the six-inch (1:10560) sheets; each is further subdivided into sixteen smaller boxes to indicate the 1:2,500 plans. Thus Cushendun will be found on OS County Antrim six-inch sheet 15, and on the OSNI County Antrim 1:2,500 plan 15.6. Each of these 1:2,500 plans (the darker boxes, numbered) covers an area of 1½ miles by one mile. But superimposed on this early county-based grid is found the new re-aligned Irish Grid (lighter shading) where each block or tile covers an area of 1km x 1km. OSNI moved entirely into a metric system based on the Irish Grid, and dispensed entirely with the county boundaries, in the 1970s. The index to the Irish Grid 1:10,000 sheets and 1:2,500 plans shows how the relationship between six-inch and twenty-five inch sheets is preserved only in the fact that it takes sixteen 1:2,500 sheets to cover the area of one 1:10,000 sheet. There is absolutely no relationship between the 'old' OS county-based sheet numbering (explained below) and the new OSNI grid numbering. The sequence moves across the province, from 1:10,000 sheet numbered one, in the north-west, through to sheet numbered 285 in the south-east. Six-figure Irish Grid references are given along the margins, allowing the reader to situate any single 1km² tile into the island as a whole.

The creation of a grid is important for national mapping; problems arise, however, when maps based on home-grown grid systems are brought together for purposes of international co-operation. Significantly NATO developed its own world-wide grid system, which is used by some but by no means all international bodies. All positions on the Irish or national grid are based on observations carried out at triangulation stations during the 1950s and 1960s, using land-based systems (such as theodolites). GPS or the Global Positioning System, first developed for military use but now widely used by civilians in surveying and orienteering, is based on a different reference frame, with positions described on different co-ordinate systems. GPS positions must be converted to Irish Grid positions and vice

15 The county origins are mapped by Richard Oliver in *Sheetlines*, Apr. 1991, no. 30, p. 17; see also no. 27, p. 7 for further notes.

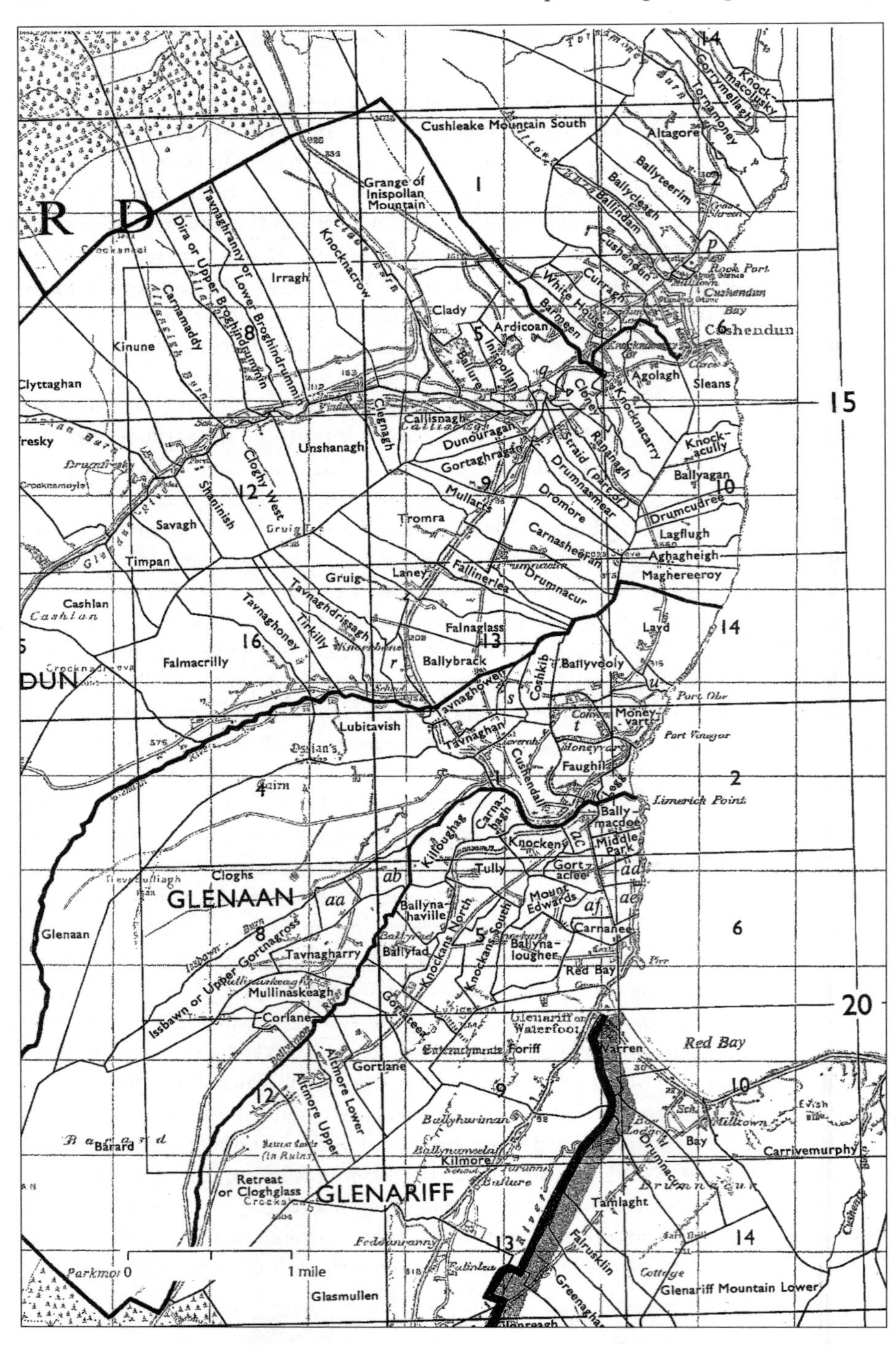

52 County Antrim index sheet, showing six-inch sheets and 1:2,500 plans

versa, a technical process that is clearly described in the OS booklet *Making maps compatible with GPS* (Dublin and Belfast, 1999). OS*i* has produced a booklet explaining the co-ordinate reference system currently used in Ireland, simply titled *The Irish Grid* and available from Map Sales in the Phoenix Park. OSNI has similarly produced explanatory notes on map projections, grid references, and other technical matters [http://www.osni.gov.uk].

The concept of 'neat lines' also needs elaboration. Neat lines enclose all the map detail, and thus define the limits of the area being mapped. In large-scale and medium-scale mapping the neat lines are based on the national or other local grid; any single place therefore can be defined by means of a grid reference (x,y co-ordinate). In small-scale maps (such as an atlas map of a country) the neat lines may be formed by two parallels of latitude and two meridians of longitude, but only where the map is on a rectangular projection. The individual sheets in a series usually join up exactly at the edges, along the neat lines, as in the one-inch OS series. However, in modern sheetline systems there is often an overlap, as in the 1:50,000 *Discovery/Discoverer* series. The sheet numbering system created for each map series therefore, must be considered in relation to the graticule (latitude and longitude) and the grid system employed, which may be the national or Irish Grid, a county-based grid, or indeed a grid system developed for that single series alone.

CALCULATING GRID REFERENCES

The advantage of a grid system is that exact location can be calculated; this is important to a wide range of map-users, including archaeologists and planners, army officers and hikers. In the case of the OS*i* *Discovery* series (fig. 47), the graticule of latitude and longitude (the outer line, in black), features alongside the national or Irish Grid (the inner line, in light blue). Firhouse is at a latitude of 53° 17' N (of the equator) and longitude of 06° 20' W (of Greenwich), as read from the outer lines. However, it is the national or Irish Grid system that is of greatest use for those seeking to locate individual places (such as a village core or cinema) simply and accurately. The national grid is marked in a larger font size at the edge of the map, along the neatlines. The side of each grid square (drawn in light blue) measures 2cm on the map, and represents 1km distance on the ground. Tallaght RTC (now the Institute of Technology, Tallaght) will be found (along with all of county Dublin and more besides) in the 1km (or 100,000 metre) national grid square which has its origin 300,000 metres (or 30km) to the east of the point 0, and 200,000 metres (or 20km) to the north of the same point 0 out in the sea beyond the coast of County Cork (see figure 51). Tallaght RTC will be found at ^{3}093 (measuring along the bottom, east of the origin) and at ^{2}279 (measuring along the side, north of the origin). The final digit in each case is calculated by estimating, in tenths, how far one has travelled within the square; if one has gone half way to the next grid line, then the final digit is 5; if one is nearly but not quite at the next grid line, then 9 may be the final digit. Using a ruler measured in centimetres makes this calculation quite simple. The reader always moves away from the bottom left corner of the map, working first eastwards (the Easting) and then northwards (the Northing).

The national grid 1km squares covering the island of Ireland have been labelled A to Z (excluding the letter I) as already explained. The 1km square into which Tallaght and county Dublin falls is square O, that is, the grid square bounded by 300,000 (Easting) and 200,000 (Northing), as in figure 51. Using the grid square letters A–Z dispenses with the need to include the small (superscript) digits. If the letter O is kept, the Easting for Tallaght RTC can be written 093 (not ^{3}093) and its Northing 279 (not ^{2}279). The final six-figure grid reference, with letter code, is written without spaces or commas as O 093279. Without the grid square letter (O) appearing in front these figures would make no sense.

In many instances the grid square letter 'short hand' is not used at all. When this is the case then the first digit (of the Easting and Northing) will refer to the 100,000m grid square in which the feature is to be found (fig. 47). *Gasaitéar na hEireann/Gazetteer of Ireland* gives the grid reference for Tamhlacht, BÁC (Tallaght) as 308227. This could be formatted as ^{3}08^{2}27; that is, found within the grid square bounded by 300,000 (Easting) and 200,000 (Northing). The number of digits in a grid reference depends on the map being used. On a small scale map, on which one can barely locate the village (such as 1:250,000), grid references will be defined to six figures (308 227); at a larger scale (such as 1:50,000) references can be defined to eight places (3083 2275); on the 1:1000 or 1:2,500 large-scale town plans, it is possible to narrow down the search to about 10 metres of the point in question, with a twelve-figure grid reference. The junction at the centre of the old village of Tallaght, can be identified to twelve places on the PLACE 1:1,000 map (fig. 53), from the grid readings offered at each corner of the print-out, and using a ruler marked with cm; it is exactly 308317 227520.

ORDNANCE SURVEY SHEET NUMBERING

An understanding of grid referencing and sheet numbering systems will serve the local historian well, not alone in identifying the required sheet or plan of their area, but also in locating individual places, such as the sites of antiquities, which are given by grid reference in the publications of the Office of Public Works. The index maps produced by the Ordnance Survey for its townland and later 1:2,500 series provide a direct entry point into this complex subject; a sample from County Dublin is reproduced as figure 54. The one-inch maps (figs. 39 and 42) ignore the county boundaries that were crucial to the organisation of the six-inch and twenty-five inch series; however, along the bottom margin of each one-inch sheet is a thumbnail map relating the one-inch coverage to the relevant six-inch sheets. Access to the 'old' twenty-five inch series is via the six-inch sheet number. Firstly, it may take several sheets to cover the entire territory being mapped at a particular scale: it takes 1,907 sheets at the scale of 1:10,560 (six-inch) or eighty-nine sheets at the scale of 1:50,000, to cover the island of Ireland.

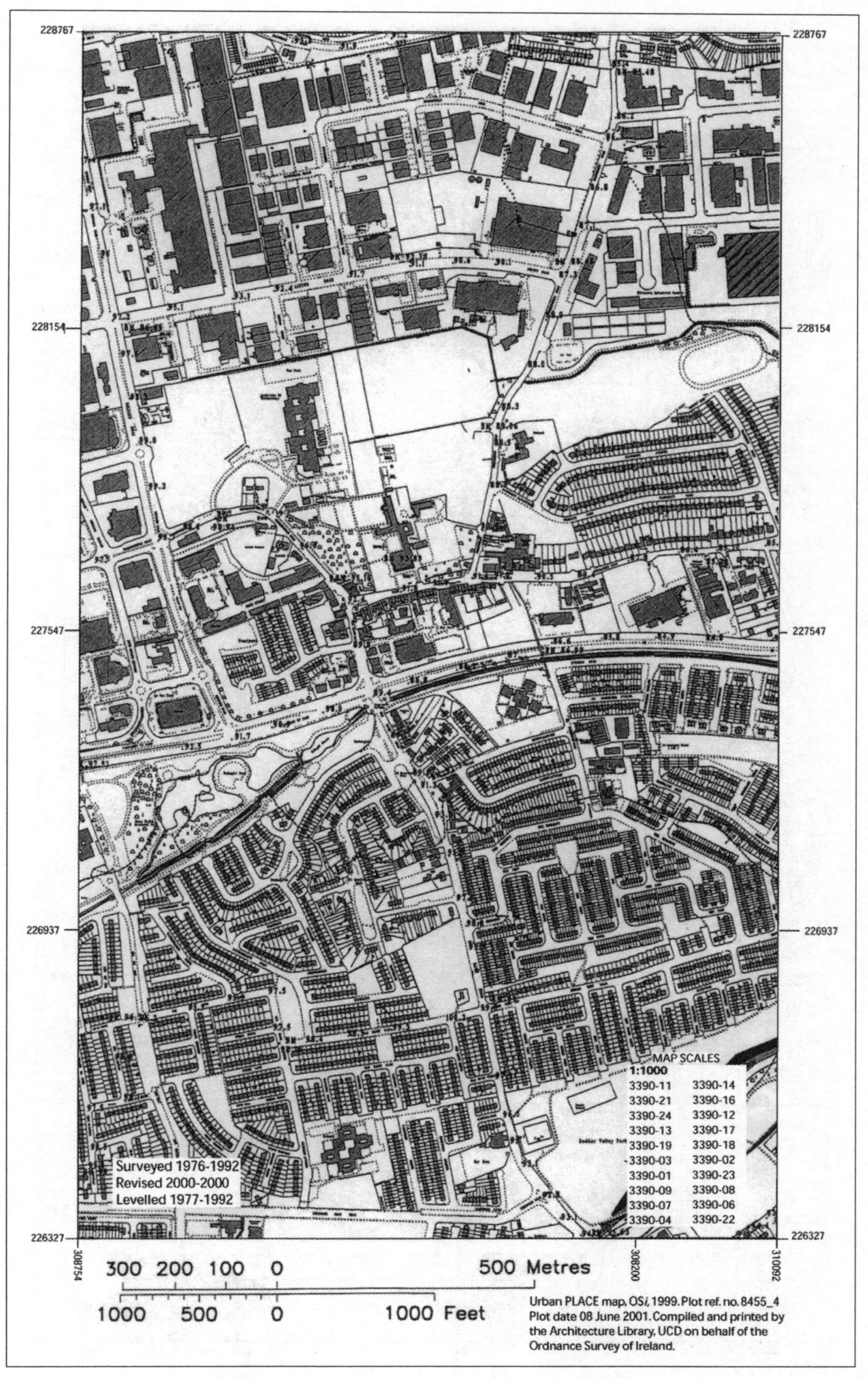

53 Tallaght, Urban PLACE map, 1:10,000, surveyed 1976–1992, revised 2000 (plot date: 8 June 2001)
Mapping by Ordnance Survey Ireland, Permit no. APL0000104 © Government of Ireland

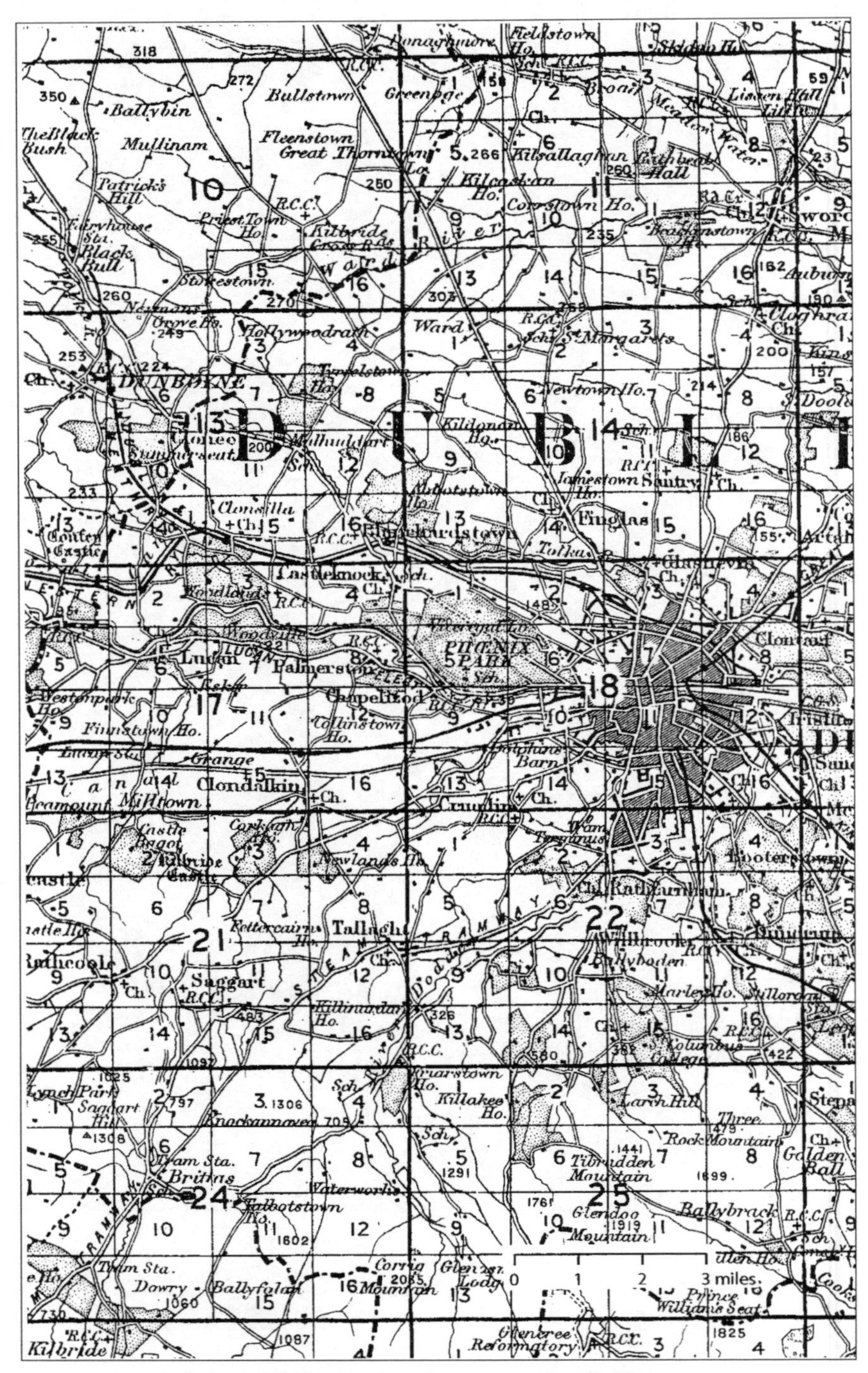

54 Index to OS six-inch and 1:2,500 series, Dublin county. Scale of original index map is one inch to one mile

In the case of the early Ordnance Survey mapping (pre-1920s), once the county is identified (remembering that there have been some name and boundary changes, such as Queen's County/Laois; King's County/Offaly), it is a simple matter to identify the number of the six-inch sheet required. This can be done through a number of routes, including the townland index[16] and the tenement valuation (the 'General' or 'Griffith's' valuation). The most direct route however, is via the index maps produced by the Ordnance Survey. The earliest series (illustrated in figure 38, Kerry) is the single-sheet county index to the original and later six-inch sheets. Because the size of each county in Ireland varies drastically, the number of six-inch sheets required to cover each county in full also differs greatly. To allow for this variation, the six-inch index maps are at different scales, allowing each county, regardless of size, to be represented on a single sheet. This difference does not matter to the researcher who is merely trying to identify the single sheet or sheets which cover their area of interest. The Louth six-inch index sheet (covering 25 six-inch maps) is at a scale of one inch to 1½ miles; the Kerry index sheet (covering 111 six-inch maps) is at a scale of one inch to three miles; the Galway index sheet (figs. 49a and 49c) is also at a scale of one inch to three miles (covering 137 six-inch maps). In the case of Tallaght, for example, the correct six-inch sheet can be identified by consulting the Dublin county index map (at a scale of one inch to one and a half miles), and finding the sheet or sheets within which Tallaght features, that is, on sheet 21, and on the adjoining sheet 22 (fig. 54). It is important to take note that the county index maps include only the six-inch numbers, as in figure 49a, which is from the 1841 townland index for County Galway. Six-inch sheet 46 includes the village of Mount Bellew (fig. 49b).

The county index maps (figs. 38, 49a, 54) are frequently used as base maps, as they provide a wonderfully clear picture of the shapes, sizes and names of the townlands. As noted elsewhere, alongside each index map is a table of parish and barony areas for the county which are invaluable for local historians trying to ascertain the administrative units into which his/her area of study falls.[17]

The introduction of the 1:2,500 maps required a new index map series known as the townland index maps, which shows both the 1:2,500 and six-inch sheet lines. These index maps (fig. 49c) are all at the same scale (one inch or 1:63,360); however, as the counties vary so much in size and shape the index sheets themselves are of different shapes and sizes, most counties needing more than one sheet. The townland index maps are also valuable maps in their own right as they are at a manageable scale, show and name the townlands clearly, and have selected a sufficient number of other features (as in villages, churches, schools, roadways and

16 *Census of Ireland 1851, general alphabetical index to the townlands and towns, parishes and baronies of Ireland* (Dublin, 1861; facsimile reprint Baltimore, 1984). **17** Andrews, *History in the Ordnance map* (1974), p. 22.

barony boundaries) to allow them to be used as base maps for local studies. Andrews observes that the townland boundaries here are not always the same as those mapped on the original first-edition six-inch maps.[18] At all stages note needs to be taken of the dates of edition and revision of indexes as well as of the maps to which they refer.

Once the six-inch sheet is known, it is possible to work out the number of the 1:2,500 sheet(s) which cover the area of especial interest, by dividing the 'six inch' sheet area into x16 equal parts, and allotting numbers 1–16 (see figure 54, Dublin index). In brief, it takes x16 sheets at the popular 25in. (1;2,500) scale to cover exactly the area mapped on a single six-inch sheet (1:10,560). The 1:2,500 sheet which covers the centre of the old village of Tallaght is titled Dublin (Uppercross) sheet 21–12. The first number refers to the area covered by the six-inch sheet, the second number gives the exact location of the desired 1:2,500 sheet within that area. Earlier editions of these townland index maps used a combination of Roman numerals and regular (Arabic) numerals to identify the 1:2,500 sheets. As the same order is used in numbering (six inch followed by 1:2,500) this should not create a problem. The 1:2,500 sheet for Mount Bellew bridge is noted as XLVI.9, where the Roman numeral XLVI refers to the six-inch sheet 46 (fig. 49c and 49d). It should also be observed that the numbers of the adjoining sheets are always included in the margin (just outside the neatlines). In the case of Mount Bellew bridge (fig. 49c) the 1:2,500 sheet directly to the west is numbered XLV.12 (remembering that each six-inch sheet is divided into 16 equal portions).

The introduction of the national or Irish grid by both survey authorities (OSNI and OS*i*), has simplified sheet numbering for some but not all map series. The six-inch (1:10,560) OSNI Irish Grid sheets (changed to 1:10,000 scale in 1968) are surveyed and published as a homogenous series with a single sheet numbering system for the six counties of Northern Ireland (that is, as a stand-alone series, ignoring county boundaries). The old OSNI 'county series' (1:2,500), as its name suggests, has a separate numbering system for each of the six counties (fig. 52). The principle of sixteen 1:2,500 sheets fitting into the area covered by a single six-inch sheet however is still valid. The townland of Ballybrack, for example, at the centre of figure 52, will be found on county Antrim six-inch sheet 15, and on the 1:2,500 plan numbered 15.13. Where the graticule extends into the sea the numbering is not interrupted; thus along the North Antrim coast there are 1:2,500 sheets numbered 15: 1, 2, 5, 6, 9, 10, 13 and 14, but none for the remainder of the area covered by sheet 15 (which would be 3, 4, 7, 8, 11, 12, 15, 16).

The 'old' 1:2,500 series has been replaced in Northern Ireland by the 1:2,500 Irish Grid series, which dispenses altogether with county boundaries. Sheet numbering in the Republic has also moved away from the unwieldy county structure. Each sheet in the OSi 1:1,000 series takes its number from the grid co-ordinates of one point of the sheet (east of the origin/north of the origin, or 'Eastings'/'Northings'), and ignoring the letter code, as already discussed. Usually

18 Ibid., p. 46.

the point taken as the sheet number is the grid reference from the SW corner, as that is where all calculations 'start'.

Digital mapping has moved the system forward considerably; in the case of the OSi PLACE series for example (fig. 53), a single grid reference can be entered in the search for a map of a particular locality, which can then be printed out at a variety of scales and paper sizes. The map can be centred on any given premises or topographical feature, and printed as a seamless whole, that is, without the neatlines or 'joinings' that traditionally mark the boundaries of individual sheets. Geographic information systems (GIS) allow the cartography, that is, locational information stored digitally, to be combined with massive databases in commercially valuable ways; the address system developed for the postal service in Northern Ireland is perhaps the most comprehensive such use to date.

This lengthy digression on graticules, grids and sheet numbering should at least imprint on the mind the importance of taking note of the full map reference. The following reference system is recommended for OS maps: the abbreviation OS Map (to indicate author), followed by the scale of the series, the county name, the sheet number and any part number, and the date of the edition. Thus figure 49d would be cited OS Map 25", Galway XLVI.13 (1930). The abbreviation OS Map is needed in a footnote or endnote, but if entering details on a map itself OS would suffice, as it is obvious that it is a map. Other published maps are referenced in the same way as printed books. The cartographer or publisher takes the place of the author, the title of the map is formatted in the same way as the title of any published book, followed by the place and date of publication: Alexander Taylor, *A map of the county of Kildare* (Dublin, 1783). An unpublished map should be cited like other manuscripts, noting the repository in which it is held and its reference number: Survey of Ringsend and Irishtown 1831, by Sherrard, Brassington and Gale, NAI, 2011/2/5. The title is given as on the original, and does not take italics or quotation marks as it is not published. If there is no title given, the researcher might devise their own title, but this is placed in square brackets [thus] to indicate that it is your own title and not that of the surveyor. Dates are always given in the format 8 June 1767 (not 8/6/1767). Most of the information required to correctly reference a map will be found in the marginal data: on OS maps between the neatline (the edge of the area mapped) and the margin of the paper, on all four sides. Here also the researcher may find the reference number of adjoining sheets, the legend (key) and a list of abbreviations employed, explanation of the grid system which has been used, an explanation of the North point, and how to calculate magnetic declination. In addition, each 1:50,000 *Discovery/Discoverer* map includes concise instructions on how to give a grid reference (Irish National Grid).

HEIGHT

Using the invisible mesh of latitude and longitude, or the (equally-invisible!) mesh of a national grid, the cartographer is provided with a horizontal framework for his or her world. But what about the vertical framework? One of the chief challenges in cartography has always been the representation of relief. How can one represent the third dimension, height or depth, on a two-dimensional surface, a flat page? Pictures of mountains is a crude but nevertheless striking mode employed in medieval and early modern cartography (figs. 55a, also 4, 7). The use of hachures (lines or brush strokes following the direction of steepest slope, their thickness or darkness relatively proportional to the gradient) is largely a mid-eighteenth-century development, suiting the half inch and larger scale 'county' maps that were then in production but continuing into the nineteenth century (fig. 55b, also figs. 8, 10, 11).[19] Daniel Beaufort, in his 1792 map of Ireland, explained how he tried to 'represent the mountains in such a manner as might nearly shew the space they occupy', and 'endeavoured to give an idea of their comparative height, by the varied strength of shading'.[20] This approach was graphically effective, and when carefully done could pick up small localised features such as gorges, gullies, dunes, corries and eskers; the Railway Commissioners' map of 1838 was outstandingly effective in this regard (fig. 43). But it was still artistic rather than scientific, 'dependent throughout its history on a field-sketcher's subjective assessment of land form'.[21] Hachuring features prominently in the maps which were part of Richard Musgrave's *Memoirs of the different rebellions in Ireland* … (Dublin, 1801). The cartographer uses hachuring to highlight the hills and cliffs, a process which makes it difficult for the reader to distinguish height, gradient and length. The overall message in this case is of a hostile landscape, where the hills and woods shelter the wild and savage, and all movement is along routeways which are overlooked or otherwise vulnerable to sudden attack. The OS characteristic sheet (fig. 56) illustrates the more strictly scientific techniques used on the six-inch and 25in. series: exact height above sea level is noted by spot heights, bench marks (engraved on a building or pillar at that point) and at trigonometrical stations (many of which were marked in the landscape by small concrete triangular tables or stone cairns, recalling the very point upon which the theodolite was set up to take horizontal and vertical readings). Contour lines are isopleths or lines of the same value joining places of the same height above a certain datum (usually taken as sea level); they can continue to show relief submerged below the surface of the sea or lake. Their value lies in the fact that they are numerical not merely impressionistic representations of changing heights (or depths), and it is possible to calculate gradients. It is not surprising that the addition of contours was championed by Thomas Larcom of the OS office in the 1830s for the civil engineering profession. The making of roads, railways and drainage schemes and the further development of agriculture were all assisted by the inclusion of contours, which were added to the Donegal six-inch map in 1839 and published from 1845 onwards.[22] Rocks, shingle, sand and marsh are represented in a stylised but effective fashion (fig. 58). A report commissioned in 1853 'to consider and report on the details of the reduced [one inch] map of Ireland now in course of publication', favoured contours over hill shading, a stand that was backed

19 J.H. Andrews, 'Paper landscapes, mapping Ireland's physical geography', p. 206. **20** Beaufort, *Memoir*, p. xvii. **21** Andrews, 'Paper landscapes, mapping Ireland's physical geography', p. 207. **22** Ibid., p. 208.

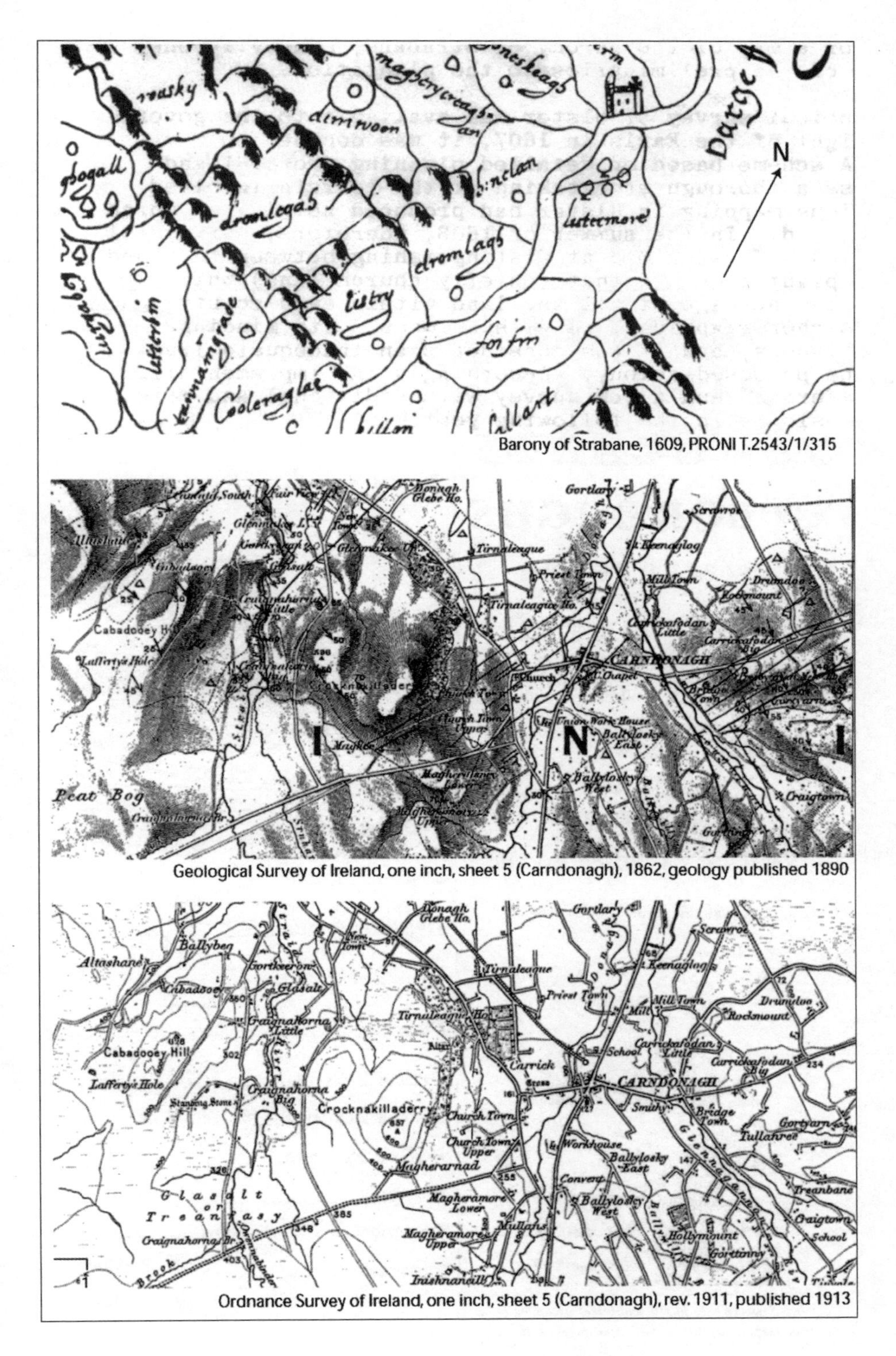

55 Representing relief by a) hill drawing b) hachures c) contours

WRITINGS

Plans published prior to 1938

*BARONIES

PARISHES

TOWNLANDS

C° BOROUGHS

MARKET TOWNS

Villages Villages *(Important)*

Workhouses

Bridges, Public Buildings, Churches.

RAILWAYS,

Locks, Manufactories, Bridges, &c.

PARKS

BAYS & HARBOURS

NAVIGABLE or TIDAL RIVERS

Smaller Rivers, Brooks.

Antiquities

Area Figures

4·37°

Bench Marks *Minor Altitudes*

B.M. 5·47 *121*

Initial Letters to Small Objects.

M.P Mile Post ***M.S Mile Stone***

S.P Signal Post ***S.B. Signal Box***

These Examples vary in Size and Extent according to the Importance of the District and Object to which they refer.

SYMBOLS

Bridges

Weirs or Dams

Ferries

Fords

Canals

Railways — *Cutting* — *Embankment*

Tunnels

Light Railways

1st Class Roads

2nd Class Roads

3rd Class Roads

Unfenced Roads

Quarries

Gravel Pits

Forts & Mounds

Wells & Springs

Pumps

Trigonometrical Station

Contours 200 *(Black)*

" 200 *(Red)*

56 OS Characteristic sheet: writings and symbols for six-inch and twenty-five inch maps.

by the 'preponderence of scientific opinion'.[23] The combination of contours (at 10m intervals) with colour (at 100m intervals) on the 1:50,000 *Discovery* maps (fig. 47) is most effective, and the result of high quality computerised mapping and printing, along with nearly two centuries of experience in surveying and map-making.

To fix heights, the first essential is to have a fixed datum or zero to which all elevations can be related. The bogs commissioners, appointed in 1809, were fully alert to the danger that their meticulous but separate district surveys would be of little use unless they were correctly levelled. They instructed the district engineers working nearest to Dublin to compare their levels to the height 'of the platform on the capital of the Column erected to the memory of Lord Nelson'; they were then to communicate these findings to the engineers further from the city, 'for the purposes of carrying forward the comparison'.[24] Standardising levels relative to this monument in Sackville street [O'Connell street], was an interim measure; the long-term ambition of the commissioners was to persuade the Ballast Office 'to mark at the Pigeon House Dock the level of high water in an ordinary spring-tide in the Bay of Dublin', which would act as a standard datum. As a temporary expedient, the engineers working on the bogs survey were instructed 'to leave permanent marks at the extremities of the several levels, and to lay down all remarkable objects which are likely to be permanent, such as raths, castles, towers, cairns, hill-tops, market-houses, &c.'; the ambition to correctly level these separate surveys was at least partly behind the richness of topographical detail found in the bogs commission maps (fig. 11).

The first all-Ireland levelling was carried out by the Ordnance Survey between 1839 and 1843, and all readings referred to a new datum: the low tide mark at Poolbeg Lighthouse, Dublin Bay, on 8 April 1837 (and engraved on its base).[25] This 1837 figure was taken as the OD (Ordnance Datum) or zero, and used as the baseline against which all subsequent OS calculations country-wide were to be fixed. Some discrepancies were found in a rigorous checking undertaken between 1887 and 1890, but most were very minor.[26] There are two other ODs also in use: Belfast and Malin Head (County Donegal). The Belfast datum, which is still used for the large-scale map of Belfast (1:1,250), is derived from readings over a six-year period 1951–6 at Clarendon Dock in Belfast, and noted on maps as MSL Belfast (mean sea level). In an entirely new geodetic levelling of Northern Ireland (1952–8) all parts of the province were levelled with reference to MSL Belfast, which is on average 2.71 metres above Poolbeg. This difference may be ignored for general local history purposes but where exact heights matter (for example, in an engineering study) OSNI can supply the necessary correction factor for any particular area. MSL Belfast was expressed in feet on OSNI maps until the late 1960s when heights began to be expressed in metres. Malin Head OD, a co-operative venture (established 1957) between the OS*i* and OSNI is the result of tidal readings gauged over ten years

23 *Report of the select committee on the map of Ireland, 1852–53* (921) xxxix. **24** *Second report of the commissioners appointed to enquire into the nature and extent of the several bogs in Ireland*, 1811 (96), pt. no 7, p. 29, appendix no 1. **25** J. Cameron, *Abstract of principal lines of spirit levelling in Ireland*, 1855; quoted in Andrews, *A paper landscape*, p. 322. **26** Andrews, *A paper landscape*, pp 110–11.

(January 1960–December 1969), and is the datum used by both surveys for small scale map series which cover the whole island of Ireland (such as the 1:50,000). MSL Malin Head is about 0.037 metres above MSL Belfast, an insignificant difference for small-scale mapping.[27] The movement from feet to metres may cause some short-term bewilderment for map users, especially where comparisons are made between a series of maps, or where manuscript data has been superimposed in a different scale to the base map; it is therefore worthy of especial note.

BOUNDARIES

The fixing of boundaries is a fundamental purpose of most if not all map-making: drawing lines to separate land from sea; recording the limits of property (for transfer, forfeiture, or taxing of land); separating one parish (and its funding circle) from its neighbour, or distinguishing 'profitable' from 'unprofitable' land. But the surveyor needs 'real' features – ditches, hedges, streams, roadways – to help make the transfer to paper. Sir William Petty longed for the day when 'quick-set Hedges, being grown up, would distinguish the Bounds of Lands, beautify the Country, shade and shelter Cattel, furnish Wood, Fuel, Timber and Fruit, in a better manner than ever was yet known in Ireland or England'.[28] He regarded the bounding of space as the essential prelude to civilisation, ridiculing the situation where territory was named after the 'Grandee or Tierne' rather than by land 'Geometrically delineated'.[29]

Boundaries such as town defences or a demesne wall may loom even larger in the landscape presented by the cartographer than they do in reality (fig. 7). Most follow the line of natural or man-made features, such as river courses, earthen banks and routeways. In fact, they can remain extant long after the river has been re-routed or built over, the roadway bypassed, and the parishes united, thereby providing the local historian with a valuable entrance point to past geographies. The 'sides' to any unit have to be defined and measurements of at least some noted before acreage can be determined; to calculate the area of a triangle, for example, we need to know the base and height. Hence the work of the Boundary Commission had to precede that of the Valuation Office, as the formula to calculate rateable valuation depended on both acreage and land quality. Along the coastline, one is reminded of the local farmer who countered the inquiry into how many acres of land he held with the question 'do you mean when the tide is in or the tide is out?' There was much serious discussion by makers and users of pre-OS estate maps as to the exact 'thread' of a boundary where roads, ditches and field banks were of appreciable width.[30] Deciding where the coastline runs is even more important for seafarers, whose lives depend on the reliability of their charts. Different definitions have been used: the 1:50,000 *Discovery* series takes as the high

27 Notes from the levelling and technical information pages, OSNI [http://www.osni.gov.uk]. **28** William Petty, *Political anatomy of Ireland* (London, 1691), p. 119. **29** Ibid., pp 104–5. **30** Andrews, *Plantation acres*, pp 120–1.

water line the start of vegetation cover. Where boundary lines overlie – such as a field edge which also marks the limit of the townland, barony and parish – the cartographer has always faced a dilemma. How to represent multiple boundaries but still along a single line, and without confounding them with linear features such as streams? Beaufort's solution (1792) prefigures that of the Ordnance Survey; he boasted how the 'boundaries of bishopricks' was 'expressed by a chain of fine pearls, and where they coincide with the bounds of counties or baronies, the pearl are intermixed with the round or long dots, by which those bounds are respectively denoted'.[31] The Ordnance Survey, less flamboyantly but no less effectively, combined dotted and dashed lines in such a way that up to four distinct boundaries could be represented by a single line (fig. 57).

The scale of the work accomplished by the Boundary Commission in the 1820s is impressive, as already discussed in chapter I. Richard Griffith set out to map the boundaries of *c.*69,000 townlands (*c.*19,000,000 statute acres), with a pitifully small staff.[32] The importance of this work cannot be overstated. It provided a framework for the collection of local taxes and was quickly adopted as the basis for organising the census of population. Most significantly, it provided an authoritative framework within which municipal and rural administration might be reformed, laying down the basis for local electoral divisions. The delimiting of poor law unions from 1838 and the drawing up of dispensary districts from 1850 provided the cartographic basis upon which the public health and social welfare systems were built in Ireland. It also made possible the creation of urban and rural districts in 1898, the foundation of our electoral geography.

GENERALISATION

Moving from the real world to the limited canvas of the map impels the cartographer to classify and conventionalise phenomena without attempting to express their individuality. In the first surviving set of *Instructions for the interior survey of Ireland* (1825), Thomas Colby declares in the opening article that the survey 'is to be performed on a scale of six inches to one English mile', and furthermore that the plans 'are to be drawn with all the accuracy and minuteness of detail which that scale admits'. The scale at which the map is to be drawn determines the amount of detail that can be shown and still remain intelligible to the human eye. But the decision about what to include and how – the level of generalisation, and questions of representation – must still be made by the cartographer, often following orders from his employers or superiors. Colby's first instructions were impossibly generous in some respects, and were inevitably followed by many more circulars.

Both natural features (most notably rivers, coastlines, lakes, bog, turloughs) and man-made features (roads, canals, housing) are represented by point, line or area

31 Beaufort, *Memoir*, p. xv. **32** Ibid., p. 348; in fact the final number of townlands is approximately 62,000, while there are 2,500 civil parishes, Michael Herity, *Ordnance Survey Letters Donegal* (Dublin, 2000), p. xi.

BOUNDARIES

	SCALES	
	1:10,560 or 6" to 1 Mile.	1:2,500 or 25·344" to 1 Mile.
Counties		
Baronies		
Parishes (only shown in County Boros. & Urban Dists.)		
Townlands		
Counties & Baronies		
Baronies & Parishes		
Parishes & Townlands		
Counties, Baronies, Parishes & Townlands		
County Boroughs, Urban Districts, & Towns (when not coincident with other Boundaries)	(Described)	(Described)
Wards (only shown in County Boros.)		
When Centre of River		C.R.
When Centre of Stream		C.S.
When Side of River		S.R. S.R.
When Side of Stream		S.S. S.S.
When Centre of Road		C.R.
When Centre of Wall or Fence		C.W. C.F.
When Face of Wall or Fence		F.W. F.F.
When 6' Root of Hedge		6' R H

57 OS Characteristic sheet: boundaries for six-inch and twenty-five inch maps

symbols (or combinations thereof). Where drawings appear to be individualised, for example walled towns and grand houses, the reader needs to exercise caution. The drawings may indeed be taken from the real world; alternatively they could be ornate but still conventional symbols. Many early maps have no key at all, expecting the reader to be so versed in the cartographic language of the time that it was not even an issue.[33] The coded language used in nautical charts required unambiguous explanation, as already introduced in the case of Murdoch Mackenzie's *Maritim survey of Ireland and the west of Great Britain* (first published 1776, figure 12 is from an 1821 edition). Mackenzie's reference list (fig. 12b) allows the sea captain to interpret the chart with confidence, trusting that the information on tidal depth, sandbanks, currents and the presence of rocks is sufficiently accurate to ensure the safe passage of his vessel. The Intelligence Division of the War Office published its own characteristic sheet (I.D., W.O., no. 1123) titled *Conventional signs and terms used in military topography* (London, 1895, 1902), simplifying some of the symbols used by the OS, and explaining how information of military interest might be packed into a sketch map at the scale of six inches to one mile or three inches to one mile. Troop numbers, composition and successive positions, 'reserves, picquets, supports, day sentries, vedettes and patrols' had their own small symbol, with Arabic numerals to show the number of companies in each. Fieldworks, clearances, demolitions, 'entanglements or abattis', intrenchments ('deliberate' or 'hasty'), batteries and redoubts, gun pits and rifle pits were symbolised. There was a distinct military slant under 'natural features, cultivation etc.' noting 'camping ground', hedgerows 'with and without trees', postal telegraph offices (*P.T.O.*), forges or smithies (*F.*), 'important public houses' (*P.H.*), and 'isolated wells or wells in districts where there is little water to be shown by *W*'. The tints to be used for natural features, roads, masonry and wooden buildings are specified, while British troops are always to be shown in 'Crimson lake'; 'Opposing force: Blue'.

Even in cases where the key is spelled out, the cartographer is quite likely to have extended creatively beyond that, as with the Taylor and Skinner *Maps of the roads of Ireland* (fig. 10). On the OS characteristic sheet (six-inch) churches, wells and springs, pumps, forts and mounds, and heights are all represented by point symbols, while forestry, shingle and sand are represented by areal symbols. Individual houses can be shown in sparsely populated areas, but cartographers face the problem of what to do elsewhere. Linear symbols bring real technical difficulties, as there are only so many thicknesses and styles which can be used. How thin a line is it possible to draw and still discern?

33 See 'Short and necessary instructions to beginners for the understanding of maps' where Moll explains the symbols used for linear features and for settlements, concluding with the note, 'these we have mentioned are the common ones, and being well known to those that are conversant with maps, the gravers omit explaining'. Herman Moll, *The compleat geographer* (London, 1723), p. xiv.

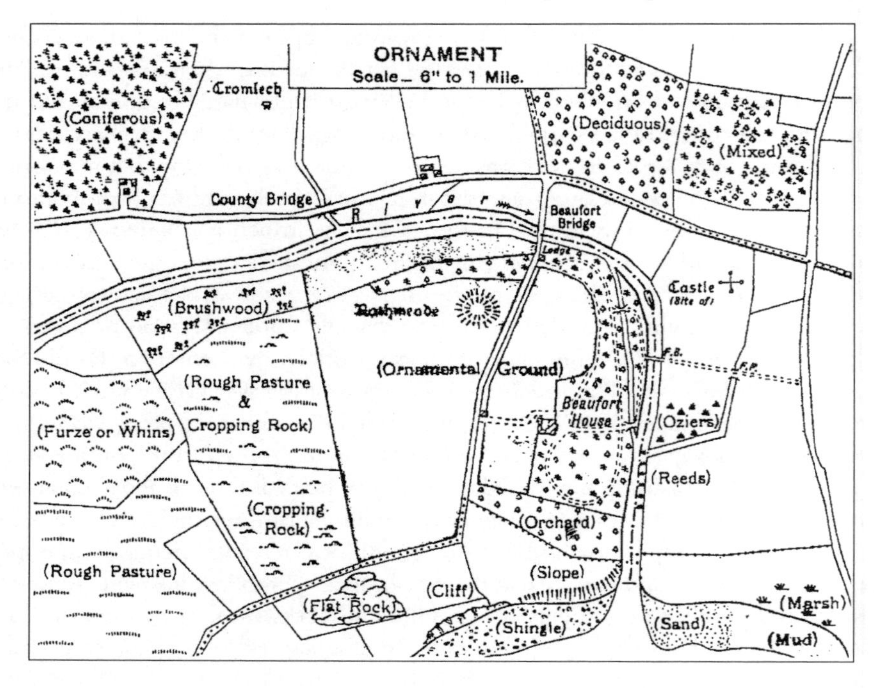

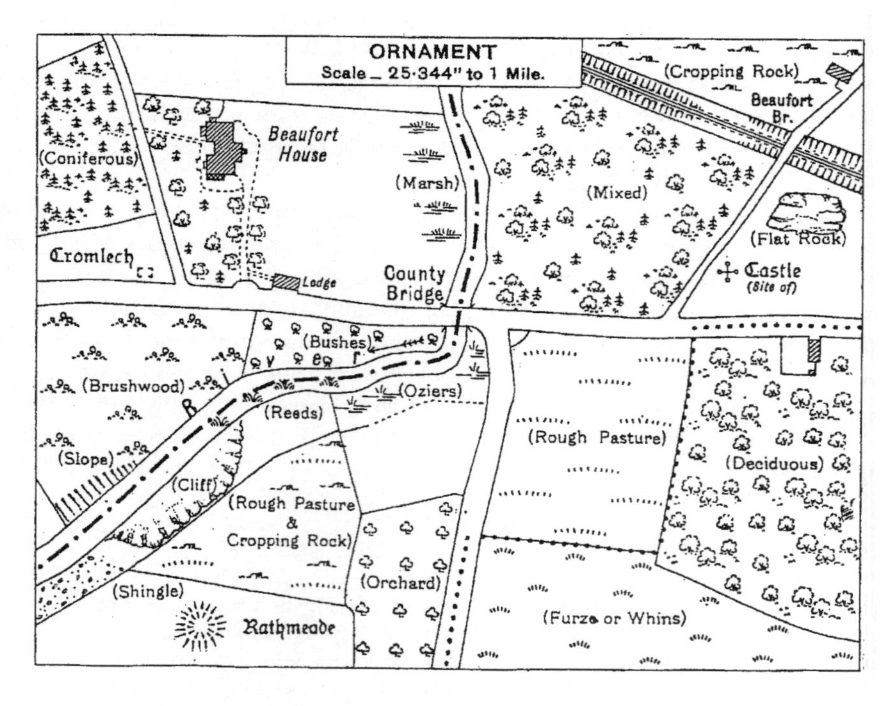

58 OS Characteristic sheet: ornament for six-inch and twenty-five inch maps

PLACENAMES

Placenames are an important part of the coded world that is the map. It is typically the placenames that distinguish a map from an aerial photograph, a satellite image, or a landscape view. This poses the challenge of what to name and how to name it. Size, font style, spacing and orientation can be used to distinguish different categories; it is difficult to surpass the ingenuity of the Ordnance Survey (fig. 56) in this regard. Filling a map with placenames that reflect an association with the 'mother country' has been practised by colonising powers in every part of the globe. But it is rarely a simple process. There is little evidence that successive central administrations troubled themselves with the standardisation of Irish placenames even after the first edition of the six-inch survey was published in 1829–42. William Petty, reflecting on the land settlement which his Down Survey (1664–9) had facilitated, urged that uniform spellings of all land grants be adopted, 'and that where the same Land hath other Names, or hath been spelled with other Conscription of Letters or Syllables, that the same be mentioned with an *alias*'. Much as he might have desired it, Petty recognised that native names would never be entirely erased from the landscape.[34]

Taylor and Skinner excused any errors in their *Maps of the roads of Ireland* (1778, figure 10) by complaining of 'the difficulty of obtaining proper information, and the arbitrary manner of writing them' (*sic*). The lengthy, transparent, scholarly and in general fair process by which the Ordnance Survey (from 1824) decided on place name spelling is dealt with in chapter I. However, central government never 'imposed' the OS spelling, allowing separate government departments such as the post office to go their own way. The population census conforms to OS nomenclature for the simple reason that in 1841 an OS officer happened to be one of the census commissioners. Placename standardisation in Ireland has been described by Andrews as 'an uncoordinated, gradual, consensual and essentially private process' which was 'exactly analogous to the standardisation of spelling in the rest of the English language that was taking place at the same period entirely without government intervention'. Much of this happened in the eighteenth century, but was accelerated by the work of the OS in the nineteenth century. The persistence of 'alternative' spellings in different government departments right through to the present is a legacy of this, while in some instances the 'official' name of the place (as decided by the local authority and mapped by the OS) is not widely used. Thus we have Droichead Nua (Newbridge), John F. Kennedy Square (Eyre Square, Galway), or the continuing confusion with Lehinch or Lahinch (County Clare) and Downies or Downings (County Donegal).

UNITS OF MEASUREMENT

The different units used in surveying may pose difficulties for the modern researcher, as indeed it vexed Sir William Petty and other state-employed cartographers who

34 Petty, *Political anatomy of Ireland*, pp 106–7.

sought to impose order on the ostensibly 'chaotic' landscape of early modern Ireland. Appendix VII is a conversion table to assist in making the jump between 'Irish', imperial and metric units. The distinction between Irish and English or statute or imperial acres derives from the difference between the Irish perch and the English perch. Petty explained the difference with admirable clarity: 'the perch in Ireland is 21 foot, that of England but 16 and a half. Where the acre of 160 perches is as 121 to 196, that is 121 Irish acres do make 196 English statute acres'.[35] The basic unit of measurement in Irish land-surveying before the Ordnance Surveyor was the perch, and the scale of 40 Irish perches (or 10 chains) to one inch (1:10,080) was commonly used in plantation and estate surveys. Alongside the Irish and statute perches (and the resulting Irish and statute acres), there was also the Cunningham perch of 18 feet nine inches which was of Scottish origin and common in parts of Ulster.[36] The OS employees were provided with a ready reckoner, or printed sheet converting Irish to statute acres and vice versa, and also for converting Cunningham acres to statute acres (NAI, OS 111/31). In 1824 (by the 5 George IV, *c.*74), statute measure was imposed on Ireland, but compliance took some time. In many parts of the country the traditional type of acre was specified alongside statute acres, while land continued to be let in many other types of locally-specific units such as ploughlands and quarters. The term 'perch' applied to both lineal and areal division; in the valuation records, for example, areas are given as ARP, or acres/roods/perches.

The OS intended to include 'area books' with the twenty-five inch maps; however, only those for the county of Dublin were published between 1868 and 1872 (as introduced in chapter I), under the heading *Book of reference to the plan of the parish of———, county of Dublin 9* (*sic*). Among the prefatory remarks at the start of each area book is an explanation of the 'decimal acreages' employed by the OS for the 1:2,500 series. How to convert decimal fractions of an acre (the style employed by the OS) to the more commonly understood acres, roods and perches, is outlined step by step, using a worked example. There are forty perches in a rood and 160 perches in one acre; in other words, one rood is ¼ part of an acre.

To convert decimal fractions of an acre to roods and perches, multiply the decimal first by 4 and then by 40, preserving the same number of decimals in the product.

Example:

633.357	Decimal acres, as noted on the OS 1:2,500 sheets
4	X *multiply the decimal only*
1.428	
40	X *multiply the decimal only*
17.120	
Answer:	633 acres, 1 rood, 17 perches

The use of decimal acreages allowed the cartographer to sum the acreages directly, and is to be recommended in electronic data processing (as with the Excel

35 Ibid., p. 51. **36** Andrews, *Plantation acres*, p. 126.

programme). With all non-OS maps, the reader needs to check against the OS record to see whether Irish, statute or another type of measure (such as Cunningham acres) is in use. In comparing acreages and distances over time, the reader should not rush to condemn too quickly, if, for example, measurements fail to meet the scientific exactitude of the Ordnance Survey. At the scale of the individual farm or townland the experienced practitioner could produce very exact results from a chained survey, but calculating acreages on the larger scale of the parish and county was a very different project. And where very small fields were involved, including or excluding hedges and ditches could materially affect the final result. Advances in technology, and trends internationally also impacted on the accuracy of measurement. In respect of the instruments available to the surveyor, the catalogue to the National Science Museum in Maynooth has already been mentioned; in addition, chapter 3 on 'surveying instruments' in G. L'E. Turner, *Antique scientific instruments* (London, 1980) should be consulted.

DATING MAPS

The question of dating maps requires care. As noted by Andrews in an appendix to the *Paper landscape*, year of publication on OS maps does not become standard until the 1880s. Various dates are employed: those of survey or revision, engraving, lithography or zincography, publication, and the printing of individual impressions. Material may be very much older than the date of publication, while a new edition may be based on a partial resurvey. In his dating of OS maps, Andrews directs attention to the names of Irish staff members found in printed footnotes or in manuscript annotations, in addition to the dates given for any step along the way from survey to final publication.[37] In a study of the Dublin six-inch sheet 18, admittedly a particularly complex map, 'the topographical content may pass through many more states than its explanatory footnote, and secondly, that there is nothing like a one-to-one correspondence between the successive stages of the map and those of the landscape it depicts'. We are left with 'a composite document embodying material from a succession of dates'.[38] The move from manuscript and hand-copied maps to 'the exactly repeatable pictorial statement of the printing press', brought enormous commercial advantages, but also attendant difficulties, with an understandable unwillingness among map publishers to update unnecessarily. The copper-plate printing system fostered conservatism, and few changes were made, particularly to hydrographic maps.[39]

37 Andrews, *Paper landscape*, pp 327–33. **38** Andrews, 'Medium and message in early six-inch Irish Ordnance maps: the case of Dublin city', p. 592. **39** Harley, *Maps, knowledge and power*, p. 61.

CHAPTER 3

Locating maps: a guide to selected archives

STARTING OUT: LOCATING YOUR MAPS

When embarking on cartographic research, the local historian will find it best to establish a number of lines of inquiry simultaneously. Place, time, theme and personalities are obvious starting ponts. The researcher must be prepared for constant reworking, backtracking, and circling, as sufficient pieces of a jigsaw puzzle come to hand to allow the construction, and reconstruction, of a likely picture. There must also be a willingness, at the outset, to move beyond the immediate townland or village that is the focus of inquiry, to work at different scales, and over a much longer period than might have been first intended. Clues about early layouts, landuses and occupations will often be found in maps of another historical period. As M.R.G. Conzen showed in his classic studies of the towns of Ludlow, Alnwick and Conway in Britain, the large-scale nineteenth century town plan can be the medieval archaeologist's most valued source and tool of analysis.[1]

The map record of individual areas varies enormously, but for different reasons: 'planted' counties are especially fortunate for the early modern period; south and east coastal areas for reasons of safe anchorage; 'scenic' districts on railway lines for nineteenth-century tourism promotion. The province of Leinster (and specifically, the county of Kildare) is exceptionally fortunate in its travel and estate mapping, due to the presence of so many wealthy and enlightened clients who subscribed in advance, ensuring that their handsome residences were included in the latest road map. But whether one's locality of interest is in Northern Ireland or the Republic, local researchers can be assured that their place of study 'features', and in more than just the Ordnance Survey record.

An essential but often neglected preliminary is to establish exactly the various ways in which a particular place might be named, and spelt, over the years. Derivation, change, current understanding, 'popular' abbreviations and local pronunciation are all part of this placename exploration. Obvious examples are Maryborough/Portlaoise; Bunclody/Newtownbarry; Lahinch/Lehinch; Innfield/Enfield; Ennistymon/Ennistimon; and Cobh/Cove/Cove of Cork/Queenstown. The *Census of Ireland, general alphabetical index to the townlands and towns, parishes and baronies of Ireland* (Dublin, 1861; facsimile reprint Baltimore, 1984), will allow the

1 M.R.G. Conzen, *Alnwick, Northumberland, a study in town plan analysis*, Institute of British Geographers, special publications no. 27, 1960; idem, 'The use of town plans in the study of urban history' in H.J. Dyos (ed.), *The study of urban history* (London, 1968), pp 113–30.

researcher to identify the civil parish, ward, poor law union, barony and county for each of the c.62,000 townlands of Ireland.[2] This index also records the final OS spelling, and most usefully names the OS six-inch map (county, sheet number) on which the placename features. The OS 'namebooks' will take the reader further, as they record the variants that the OS officers faced in the field, the derivation they judged most likely, and show the basis for what was for each small place indeed a landmark decision, as its name became largely standardised for posterity (see fig. 28). It is crucial to work out where each named place stands in the administrative hierarchy over time; the name of the barony, for example, is required before progressing to Petty's Down Survey maps, while each 'level' of the address is required before one can successfully access the 1901 and 1911 household census returns in the National Archives of Ireland. Samuel Lewis' *Topographical dictionary* (London, 1837) can also be used to determine the parish and barony within which a settlement stands, while it also includes variants on the placename spelling. Current or recent administrative boundaries can provide entry points to earlier district divisions. It can be problematic to assume that one knows these details and proceed with one's research; the case of Rosbercon, outside New Ross, illustrates this exactly. A medieval settlement on the west bank of the Barrow-Nore, it is within the Roman Catholic diocese of Ossory (Kilkenny) and as part of the parish of Tullogher is most certainly a stronghold of Kilkenny GAA; however, the municipal boundaries report of 1837 included it within the urban district of New Ross, County Wexford, and it was also part of the New Ross poor law union. The electoral boundary divides neighbouring Rosbercon houses between constituencies in Kilkenny and Wexford; citizens cast their votes for Kilkenny candidates in their local national school, which is within County Wexford. *Townlands in Poor Law unions: a reprint of Poor Law Union pamphlets of the General Registrar's Office* (Salem, Mass., 1997), by George B. Handron, is a very useful tool in anomalous situations such as Rosbercon. This facsimile reprint lists the townlands in each of 163 unions *c.*1885–86, with the area and population. Supplementary information is given for Belfast Union (1911, showing the townlands and streets in each district electoral division), Cork Union (1901, showing 'component parts' of each registrar's district), and for both Dublin city unions (1906, also showing 'component parts' of each registrar's district). This guide can be used in conjunction with the one inch to ten mile map of rural and urban districts produced by the OS for the Local Government Board for Ireland in 1899 and subsequently revised. The Handron guide illustrates exactly how the union disregarded county boundaries, and how the system coped with boundary changes, as in the extension of the Dublin city boundary in Jan. 1903. Celbridge Union, for example, though based largely in County Kildare also included townlands in counties Dublin (Rathcoole, Lucan) and Meath (Rodanstown). This reprint allows the reader to trace the ways in which the Poor Law union became the basis for administration and local elections in Ireland. From 1864, with the compulsory registration of births and deaths, the union became the superintendent registrar's

2 Herity, *Ordnance Survey letters Donegal*, p. xi.

district; within each a dispensary district or registrar's district was created. The connection between townland and DED (district electoral division) can be established. The Ordnance Survey *Gasaitéar na hÉireann/Gazetteer of Ireland* (Dublin, 1989), available in public and also second level school libraries, provides a concise but rather bald listing, in Irish and in English, of the 'official' spelling of each centre of population (over 1,500 population in 1971, or with its own post office), and major geographic feature (river, mountain, and so on), in the Republic; grid references and a guide to pronunciation are included. John Bartholomew's *The survey gazetteer of the British Isles* (Edinburgh, various editions) is especially useful where the researcher is trying to separate out places in Ireland (such as Newport) which share the same placename with places in Britain or elsewhere in Ireland. It also includes a useful etymology of placenames. Population numbers are given according to the most recent census, and abbreviations such as TO (telegraph office) are listed at the outset. The entries though brief are informative; the entry for Maynooth (ninth edition) reads:

> Maynooth, town, with rwy st., G.S.W.R., co. Kildare, on Royal Canal, 15 m. W. of Dublin; pop. 886 [1936 census], P.O. T.O. Contains the Royal College of St. Patrick, the principal educational establishment of the Roman Catholic Church in Ireland.

Éire thuaidh/Ireland north, a cultural map and gazetteer of Irish placenames (Belfast, 1988), produced by the Ordnance Survey of Northern Ireland and Patrick McKay's *Dictionary of Ulster placenames* (Belfast, 1999), are invaluable resources for locating places throughout the nine Ulster counties. The researcher could also usefully consult commercial county and city directories for parish, barony and county details of the place in question; *Thom's*, for example, gives these administrative details for each entry, and in the Dublin city section notes where each named street begins and ends. The larger urban areas are particularly well served; Dublin is fortunate in having C.T. McCready's *Dublin streetnames*, 1898 (facsimile reprint 1985) which notes the earliest appearance of each name on a particular map, and is methodologically a model of how a local historian might proceed, albeit it on a more modest scale. The 'topographical information' in each fascicle of the *Irish Historic Towns Atlas* can be consulted for an example of how changes in placenames might be recorded, but also for direction on cartographic and other sources for placename study.

The need to operate over several levels of the spatial hierarchy is self-evident; in the New Ross suburb of Rosbercon, for example, the researcher might consider both counties of Wexford and Kilkenny (county maps, road maps, grand jury maps), the baronies of Bantry (Wexford) and Ida (Kilkenny). They should also search under south-east Ireland (directories), Barrow and Nore (navigation studies), Old Ross (the earlier unsuccessful Anglo-Norman settlement), as well as themes such as 1798 (maps to illustrate memoirs on the rebellion), river navigation (portolan, admiralty and harbour commissioner maps), antiquities (the OPW record), railways (guide books to the South East line, as well as formal planning maps), and estate mapping (the Tottenhams were landlords of New Ross). And this is still in advance of accessing the standard OS six-inch, 25in. and 5ft. sheets, and the annotated maps of the Valuation Office, all of which can be searched under parish name and OS district (explained below).

LOCAL RESPONSIBILITIES

Establishing who might have held responsibility, over time, for the area under consideration is one of the chief challenges facing the local historian in search of maps. The Dublin Port company, for example, is the successor to the 1708 Ballast Office committee of Dublin city council (known as the Ballast Committee), followed in 1786 by the 'Corporation for preserving and improving the port of Dublin' (the Ballast Board), reconstituted in 1867 as the Dublin Port and Docks board, which was again reconstructed under the 1946 Harbours Act. A commercial harbour company, operating under company law, was established under the 1996 Harbours Act, this time to be known simply as Dublin Port.[3] With each structural change there was some change to its sphere of activities. For example, responsibility for the erection and maintenance of bridges right along the Liffey was held by the harbour authority dating back to its earliest years as a committee (albeit with a lot of autonomy) of the city corporation. It also held responsibility for several smaller ports, including Kingstown/Dún Laoghaire, Skerries, and Balbriggan. From 1810 to 1867, the Dublin Ballast Board was responsible for the erection and maintenance of all lighthouses on the coasts of Ireland; only in 1867 was this task given over by act of parliament to a dedicated body, the Commissioners of Irish Lights. The cartographic and other records held in the archive of Dublin Port (at Port Centre, Alexandra Road, Dublin 1) therefore encompass those produced by its forerunners, whose functions, range of duties and priorities differed in important respects from those of today's harbour authority.

Posing the question 'who was in charge of what and when?' will lead to unexpected riches. Perhaps one of the collections of most importance to local studies is that of the Office of Public Works (OPW), now in the National Archives of Ireland. Its annual reports 1831/2 onwards (which include maps) were published as sessional papers and so are widely available in research libraries; they can also be consulted in the OPW head office and library in St Stephen's Green, Dublin 2. The OPW library (open to researchers and students with letter from their college library explaining what they need to consult) also holds copies of Irish statutes 1310–1800, general and local British legislation 1828–1939, Irish legislation 1922–, parliamentary debates, government and commission reports, copies of all local authority development plans, and a range of journals and reference texts. The Barrack Board and Board of Works was established in 1759 to control and supervise the construction and repair of all public buildings in Ireland, following on several

3 *Dublin Port company yearbook 2000*, pp 36–41.

decades of outrageous abuse of public monies by the existing barrack establishment.[4] By an act of 1793 (33 Geo. III c.34) all buildings which had been erected solely at public expense, both civil and military, were placed under this Board, but in 1802 the civil and military functions were separated, with the establishment of a barrack inspectorate and Commissioners for Civil Buildings.[5] The Irish Barrack Department was transferred to His Majesty's Ordnance in London in 1823, but the OPW continued to play a very practical role in the management of military buildings. It had responsibility for refurbishing military barracks for constabulary use as well as designing and constructing numerous new police stations around the country. While plans for the refurbishment and expansion of military barracks were drawn up by Royal Engineers, it was the Board of Works which most usually implemented them. Plans from the late 1860s onwards to upgrade sanitary arrangements and to develop proper married quarters at military barracks were also implemented by the OPW. To carry out such projects, both record maps (showing the inherited fabric) and planning maps produced by engineers and architects were vital; the OPW files (NAI, OPW 5HC) hold several hundred sets of such maps.[6] The *Public Works (Ireland) Act* (1 & 2 Will. IV c.33) marks the handing over of the functions of the Civil Buildings Commissioners to the Board of Works. This explains why some of the OPW archive materials (including maps) pre-date the 1830s: they have come from various bodies which were centralised in the OPW, including the Commissioners of Inland Navigation (founded 1730), and the Barrack Board (founded 1759). Among the multiple responsibilities of the OPW were railway development, inland navigation and protection of archaeologically important buildings. The care of historic monuments vested in the Board of Works generated 'registers of inspection' and surveys including drawings, plans, sections, photos, OS extracts, and sketch maps showing the layout of monuments within individual cemeteries or enclosures. In practice, the OPW was used to implement government policy in the absence of other obvious mechanisms. Many of these functions have since been taken over by state, semi-state and private bodies, while the OPW takes on new and as always very practical challenges, including most recently the housing of refugees and asylum-seekers. Royal harbours, piers, fisheries, arterial drainage, roads and famine-relief schemes came within its remit. Its earliest pre-occupation is with state buildings in Dublin, such as Dublin Castle, the Four Courts, and the vice regal lodge in the Phoenix Park. Put most crudely, wherever state money was spent on buildings the OPW was almost certainly involved, and in the course of fulfilling its obligations required maps, plans and architectural drawings covering an immensely wide variety of districts and premises.

The context within which each map in the OPW archive was created and the purpose for which it was intended is made clear from the material with which it is

4 *The secret history and memoirs of the barracks of Ireland*, 3rd edn (London, 1747). **5** Lohan, *Guide to the archives of the Office of Public Works*, p. 37. **6** The War Office files in the National Archives (Kew) should also be consulted for maps of barracks throughout Ireland and also the archive of the Department of Defence.

lodged. Memorials or applications to the OPW for a loan to erect a harbour, pier and quays in Ballywalter, County Down are accompanied by a map of the district, but also by correspondence, public notices, draft and final declarations, details of loan charges, and notes on the district to be assessed.[7] The annual *Public Works (Ireland)* reports bring map and text together, though obviously without the correspondence, gossip and trivia that lighten historical research. In its first official report to parliament (1832–3) the board included a 'Sketch of the town and harbour of Galway showing the improvements in progress' by a civil engineer named H.H. Killaly, at the very useful scale of five feet to one mile.[8] Additional responsibilities, such as the famine relief schemes, generated new cartographic materials. The *Public Works (Ireland)* report for 1847–8 includes a map 'exhibiting the numbers employed … during the relief operations of 1846–7' while the report for 1880–81, when another catastrophic famine was widely feared, also includes a distribution map of the relief works in hand.[9] That the OS co-operated with the OPW at this time of national calamity is evident in the willingness of Captain Larcom to supply 'office proofs' of OS maps of Kerry in January 1846 'to enable the county surveyor to lay down public works projects for the county', though insisting that these as yet unpublished OS maps be returned to headquarters when the county surveyor has completed his own maps.[10]

LOCAL HISTORY COLLECTIONS: A GUIDE TO THE MAIN ARCHIVES

Having determined the administrative units within which the place under investigation might be situated, the various ways in which the placename might be spelt, and the persons or organisations that might be involved in mapping that locality, the challenge is to find pre-OS maps of the area under scrutiny. Appendix 8 lists some of the questions that might be posed at what can be a rather daunting stage of research.

A good place to begin the quest for early maps is the local history journal, as historic maps are often noted here, either as the subject of a study in their own right, or utilised as sources or illustrations in other studies. Local journals have long been a significant forum for articles on archaeological excavations, and for most, maps are central to both the research and presentation of findings. The *Journal of the Cork Historical and Archaeological Society* is an excellent example of this, but the same case can also be made for the journals of societies in Louth, Tipperary, Galway, Kildare, Kilkenny, Waterford and elsewhere. Some local journals have published their own map lists, as in the case of county Meath.[11] An article by Christopher J.

7 From 'pier and harbour structures' index, NAI, OPW 8/30, 1846–54. **8** *Public Works (Ireland), first report,* 1832–3, NLI, 3525. **9** *Public Works (Ireland), sixteenth report*, 1847–8; *Public Works (Ireland) forty-ninth report*, 1880–1. **10** NAI, Relief commission papers, distress reports Z series, RLFC2/Z762, nos. Z762, Z974, Z1514, January 1846; Kerry first edition OS six-inch was published in November 1846; Andrews, *A paper landscape*, p. 333. **11** Maeve Reilly, 'Meath maps in the National Library' in *Old Drogheda Society Journal,* 8 (1992),

Woods titled 'Maps as a source for Ulster local studies', in *Ulster Local Studies*, ix, 21 (1985) pp 174–7 will direct the reader to papers on maps of the province carried by a number of journals. Many local scholars of the nineteenth century had a particular regard for old maps, and both printing and later photographic advances enabled them to share this with a select readership. In many cases the commentary is more antiquarian than historical, and the maps taken as incontestable documents; however, even when this is the case, the new researcher is still indebted to these scholars for the quality map reproductions, whether as hand-drawn copies or, in the later decades of the nineteenth century, as photographic plates. Early volumes of the *Ulster Journal of Archaeology* were generous with map reproductions, including several on the destruction of the Spanish Armada off the Ulster coast.[12] An 1710 map of Gowran will be found in the 1856 *Journal of Kilkenny and South East Ireland Archaeological Society*.[13] In an 1896 pamphlet on Limerick, a map of 'the siege of Glin castle 1600' is included; it is a typical instance of how local maps can be taken completely out of their creative context and used for purely artistic reasons, or worse, as documents whose antiquity is sufficient guard against close interrogation.[14]

More usually, it is the footnotes in articles which will direct the reader to relevant collections in the National Library of Ireland (Dublin), Public Record Office of Northern Ireland (Belfast), or other major repository, and may also alert the reader to maps held in private hands for which there is otherwise no record. As always, the researcher needs to think laterally; an essay on placenames, for example, is likely to be map-based. This writer has located essays on maps in practically all the major, long-running local history journals, and many of these have also been noted in the Ferguson bibliography already discussed.[15] A useful listing of local history journals (provinces of Leinster, Munster and Connacht), and new monographs will be found in the *Local History Review* (Maynooth, 2002), edited by Maeve Mulryan-Moloney. This provides contact names and addresses of all active societies in the Republic affiliated to the Federation of Local History Societies (numbering, in the 2002 edition, about 130 societies), and notes lecture courses, fieldtrips, exhibitions,

pp 34–5; this lists 26 collections of maps in the NLI, some with shelfmarks. **12** *Ulster Journal of Archaeology*, 1859; also 'Map of 1609, Spanish armada, of W and NW coast of Ireland, drawn in 1609, from the original in BM Allingham', 1894, vol. I, p. 180. **13** Rev J. Graves, 'The records of the ancient borough town of the county of Kilkenny' in *Journal of Kilkenny and SE Ireland Archaeological Society* 1856 (N.S), vol. 1, p. 92. **14** Dowd, *Round about the city of Limerick* (Limerick, 1896), NLI Ir 94144 d 1. **15** For example, see: H.W. Lett, 'Maps of the mountains of Mourne in the county of Down' in *Ulster Journal of Archaeology*, 2nd series, 8 (1902), pp 133–7; Patrick S. O'Sullivan, 'Land surveys and mapping of eighteenth-century Kanturk' in *Journal of the Cork Historical and Archaeological Society*, vol. 95, no. 254 (1990), pp 88–106; Deirdre Morton, 'Some early maps of County Antrim' in *Ulster Placename Society Bulletin*, 2 (1954), pp 56–60; Reilly, 'Meath maps in the National Library', pp 34–5; Anon., 'The Down Survey maps of County Waterford' in *Decies*, 44 (1991), pp 23–38; B.H. St. J. O'Neill, 'Notes on the fortification of Kinsale harbour' in *Cork Historical Society Journal*, 2nd series, 45 (1940), pp 110–16; Patrick J. Duffy, 'Maps of Farney in Longleat, Wiltshire, and Warwick County Record Office' in *Clogher Record*, 12 (1987), pp 369–72.

heritage centres, and current research projects; the Federation for Ulster Local Studies provides a similar service in their guide to societies in Northern Ireland. The value of the local journal or local society resource cannot be overemphasised. Some local history societies from time to time have produced facsimile reprints of old maps and there can be few more worthwhile ways of investing resources. The Ballymena Local History group, for example, has reproduced an 1840s 'tourist' map of the town for today's visitors and residents, putting back into general circulation a heritage document that is immensely important in making sense of the shape of the present town, as it stands poised on the brink of substantial new development.

County or regional libraries, particularly those which have invested seriously in developing 'local studies' sections, are ideally placed to assist the researcher in pursuit of maps and map-related sources. Longford county library, for example, under its librarian Mary Carleton, has assembled both originals and copies of many maps that relate to the county over time and at as many different scales as possible. Monaghan County Museum and the North Down Heritage Centre Bangor hold important estate collections. At the very least, practically every public library in the country will hold some recent or current OS maps of their immediate area. These in themselves can provide excellent starting points, allowing one to 'walk the land', large-scale map in hand, seeing the familiar from another perspective, noticing field or plot boundaries, placenames, built fabric, landuse, and overall patterning, and in the process generating new questions and fostering the curiosity and absorption that are part of the satisfaction of local history. Along with their journal collections the library is likely to hold, or be prepared to acquire, most published local history works relating to their catchment area. Very new libraries can be treasure houses of the old as well, as in the case of the county library (South Dublin County Council) at Tallaght, which has already built up an exceptional local history reference section and map collection carefully catalogued, and welcomes deposits of original material related to its area.

Many libraries have produced guides to local history sources which also incorporate reference to maps; guides produced by libraries or county councils for Cavan, Kildare, Sligo, and Westmeath have been identified,[16] and there are certain to be others also. PRONI has produced guides to the pre-Ordnance Survey maps of the six Northern Ireland counties, one for Belfast and another for the island of Ireland. Although these booklets have been superseded by computer cataloguing and more recent guides (notably the county guides, discussed below), they are still a useful entry point to at least some of the early maps of the northern counties held

16 Sarah Cullen, *Books and authors of County Cavan: a bibliography and an essay* (Cavan, 1965), pp 81–3; Michael V. Kavanagh, *A contribution towards a bibliography of the history of County Kildare in printed books* (Naas, 1977), pp 104–13; John McTernan, *Sligo sources for local history: a catalogue of the local history collection, with an introduction and guide to sources* (Sligo, 1988), pp 104–16; Marian Keaney, *Westmeath local studies: a guide to sources* (Mullingar, 1982).

in the National Archives (Kew), NLI, TCD and PRONI. The work of PRONI staff in acquiring good quality photocopies of Ulster maps held in other repositories should be acknowledged; in PRONI notation, the prefix 'T' means a transcript while 'D' means a privately deposited record. Records that come from government or public bodies have their own prefix, eg. VAL (Valuation), OS (Ordnance Survey). *A union list of Belfast maps to 1900* (Belfast, 1998) excludes OS maps, as these are covered in the PRONI published catalogue *Northern Ireland town plans 1828–1966* (Belfast, 1981). The *Union list of Belfast maps* covers both manuscript and printed versions held by Belfast Library and Education Board at Belfast Central Library, by the Linen Hall Library, by the Ulster Museum, and by PRONI, with notes on size and where already published. The most impressive guide yet encountered is that by Mary Kavanagh, former county librarian of Galway, whose millennium text *Galway/Gaillimh, a bibliography* (Galway, 2000) has identified 279 maps of Galway city and county, deposited in archives throughout the British Isles. Many of these can be consulted in Island House, Galway, in photocopy or facsimile form, while the library headquarters also holds its own collection. The production of guides is not confined to the libraries; a bibliographic guide titled *Historic Sligo: a bibliographic introduction to the antiquities and history, maps and surveys … of County Sligo* (Sligo, 1965) by John McTernan covers maps in the county library, NLI, British Library, and University College London. PRONI has produced its own guides for counties Fermanagh, Tyrone and Armagh. Its 'Guide to the records of the Irish Society and the London Companies' acts as a guide to county Londonderry. Peter Collins' guidebook *County Monaghan sources in the Public Record Office of Northern Ireland* (Belfast, 1998), has already been introduced under 'estate mapping'. The best place to check PRONI map holdings is the 2 vol. 'Guide to landed estate records' which lists all maps held for each county, excluding OS and VO. The staff at the local or county library are best placed in the first instance to advise if there is indeed a guide to map resources, either in published or typescript form.

County histories or essay collections, also likely to be stocked by the local library, may include essays on maps, or (more likely) historic map extracts used as illustrations. Hardiman's *History of Galway* (Galway, 1820), includes maps from 1610 (Speed) and 1691, while Hore's multi-volume *History of the town and county of Wexford* (Wexford, 1904) includes early plans of Duncannon Fort (1624, 1645), the Down Survey map of the parish of Hook (1655–6) and the 1771 Housland map of the town of Wexford (vol. iv). *Dublin through space and time* (Dublin, 2001), edited by Joseph Brady and Anngret Simms, includes an explanatory note on published city maps from 1610 (pp 341–6), while various essays in the volume itself are generously illustrated with extracts from contemporary maps and maps constructed by the contributors from primary data. This is supplemented by Ruth McManus' *Dublin 1910–1940: shaping the city and suburbs* (Dublin, 2002) which is also most generously illustrated with original maps and reconstructions, and includes as an appendix a note on Dublin maps 1900–1930s. Belfast city is well served by a succession of local histories which incorporated earlier maps as illustrations, including George Benn, *A history of the town of Belfast* (Belfast, 1877) and D.J. Owen, *A short history of the port of Belfast* (Belfast, 1917). In the ongoing 'history and society' series published by William Nolan, the volumes on Wexford, Wicklow, Offaly and Armagh feature landmark essays on the mapping history of the county in question.[17] John Feehan's *Laois: an environmental history* (Ballykilcavan, 1983), includes a note on maps of the county. And in more general essays on mapping in Ireland, there is always a chance that one's county or a cartographer who worked locally will feature in the examples selected.[18]

Complementary to the local journals, the journals of scientific societies or professional bodies should also be consulted; back issues of most of the major series are held by the NLI as well as by university libraries, while several of the societies also maintain their own libraries but restrict access to members. The 'Dublin society for improving husbandry, manufactures, and other useful arts and sciences' (founded 1731), through its county series of statistical surveys, dated 1797–1805, includes (in almost all cases) county maps that are of significance as they pre-date the OS. The Carlow volume, for example, boasts maps of the Barrow navigation. Through its 'Committee of chemistry and mineralogy', the society sponsored geological surveying, and both its *Journal* and later *Scientific Proceedings* acted as a major vehicle for publishing research findings, such as Griffith's 1821 study of the Kilmaleady 'moving bog' in King's County (Offaly), or Kinahan's 'map of Croghan Kinshelagh gold district' (1882).[19] Known since 1820 as the Royal Dublin Society, the NLI, RIA and university libraries hold collections of its journals. The members' library of the RDS is at Merrion Road, Ballsbridge, Dublin 4; access for bona fide researchers is generally allowed, written application should be made to the Registrar of the RDS in the first instance. Geological surveying was also encouraged by the Belfast Natural History and Philosophical Society, by the Royal Irish Academy, by the TCD-based Geological Society of Dublin (founded 1831, from 1864 the Royal Geological Society of Ireland), and by the Institution of Engineers of Ireland (22 Clyde Road, Ballsbridge, Dublin 4); published maps will be found in all of their

17 Patrick J. Power, 'A survey of some Wicklow maps, 1500–1990' in Ken Hannigan and William Nolan (eds), *Wicklow history and society: interdisciplinary essays on the history of an Irish county* (Dublin, 1994), pp 723–60; J.H. Andrews, 'Landmarks in early Wexford cartography' in Kevin Whelan (ed.), *Wexford history and society* (Dublin, 1987) pp 447–66; J.H. Andrews and Rolf Loeber, 'An Elizabethan map of Leix and Offaly: cartography, topography and architecture' in William Nolan and Timothy P. O'Neill (eds), *Offaly, history and society* (Dublin, 1998), pp 243–86; R.J. Hunter, 'County Armagh: a map of plantations, *c.*1610' in A.J. Hughes and William Nolan (eds), *Armagh history and society* (Dublin, 2001), pp 264–94. **18** For example, see J.H. Andrews, 'Map and mapmakers' in W. Nolan (ed.), *The shaping of Ireland: the geographical perspective* (Cork, 1986), pp 99–110, concerned with John Thomas' map of Fermanagh, 1594; Josias Bateman, early eighteenth-century surveyor in counties Waterford and Cork; and John Hampton, early nineteenth-century surveyor. **19** Richard Griffith, 'Sketch of the bog of Kilmaleady in the King's county showing the moving bog' in *Journal of the Royal Dublin Society*, i (1821), p. 144; G.H. Kinahan, 'Map of Croghan Kinshelagh gold district', in *Scientific Proceedings of the Royal Dublin Society*, NS III (1882), pl. 21. **20** For example, F.T. Hardman, 'Age and mode of formation of Lough Neagh' in *Journal of the Royal Geological Society of Ireland,* iv (1875), pl. XI.

journals.[20] *The Proceedings of the Royal Irish Academy* has a long tradition of contributing to map studies, as in the work of T.J. Westropp on early Italian maps of Ireland.[21] Back runs of the *Proceedings* are held by most academic libraries throughout Ireland and the UK. David C. Jolly has produced a useful index titled *Maps in British periodicals: major monthlies before 1800* (part 1) and *annual scientific periodicals, miscellaneous magazines 1780–91* (part 2) (Brookline, Mass., 1990). Jolly's text reveals how subscribers to the *Gentleman's Magazine* in 1779 (vol. 49) could study a map of the Grand Canal system 'with the collateral canals to the Rivers Barrow and Boyne'. Those taking the *London Magazine or Gentleman's Monthly Intelligencer* in 1765–6 were treated to a series of maps of the four provinces 'into which Ireland is divided by geographers'.[22] Those who took the *Political Magazine* in April 1782 (vol. iii) could study 'a new and accurate map of the kingdom of Ireland with the roads, from the latest survey'. Maps were an important tool in the scholarly exchanges of the eighteenth century, a role that was to be further developed in the century which followed.

MAJOR LIBRARIES AND ARCHIVES

For the names, addresses, opening hours and brief introduction to the holdings of the major repositories throughout Ireland the standard reference is Séamus Helferty and Raymond Refaussé (eds), *Directory of Irish archives*, 4th edn (Dublin, 2003). Researchers should be mindful of the fact that the presence of maps in archives may not be indicated explicitly but may still be found under headings such as estate papers and diocesan records, and among the records of bodies which routinely employed maps in the management of their business, as with the War Office maps of barrack properties throughout Ireland. An explicitly cartographic guide is Helen Wallis and Anita McConnell, *Historians' guide to early British maps: a guide to the location of pre-1900 maps of the British Isles preserved in the United Kingdom and Ireland* (London, 1994). Familiarly known as 'the Hayes catalogue', the eleven-volume *Manuscript sources for the study of Irish civilization* (Boston, Mass., 1965; three volume supplement 1965–79, published 1979) by Richard Hayes is a 'union catalogue', that is, it refers to material held in the National Library of Ireland, and in a number of major Irish and British libraries. Researchers need to look under subject: maps and survey, and under places: [county]: maps. They might also search under surveyor's name (under people). A union list designed in a very different way is Sarah Bendall's *Dictionary of land surveyors and local map-makers of Great Britain and Ireland, 1530–1850* (London, 1997). It aims to include all surveyors who made maps of areas less than one county in size over this period; holdings in approximately 400 repositories have been traced, which makes it the most comprehensive union

21 For example, T.J. Westropp, 'Early Italian maps of Ireland from 1300 to 1600, with notes on foreign settlers and trade' in *RIA Proc.*, xxxi, C (1913), pp 361–428. **22** Leinster (vol. 34, Aug. 1765); Ulster (vol. 35, Jan. 1766); Munster (vol. 35, Mar. 1766); Connaught (vol. 35, May 1766).

catalogue in the field of cartography yet produced. Map collections in Northern Ireland are well served by Joan Chibnall's *A directory of UK map collections*, 4th edn (2000). While earlier editions of this text were available in printed form, this latest update has been published online only, by the British Cartographic Society [http://www.cartography.org.uk/]. It covers *c.*400 institutions in the UK holding maps, including the three national libraries (British, Scottish, Welsh), museums and societies, university libraries and local authority libraries. Kate McAllister's 'Maps in local studies libraries' (*Ulster Local Studies*, xiii, no. 1 (1991), pp 48–54) will also be useful to those relying on collections in Northern Ireland. The *Guide to collections in the Dublin libraries, printed books to 1850 and special collections* (Dublin, 1982) refers to map collections in the City Council's network of libraries, including the Dublin City Library and Archive, 138–144 Pearse Street. As the principal centre for research on the city, the Pearse Street premises now holds on a single site the original Gilbert collection, the Dublin City Libraries' Dublin and Irish collection, and the city archives; its map resources include first edition six-inch maps of Dublin, selections from various other OS series, maps which were part of original archive collections relating to Dublin (as in the plans of the Wide Streets Commissioners) and maps from commercial directories [http://www.iol.ie/dublincitylibrary]. The *Reports and printed documents of the Corporation of Dublin* (*RPDCD*), held at Pearse Street, include many maps and plans, illustrating slum clearance, housing projects and 'municipal improvements'. To mark the Tenth International Conference on the History of Cartography held in Dublin in 1983, Mary Clark of the Dublin City Archive produced a guide to the Corporation's map collection, titled *The book of maps of the Dublin city surveyors, 1695–1827* (Dublin, 1983), though the maps were not reproduced in this instance.

Many guides are limited to specific chronological periods or classes of maps (military, plantation, OPW); some of these have already been noted in the discussion on historical background to map creation in Ireland. The periodical *Irish Archives* includes occasional reference to map sources; the issue devoted to sources for Irish maritime history (vol. ii, no. 1 (1992), pp 1–98), for example, includes frequent reference to collections of charts and maps.

Where maps are concerned, there is a strong case to be made for approaching a local history research project on a British Isles basis. The British Library, the National Archives (Kew), the National Maritime Museum at Greenwich, Cambridge University Library and the Bodleian and other libraries at Oxford all have extensive holdings of Irish maps, while a trawl through the general guides to archives in Britain will identify other repositories that are worth exploring. The directories edited by Joan Chibnall (4th edn , 2000, online) and by Helen Wallis (1994), already introduced, are essential references. The online search tool, the National Register of Archives website [http://www.hmc.gov.uk/nra], covering both Ireland and the UK, will also identify locations that hold cartographic materials (and in this case, also specific items, by catalogue number). Irish maps may of course be found in archives well beyond these islands; for a routeway into other library possibilities, Internet sites devoted to map history are worth exploring, reminding oneself all the time to search under 'British Isles' as well as 'Ireland', using search terms such as 'chart', 'survey' and 'plan' as well as 'map'. The University of Edinburgh project site titled 'Charting the nation' has already been introduced [http://www.chartingthe nation.lib.ed.ac.uk]. The Geography and Map Division of the Library of Congress, Washington D.C. holds over 12,000 maps of Ireland [http://catalog.loc.gov/]. These are now fully catalogued under the heading: single sheet maps and maps bound in atlases; medium and large scale multiple-sheet series of maps; British Admiralty Charts. There is both a general index and a title index. The Map Room at the Sterling Library, Yale University has a small but significant Irish collection [http://orbis.library.yale.edu], including early road and administrative maps and maps published by the War Office (1942). The Bibliothèque Nationale de France, Paris [http://www.bnf.fr/] is another major international collection with some maps of Irish interest. The map room of the Bodleian Library, Oxford, has an impressive on-line site [http://www.bodley.ox.ac.uk/] and useful map links. The British Library Map Room maintains its own 'useful links' page [http://www. bl.uk/catalogues/listings.htm]. Oddens' bookmarks [http://oddens.geog.uu.nl/index.html] is a long-established site created by Roelof Oddens, map curator at Utrecht University, with links to many thousands of web sites of cartographic interest.

The 'Map history/History of Cartography' gateway site is part of the WWW-Virtual Library: History. It is hosted by the Institute for Historical Research in London [http://ihr.sas.ac.uk/maps/] and is perhaps the best entry point to cartographic history and related resources on the web. It also provides the most up-to-date introduction to events of cartographic interest, including the long-running lecture series titled 'Maps and society' held between October and May each year at the Warburg Institute, University of London; the history of cartography seminar series held by Cambridge; the Oxford seminar in cartography series, and public talks relating to maps held at the British Library.

FINDING AIDS: WITHIN THE LIBRARY

An introduction to the maps held by the National Library of Ireland, Kildare Street, Dublin 2, is to be found online at http://www.nli.ie. There is an unpublished handlist of both general OS and manuscript NLI maps at the duty desks in both the main reading room and Manuscripts Department reading room. An incomplete but useful card index is held in the catalogues room of the main library. The subdivisions of this card index include county or placename, personal name (of cartographer, commentator, client), theme (railways, canals, named rivers, inquiries), and title of map. It provides an unrivalled introduction to the way in which maps are entwined with history at both national and local level. It lists maps that are part of books, parliamentary reports and other collections which it would be difficult to locate without prior knowledge of their existence. Under place name (for example, Galway city) the maps are listed chronologically. Within each county there are also thematic divisions, including estates and roads. The headings vary from county to county, as not every heading is generally applicable. Londonderry, for example, has a subsection on maps of the siege. Under the general heading 'Ireland: administration', there are numerous subdivisions, including maps on agriculture, banks and bogs. And wherever there is a map, the local historian can be certain that there is a good story: ambitions to improve the local harbour, to reclaim land, to promote tourism. Dublin is far better mapped than anywhere else, but an imaginative approach is likely to yield results for practically everywhere. Most early printed maps, estate maps, marked-up OS maps, and surveys are in the Manuscripts Department (2–3 Kildare Street, http://www.nli.ie). Hayes' catalogue and supplement, available in the NLI catalogues room and in university libraries, should be searched. Additional map catalogues in the Manuscripts reading room include the Longfield manuscripts maps list, and an unpublished manuscripts maps card index. Some of the 'collection lists' cite maps pertinent to these collections.[23] Some original maps will not be available to readers due to their fragility or because they are undergoing conservation; but may be consulted in an alternative format, such as microfilm. A major project to rehouse the NLI's collection of over 47,000 OS maps commenced 1 Feb. 2002; and a major digitising project by the Ordnance Survey of Ireland will allow online access to the OS historic collection held by the NLI. Selected maps are reproduced in Noel Kissane (ed.), *Treasures from the National Library of Ireland* (Drogheda, 1994), pp 175–201 and in the facsimile collections *Ireland from maps* and *Historic Dublin maps* already introduced. Photocopies of Manuscript materials are never provided but photographs, slides and transparencies may be ordered; notice of current regulations and order forms will be found at the duty desks.

The National Archives of Ireland has a large collection of maps. As the NAI incorporates the former Public Record Office of Ireland (PROI), readers may follow up 'old' PROI references in the NAI. The holdings of the NAI are immense and varied. Some map series were created or collected by departments in the exercise of their statutory functions, most obviously with the Ordnance Survey, but also with the Valuation Office, the Quit Rent Office, the Incumbered Estates Court, the Office of Public Works and the Records of the Clerks of the Crown and Peace. However, maps and map matters feature in the records of many other departments, including the Relief Commissioners (1846) and among what are classed as 'business records'. The allocation of OS map plates 1922–3 to the newly-established Ordnance Survey of Northern Ireland is covered in the records of the Department of the Taoiseach (NAI, S 2137) and of the Department of Finance (NAI, FIN 1/2169). Maps were important in the making of *Saorstát Éireann*; the Compensation (Ireland) Commission (1922) and the North East Boundary Bureau (1922–3) were only two of the many organisations whose mapping interests are recorded in the National Archives of Ireland files (NAI, FIN 1/319; S33; FIN 1/2167). The Quit Rent Office (QRO) papers which are held in the NAI include a wide variety of

23 R.J. Hayes (ed.) *Manuscript sources for the history of Irish civlisation* (Boston, Mass., 1965; 3 vols supplement 1965–79).

map types at different scales: Down Survey maps, grand jury maps, private estate maps, and Ordnance Survey extracts. The printed QRO catalogue should be searched under place as well as family name, as only a small percentage of its map holdings are listed separately under 'maps'. A search of the card index which lists holdings by county is another fruitful entrance point; here placenames are noted, which leads in many instances to the immense collection of private estate papers held by the NAI. Once the key family names have been identified – Trench (King's County and Tipperary), Palmer (Mayo, Sligo), Colclough (Wexford) – the researcher can go to the dedicated estate catalogue for further details. Some estates are very well represented in the national repositories, but the records of others, which were equally significant and carefully managed in their time, have not survived or are otherwise beyond public access.

Other key cartographic holdings in the National Archives of Ireland are the 'Landed estates court rentals' 1848–90; many of those relating to sales in Northern Ireland are held in PRONI. Popularly known as the 'O'Brien rentals' and covering conveyancing up to 1881, this is the most complete sequence of the sets of surviving rentals. The other set held by the NAI is the Quit Rent Office collection (which though small has the advantage of including late nineteenth- and early twentieth-century rentals), while the NLI also holds two sequences, that compiled for the Land Commission as a reference set (to the mid-1890s), and a set compiled between 1850 and 1864 by a Tipperary MP, Joseph Burke which was bequeathed to the RDS Library and hence to the NLI.[24] While there is much overlap, each set nevertheless contains material unique to itself and is therefore worthy of attention. These are descriptive rentals of estates issued by the Incumbered Estates Court prior to auction. They are essentially advertising materials or auctioneers' brochures, of lands and other property held by landlords bankrupted during the famine, on whose behalf the state now sought to find new owners. The index to the 'Landed estates court rentals' may be searched under owner's name, date of transfer, county, civil parish and townland where land was held, though there are some inadequacies in the index particularly relating to counties Dublin and Kildare, and to the cities of Kilkenny and Dublin.[25] City estate properties are listed by street, a reminder that many of the property parcels transferred to new owners under the land acts were relatively small. Lists of tenants, the land they held and their conditions of tenure are accompanied by handsome coloured maps of the lots to be auctioned. From 1862 the OS was required to provide the court with countless six-inch extracts, which formed the basis for these rental maps. Urban as well as rural property was disposed of through these courts. The lists are arranged by parish and county, and for larger urban areas, by street name. The National Archives of Ireland has a

24 Mary Cecilia Lyons, *Illustrated incumbered estates Ireland, 1850–1905* (Whitegate, County Clare, 1993), pp xv–xvi. **25** Not all the rentals recorded in these volumes are listed in the index, with townlands in the baronies of Carbury, Kilkea and Moone (Kildare), Coolock, Nethercross, Newcastle and Rathdown (Dublin) and Kilkenny city not indexed at all; for a full explanation see Andrés Eríksson and Cormac Ó Gráda, *Estate records of the Irish famine* (Dublin, 1995), p. 50.

substantial card 'Index to maps and rentals' which covers all parts of the country. However, it cannot cover all its holdings, as maps feature in so many different collections. The card catalogue titled 'miscellaneous index' also reveals map holdings.[26] The records of the Land Commission, another key collection for local history research, are still held in the Irish Land Commission (Records Branch) adjoining the National Archives of Ireland in Bishop Street, but regrettably access for research is severely restricted, at least at the time of writing. This archive holds the title documents and associated documents, including maps of estates acquired by the Land Commission and the Congested Districts Board, under a series of Land Acts dating from 1881. Estate maps and resale maps, as the ownership of hundreds of thousands of acres of land was transferred to small farmers, makes this one of the most important map archives in the country, which it is to be hoped will soon become more accessible. The Land Commission also holds the maps of glebe properties (Established Church) for the period 1835 to 1877.

The map archive of the Clerks of the Crown and Peace, also in the NAI, contains copies of some of the maps which were deposited in the House of Lords Record Office, Westminster, in the process of securing a private act of parliament to enable municipal or other 'improvements'. The records of the Circuit Court, Dublin, for example, include railway plans and books of reference (NAI, IC/14/23–26), similar records for tramways (NAI, IC/14/30/68–69); and the 'county and city electric plans' for Dublin 1900 (NAI, IC/72/41).

The most important cartographic collection in the National Archives of Ireland is that of the Ordnance Survey. The on-line introduction to the Ordnance Survey archive [http://www.nationalarchives.ie/] is the most direct entry point to what is without exaggeration an awesome scientific, artistic and organisational achievement. In size, completeness, content, and variety of formats, this is an extraordinary record. It covers all Ordnance Survey procedures, from the measurements taken in the field with theodolites and chains through to the engraving on copper plates. Record keeping was of prime concern, and much thought was invested in setting up a system by which every single document could be tracked at any point in time. Every notebook, tracing, fair plan and instrument, for example, could be followed from the day it was issued from central stores to staff until the day it was dispensed with and returned to the fireproof or other stores in the Phoenix Park. In each case the date of return and code numbers of the item were duly entered in a register. To understand the system employed by the Survey, and used in the NAI in its on-line computer catalogue, requires some application; however, the extra effort will be more than amply repaid. The transfer of administrative records from the Ordnance Survey headquarters in the Phoenix Park to Bishop Street is an ongoing project; at the time of writing most of the records relating to the six-inch townland survey and the large-scale town plans are computer-catalogued and accessible in Bishop Street (Appendix III), with the records relating to the one-inch, twenty-five-inch and other surveys yet to be transferred. The notes which introduce each series

26 For example, the Gaussen family, estate maps *c.*1863, NAI, M7049.

explain not alone the nature of the material but how it is organised, and is essential reading before attempting to use the computer search facility; as this is on-line, preliminary searching can be done before travelling to view the documents themselves. A print-out of the full catalogue is on open access in the NAI reading room; this includes the introductory notes to each class of material. Notes on accessing OS manuscript maps and allied materials (plots, examination traces, sketch maps, town name books) of urban areas and villages has already been introduced under 'large scale OS and other maps' in chapter 1.

The local 'hierarchy' of names must first be ascertained – townland(s), civil parish, barony, county – followed by determining the district division (A, B, C, D, E) within which the parish was surveyed, and the unique reference number given by the OS to each parish, as these numbers were vital to the survey's own record-keeping. The district divisions are not on a county basis, as work began in the north of each district and moved south, ignoring the county boundaries. C district, for example, stretched from County Donegal southwards through the central lowlands to county Cork; C 5 is the parish of Burt in Donegal, C 754 the parish of Inch, County Cork. The townland of Tallaght, County Dublin, is coded E 289; Maynooth, which is within the parish of Laraghbrien (Laraghbryan, Laraghbrian), barony of Salt North, is coded A 56.K.[27] There are ninety-nine parishes within the county of Dublin, each with the prefix E followed by its own number, as in Grangegorman, E 278 and St Werburgh E 299. These codes will be found in the 'List of parishes and reference numbers' produced by the National Archives of Ireland as its own entry point to the Ordnance Survey records in its care. It is bound as a separate volume in the reading room, alongside the OS and VO catalogues. The OS 'List of parishes …' held in the NAI is prefaced by a sketch map illustrating the OS districts and sub-districts 1831–46; these are also mapped in Andrews, *A paper landscape*, p. 53. Researchers should be conscious that the final 'authoritative' spelling of the parish name, as engraved on the published OS maps, may not exactly match the spelling in these indexes; all variations should be used until the OS reference code is ascertained, and then this code should be checked against entries in each series. Where difficulties in identifying the parish code are encountered (due to a multiplicity of spellings), entering the county name and scrolling down through the county list until the correct parish and barony is reached will usually work.

The registers of documents, created by the Ordnance Survey itself (NAI, OS 51) is a good place to begin inquiries, as it provides a list of documents submitted to Mountjoy House, and the dates forwarded, on a parish by parish basis. This is not a comprehensive listing of all OS files (there are other registers to consult) but will provide the parish number for OS archive purposes, information on what records were not created for a particular parish thus saving fruitless searches, the numbers of volumes or notebooks produced in each record category and their code

27 The letter K was allotted by the NAI to distinguish it from the parish of Drumsnat, county Monaghan which was given the same number (A 56) by the OS, in error.

numbers. For the parish of Laraghbrien, county of Kildare for example, coded A 56K, there are no road plots or index diagram, but all other categories of records[28] were safely deposited 'in fire proof chamber' on 17 September 1839.

Another routeway directly into the 'fieldbooks' is to consult the index diagram for a particular parish (NAI, OS 104); this is the sheet onto which the chain lines connecting the trigonometrical stations are plotted, and dated. These index diagrams are also known as 'line plots', 'outline plots' and 'skeleton plots', appropriate terms as there is no topographical detail. The information is drawn from the 'content field books' (NAI, OS 58) and the 'levelling registers' (NAI, OS 65); the numbers of the books used in its construction is entered directly on the index sheet, acting as an effective finding aid to the fieldbooks for that parish. There are no index diagrams for C district, but parishes in the other four districts may be approached in this way. In the Maynooth sample, for example (fig. 23), within the polygon created by lines joining the secondary triangulation stations T[I] (Mariavilla), C[I] (Real Park), N[I] (Greenfield) and R[I] (Maynooth Tree), the four-digit numbers refer to the individual content field books in which the survey detail for this parish of Laraghbrien will be found. Book 3186 (OS 58A/56, fig. 23) covers the entrance to the college with nos. 3163 and 3165 (OS 58A/56) continuing the area directly to the north and south respectively. The researcher is strongly advised to arm themselves with a six-inch extract covering their area of study before trying to work with the index diagrams in the National Archives of Ireland. To state it most simply, the trigonometrical stations (marked with a small triangle with the height in feet) on the OS published sheets correspond exactly with those on the parish index sheets. If coloured lines are drawn joining up each station with its nearest neighbour on the OS published sheet (or sheets), the researcher will have a rough framework to hand to assist in making sense of the index maps. They will also have a ready-made guide to exactly which content field book(s) deal with that area, an advantage which will only become apparent when the researcher discovers how many notebooks were used per parish.

To follow the progress of the Ordnance Survey in any particular area, the researcher needs to know the month and year when surveying was underway, and also which section is likely to be involved at any particular time. The 'progress reports and returns' 1826–1943 (NAI, OS 1) account for all aspects of the work, and the activities of all persons employed by the Ordnance Survey, on a month-by-month basis. While the names of some occasional day labourers are excluded, the vast bulk of those in the Survey's employ are named. These reports and returns (OS 1) are organised by department, date and district, the early files (1826–47) bound into volumes, and the later material (to 1943) consisting of monthly files. While it is possible to identify the names of those officers responsible for directing local operations, it is difficult for the local historian to go too far beyond this, as most disappointingly, little of the official correspondence has survived beyond the letter registers and indexes (OS 2).

28 Fair plans, content plots, boundary remark books, boundary surveyor's sketches, boundary register, original content register and duplicate, original levelling register and duplicate, levelling books, content books, name books, field sketches.

The six-inch published maps can be identified by county name and sheet number, but the pre-publication technical plots, such as the index diagrams (NAI, OS 104) and the plots reconciling the boundaries common to adjoining parishes (NAI, OS 103), are listed under the unique parish code. So too are the 'fair plans' or final manuscript drawings (NAI, OS 105), and the proof impressions (NAI, OS 107). Full sets of the first edition six-inch OS sheets are widely available, due to the generosity of the lord lieutenant of the time who donated copies *gratis* to a range of institutions (up to 1846), including Trinity College Dublin, the RDS library, the Royal Irish Academy, the Russell Library, Maynooth, NUI Cork and NUI Galway, Queen's University Belfast, and the Armagh Observatory.[29] The manuscript town plans 1830–48, drawn at a variety of scales until the five feet: one mile (1:1056) was decided upon as the standard scale, were produced for Valuation Office (VO) purposes, and thus were never intended for publication. Only the Dublin sheets were published. As this series (NAI, OS 140) covers each village and small town as well as the larger places, they are the equivalent of the six-inch in the urban areas. It is worth recalling that not all the information gathered by the OS was or could possibly be entered on the final map, so that it may be instructive to compare the published product with its draft materials, as was illustrated in the case of Maynooth (figs. 20–21, 23–28).

Other manuscript Valuation Office materials held by the National Archives of Ireland are the fieldbooks (land), the housebooks (houses) and the mills books; a guide to the latter, titled *The millers and the mills of Ireland of about 1850*, has been compiled by William E. Hogg (Dublin, 1997).[30] The annotated maps which are no longer current have been deposited in the NAI but will require substantial conservation work before being made available to researchers. Copies of the valuation maps are now held on CD ROM at the VO headquarters in the Irish Life Centre in Abbey Street Lower, Dublin 1; printouts of sections (A4 and A3 paper) and of entire sheets (AO) may be purchased. While the originals are in colour, the printout is merely black and white, and some of the manuscript additions may not be legible; the researcher ought to build in study time to view the original once more on-screen and annotate his or her own copy. The 'cancellation books' (still held by the Valuation Office) are vital to the interpretation of these maps; the researcher is encouraged to prepare blank master sheets in advance (as illustrated in Appendix IX) so that information from a series of cancellation books can be entered efficiently. The typescript catalogues to the Valuation Office materials held by the National Archives of Ireland are shelved in the reading room (alongside the OS catalogues). While the OS catalogues (online and hardcopy) are preceded by explanatory notes on their origin and form, the VO catalogues are still merely listings. PRONI holds the original maps (VAL.2D), notebooks (VAL.2B) and the

29 The Geography and Map Division, Library of Congress, Washington, has produced a list of institutions holding first edition printed six-inch maps of Ireland. This is an appendix to its Ireland maps catalogue, which is to be launched online. **30** Counties for which mill books do not survive are Antrim, Carlow, Donegal, Dublin, Fermanagh, Kildare, Londonderry, Meath, Monaghan, Roscommon, and Tyrone.

revision lists 1865–1929 (VAL.12B) for the six Ulster counties in Northern Ireland; readers are directed to the guide to PRONI's valuation records produced by Trevor Parkhill (*Ulster Local Studies*, xvi, no. 2 (1994), pp 45–58). A computerised geographical index makes the valuation and indeed other records held by PRONI readily accessible.

The map collections of the Public Record Office of Northern Ireland are well served by catalogues and 'finding aids'. Jonathan Bardon, *A guide to local history sources in the Public Record Office of Northern Ireland* (Belfast, 2000) includes a substantial section on maps, how they might be used in local history, and a chronological overview of map-making in Ireland. Its summary of the work of the Ordnance Survey in Ireland, and a guide to the holdings of the OS in PRONI, from its establishment in the 1920s, is especially useful. PRONI has an excellent collection, as one would expect, of early modern and plantation maps. Its education facsimile volume *Plantations in Ulster,* c.*1600–41* (1975, 1989) is a valuable source of maps, associated documents and explanatory notes. But is also has an outstanding collection of estate papers including maps of landlord towns, demesnes and the estate at large, covering areas in the Republic as well as in the six Ulster counties. In the *How to use the Record Office* series PRONI has produced useful guides to maps and plans on a county basis. The six counties of Northern Ireland are covered individually, the date of commencement varying between 1570 and 1608 and continuing to the eve of the Ordnance Survey, *c.*1830, ranging from maps of the Ulster plantation period through manuscript estate maps and plans. Volume 18 is titled *General maps of Ireland and Ulster,* c.*1538*–c.*1860*. PRONI holds a full set, in bound volumes, of the Irish Encumbered Estates Rentals, available for the whole of Ireland, and already introduced under the NAI heading. The index to the Encumbered Estates Court in PRONI is referenced MIC 80/2.

Perhaps the largest collection of printed maps in Ireland is that of the Glucksman Map Library, part of Trinity College Library, Dublin [http://www.tcd.ie/]. An introduction to both printed and manuscript collections is provided by John Andrews, 'Maps and atlases', in P.K. Fox, *Treasures of the Library, TCD* (Dublin, 1986) pp 170–83. As a legal deposit library (since the Act of Union), it was in an excellent position to build up a collection, and has pursued an enlightened acquisitions policy in recent decades. A visitor to the Glucksman Map Library (and indeed to any map archive) would be well advised to have formulated a number of specific lines of inquiry in advance, as discussed above; searching under 'Dublin' for example, would be inadequate as the collection listed here is immense. Requests are directed to the map librarian who will identify items in the map catalogue, as readers (at present) do not have direct access to this. The Map Library also boasts the country's largest reference section on cartographic history and Irish maps, including specialist journals, and guides to British map collections and individual cartographers (as listed below). While the library is primarily to serve the needs of students and staff members of the University of Dublin, it has a long tradition of facilitating individual students registered in other colleges who have specific enquiries. Researchers should always make inquiries of the TCD map librarian in advance, by telephone, in writing or by email [map.library@tcd.ie]. In addition, TCD has set up a fee-based service for business and professional users, whereby its staff search, retrieve and photocopy all historical maps covering a particular site. There is a minimum cost per site, and only out-of-copyright maps can be copied (except where the customer already has an appropriate licence). Full details of that service can be obtained directly from the Map Library.

While the Glucksman Map Library houses the bulk of TCD's collection of published maps, some early printed atlases and maps are in the Department of Early Printed Books in TCD Library. Some important early manuscript maps are also to be found in the Hardiman atlas in the Department of Manuscripts. ALCID readers (as explained below) have ready access to the library but a letter of introduction is required for Manuscript readers. At the time of writing the quickest routeway into the collection of early published maps, maps in books and books dealing with map-making (on the principles of a trigonometrical survey, for example) is via the 1870s list which has been computerised and is available (in unedited format) in the Early Printed Books section, with the published volumes on open access in the main reference section of the Library. The catalogue for manuscript maps can be consulted only in the Manuscripts Department.

A full listing of Irish repositories with map holdings would be merely to rehearse most of the entries in Helferty and Refaussé's *Directory of Irish archives*. The ALCID system (Academic Libraries Co-operating in Dublin)[31] which provides registered post-graduate students and academic staff with access to the libraries of all the institutions in the network, on a reference basis, has greatly widened the circle of resources available. Other libraries should be contacted directly; while some require evidence of serious research, postgraduate registration is not always required. Marsh's Library, St Patrick's Close, Dublin 8 has an exceptionally fine pamphlet collection, many of which include maps, while it also holds treatises on scientific subjects, including mapping, surveying and geography. Its full catalogue is online [http://www.marshlibrary.ie]; most of its holdings date from the late seventeenth and eighteenth centuries, and it has a long tradition of public access.[32] The RIA holds some important cartographic materials. Among these are the earliest extant manuscript copies of the 'Books of survey and distribution' (1664–88), manuscript and typescript OS letterbook and memoir material, and original watercolours and sketches of a geological, antiquarian and geographical interest, produced either directly for the OS, or by persons employed for a period by the OS, including topographical drawings by George Victor du Noyer and George Petrie, and (*c*.1914–30) by T.J. Westropp. It also holds the scientific notebooks of the

31 At the time of writing the co-operating academic libraries are NUI Maynooth, TCD, UCD, DCU, RCSI, NUI Galway, NUI Cork, Mater Dei Institute, St Patrick's College, Drumcondra, UL, DIT, Mary Immaculate and the Royal Irish Academy. **32** For example, see *A more exact way to delineate the plot of any spactious parcel of land as baronies, parishes and townlands, as also of rivers, harbours and loughs, &c., than is as yet in practice, also a method or form of keeping the field-book, and how to cast up the superficial content of a plot most exactly*, William Bladen (printer), (Dublin, 1654).

military surveyor, Charles Vallancey (whose manuscript survey is in the British Library). The RIA has a small but fine collection of travel maps, including examples by most of the 'masters' in the field; the separate card catalogue should be searched under cartographer and place. Access to the RIA library is open to members and holders of an ALCID card; non-members and visitors should make application in advance of their visit to the librarian, with details of the research project in hand. The Representative Church Body Library holds the archives of many Church of Ireland dioceses; several of these collections, including Dublin and Meath, Ossory and Tuam, include important pre-1870 materials. It also holds over 850 collections of parish registers, principally from the Republic of Ireland, some of which include maps of parochial boundaries and parish properties. Maps were important tools in the management of estate property but also for parish and diocesan administration; surviving Church maps are especially valuable sources for the period prior to Disestablishment in 1870. It is best to contact the librarian and archivist in advance of one's visit.

An introduction to the Land Registry can be found via the Government of Ireland webpage [http://www.irlgov.ie/]. Its primary function is to provide a system of compulsory registration of property, for which purpose maps are essential. Its practice directions (1998, published online) provide a useful introduction to issues in map-making, in this case from the perspective of land registration and the legal transfer of title. The Land Register is divided on a county basis. Before approaching the Land Registry, the researcher needs to have full address (street name and number, townland name, parish, barony, county), the names of previous owners (where possible) and names of current owners. A separate Land Commission for Northern Ireland was set up in 1922; the transfer of maps and other documents from the Quit Rent Office to this newly-established office (1923–4) is covered in Department of Finance Records (NAI, FIN 1/2309).

The Geological Survey of Ireland (GSI) Beggars Bush (initially part of the Ordnance Survey and, unofficially, of the Valuation Office) holds the original six-inch geological survey field sheets of the 26 counties (1845–87), manuscript OS memoir materials (the Portlock survey, 1820s–40s) and exquisite water drawings of rock outcrops and mineral formations, by du Noyer (1836–69) and others. Part of the GSI's remit has been to build up a public database of mineral resources in Ireland. Under 'historical mine records' it holds closure plans and sections for mines, and under 'open files' holds more recent mining records, most of which are available to researchers. County-based guides have been produced by the GSI: *An index to mineral and mining records and manuscripts in Ireland* (Dublin, 1988) by C. Nolan, and *An index to references to Irish mineral resources and mining activities in the Mining Journal 1835–1920* (Dublin, 1989) by P. Jeffcock. The website [http://www.mhti.com], newsletter and publications of the Mining Heritage Trust of Ireland (headquarters at the GSI, Dublin) is a good starting point for research in this area. Among practical aids to researchers is a reprint of Grenville Cole's *Memoir of localities of minerals of economic importance and metalliferous mines in Ireland* (Dublin, 1922, republished by MHTI 1998), with biography by Patrick Wyse Jackson. The Trust also maintains an online listing *Irish mining in periodicals and anthologies 1955–95*. The GSI library is open to the public during weekday office hours. Its Ulster equivalent, the Geological Survey of Northern Ireland (20 College Gardens, Belfast BT9 6BS), holds the first edition geological field maps which relate to Northern Ireland, as well as the maps of its own 1947 six-county survey, and a reference map collection; the GSNI is open to researchers on weekdays.

The published catalogues for the maps held by the British Library (originally housed in part of the British Museum) are available outside of St Pancras, a major asset for researchers operating at a distance from London. University libraries in Ireland hold at least some of the catalogues. These map catalogues, and descriptions of manuscript maps acquired before *c.*1850, have been brought together as *The British Library map catalogue on CD–ROM format* (London, 1998). A continuously updated version of this map catalogue is now available free on the internet via COPAC (http://copac.ac.uk/). The *Catalogue of the manuscript maps, charts and plans, and of the topographical drawings in the British Museum* (3 vols, London, 1844, reprinted 1862) covers manuscript material received up to 1844, and is geographically arranged. *Catalogue of the printed maps, charts and plans* (15 vols, London, 1967) and *Ten year supplement, 1965–74* (1978) is arranged by author and geographical heading, and is an outstanding guide to the rich cartographic holdings of the British Library. This is further supplemented by the *Catalogue of cartographic materials in the British Library, 1975–1988* (3 vols, London, 1989), and later microfiche updated versions. The sales catalogue (Maps 135–137) directs the reader to a large collection of nineteenth- and twentieth-century printed estate plans, arranged alphabetically by county and then by town. The updated online COPAC version, as stated already, brings together these various catalogues into one master search tool. Search results (under place/area covered or place/sub-regions; name, date, and subject) may be re-oganised, saved and printed out, making it the essential first step to work in the British Library Map Room. The Map Room also produces a series of short but helpful guides to assist readers in their searches, such as 'Atlas maps and how to find them' and 'How to find estate maps in the British Library'. Ongoing digitisation (as part of the 'Collect Britain' project) will make available many maps and views of Ireland (and G.B.) *c.*800–1620; researchers should visit the BL map webpages for up-to-date information and access to these and other digital collections (http://www.bl.uk).

The cartographic holdings of the British Library are not confined to the Map Library. A free twenty-page leaflet titled *The map collections of the British Library* can be requested in advance, and provides information on the other catalogues containing maps, such as the Oriental and India Office collections. Most of the estate maps held by the Department of Manuscripts, for example, do not feature in the Map Library catalogue. These may be identified through the three volume *Indexes to material of cartographic interest in the Department of Manuscripts and to manuscript cartographic items elsewhere in the British Library*, held in the Map Library at Maps Ref.Z.2(1) and also available online via MOLCAT. Geographical sequence (vol. i), subject sequence (vol. iii) and nominal index (vol. ii) make this a particularly 'searchable' tool. Many books containing maps are held in the general library or rare books collection, and so these catalogues need to be considered also. The process of combining the map catalogue with the catalogues of most of the other BL collections on a joint Open Access Catalogue is well advanced; it should be available in 2005. Before a British Library reader's ticket is issued evidence of post-graduate student registration is generally required; researchers should inform themselves of the regulations governing admission before travelling to St Pancras [http://www.bl.uk/].

As with the British Museum/British Library catalogues, many university and other libraries in Ireland hold copies of various guides to the Public Record Office, now part of the National Archives at Kew. References to Irish maps and collections of maps may be found in *Maps and plans in the Public Record Office, London, vol. 1, British Isles, c.1410–1860* (London, 1967), pp 547–93. However, this has been overtaken by more recent computer cataloguing [http://www.pro.gov.uk/]. The National Archives has placed its massive catalogue online, a tremendous service to researchers overseas. It has also produced information leaflets titled *Maps in the Public Record Office* (no. 91, thirty-nine pages) and *Architectural drawings in the Public Record Office* (no. 109, thirteen pages), which may be downloaded from the website; these are essential reading, as 'some millions' of the maps and plans held by the National Archives (according to their own estimate) do not feature in the published or online catalogues at all. Most of the maps held by the National Archives were produced in the course of official departmental business and are still enclosed with the correspondence, reports, and files with which they were originally associated. This greatly increases their value to the historian who can see at first hand the larger context within which they were created. But the problem is that there is usually no mention in the series lists (which are online) that maps are present, so that the researcher must use their own ingenuity in assessing which files are likely to contain maps. A few departments did create their own map libraries which were transferred to the then PRO as discrete map collections and these can be readily identified, such as the Foreign Office, the Colonial Office and the War Office. But even these contain only a fraction of the maps produced or held by these offices, as maps continued to be part of other record sets. A system was introduced by the PRO about 1926 whereby maps and plans were removed from volumes of files and given separate map room references and a 'dummy' sheet noting the map's new location was inserted in the original file. The link between the original setting and newer map room location is maintained; these extracted maps should be pursued under both the original reference (for context) and the newer reference (when ordering the map itself). Maps come in all sizes and shapes, rolled, flat, as bound books and as loose enclosures, creating myriad difficulties for archivists; this is also reflected in the National Archives cataloguing, as extracted maps of an awkward size were catalogued according to their format (rolled or flat, atlases and maps 'of unusual form') rather than provenance (such as from the Chancery, the Exchequer, the State Paper Office, the Court of King's Bench). Along with the printed and online catalogues already mentioned, researchers can refer to the supplementary card catalogue in the Map and Large Document Room, and the 'Summary calendar of unextracted maps' also held in the Map Room. The National Archives contains a wealth of manuscript maps covering every part of the island of Ireland, and most valuably, the associated files which allow a thorough exploration of these maps. It also contains records relating to the practice of map-making, as in the papers of the Ordnance Survey and of the Barrack Board (part of the War Office). It is undoubtedly deserving of the attention of local historians working on any part of Ireland. G. Beech and R. Mitchell's *Maps for family and local history* (Kew, 2004) is part of the readers' guide series to the National Archives and most user-friendly. For hydrographic charts relating to Ireland, the *Catalogues of Charts, plans, views and sailing directions … of the British Admiralty from 1825* lists those held in Kew.

The House of Lords Record Office, Westminster (HLRO) holds a large number of plans and maps of compelling interest to local historians in Ireland. This author has identified over 1,800 separate files of Irish interest covering harbour development, docks, canals, railway, tramways, drainage and embankments, reclamation schemes, commons enclosure, markets, sewage schemes, burial grounds, reservoirs and water-works, promenades and piers, requests for boundary extensions, 'new streets', parks, slum clearance schemes and labourers' dwellings, gasworks and electric lighting. The manuscript indexes to what are termed the 'HL deposited plans' date from 1794; the last entries for the Republic of Ireland are for 1920. The researcher must firstly know the year or parliamentary session during which the submission was made as the HLRO indexes are in chronological order. Within each year the submissions are divided alphabetically, and a letter and code number ascribed. The *Chronological table of local legislation, local and personal acts* (4 vols, London, 1996) along with Rosemary Devine's *Index to the local and personal acts, 1850–1995* (4 vols, London, 1996) will assist in this process, but it must be noted that not every submission did in fact result in the passing of a private or public bill. The 'List of Private Bill evidence', a typescript in the HLRO, shows the number of volumes of printed evidence per session, with the volume number and microfilm number, and then on a separate page the manuscript evidence amongst the main papers. For example, the heading '1836 Municipal Corporations (Ireland) bill, report of the Lords Committee, vol. 2 mf7' is followed by the manuscript evidence submitted during that session. Maurice F. Bond's *Guide to the records of Parliament* (London, 1971), is the standard reference to the classes of material in the HLRO. In addition, a list of Irish plans 1794–1850 in the HLRO was published in 1966 by Brian Dietz; this does not include all the cases (as explained below) but gives the year, project name ('Belfast and Cavehill railway'), Act under which it was submitted, and notes where committee evidence exists, which is for a small percentage of cases only.[33]

A good grasp of Irish placenames is required to identify Irish material in the index to the deposited plans in the HLRO, as the suffix 'Ireland' is included in only a fraction of entries, and there are many placenames common to Britain and

33 Brian Dietz, 'A survey of manuscripts of Irish interest for the period 1715–1850 in the House of Lords Record Office' in *Analecta Hibernica*, xxiii (1966) pp 225–43.

Ireland, as in Newport, Hanover Square, Bangor, Sutton, Louth, while 'Great Western Railway' and 'Great Northern Railway' companies are found on both islands. There are some misspellings in the handwritten catalogue, as in the Kilbrush and Kilder (Kilrush and Kilkee) and Poulnashery Reclamation (1865 K3) but this will not deter the local expert. More serious are irregularities in alphabetical order. In some cases the plans are filed under the title of the relevant legislation rather than under the placename, thus there are files under L (Labourers' Dwellings Act) (1884 L34–59) and Local Government (Ireland) Act (1875 L22), and under D, Drainage and Improvement of Lands (Ireland), (1885 D7, D10). Dublin can be taken as an example of how irregularities in the filing system here need to be overcome, pending computerisation; while Dublin material is filed for the most part each year under D, relevant material will also be found under other headings.[34] However, the sheer volume and richness of the archive for students of Irish local history make the initial effort at locating the material more than worth the effort. And once the bill or project title is identified, the index lists the type of material available, in an abbreviated form (as explained under railway mapping); PSR for example stands for plan, section and book of reference; PO for provisional order.

The Caird Library at the National Maritime Museum, Greenwich, is a specialist maritime repository whose holdings include pamphlets, journals and all manner of books relating to maritime matters (including chart-making) dating from 1474. The book catalogue is available online [http://www.nmm.ac.uk]. The reader is also directed to the published catalogues: section 3 of the library catalogue *Atlases and cartography* (2 vols, London, 1971) and to R.J.B. Knight (ed.) *Guide to the manuscripts in the National Maritime Museum* (2 vols, London, 1977–80). The collection of charts, atlases and pilot books (manuscript and printed) at the NMM date from the 1400s and cover all parts of the globe; at the time of writing volume iii of the catalogue, titled *Atlases and cartography*, was not available online. It must be searched onsite under placename as well as cartographer. Most of the charts of Irish interest in the NMM are to be found bound in atlases and pilot books; charts of the Irish coast and insets of Irish harbours have been located in German, French, Dutch and English atlases published 1596–1835, and further inquiry is certain to yield even more results. Some of the internationally-renowned chartmakers and publishers who turned their attention to Irish waters and feature in the NMM collections are Lucas Waghaener (1584); Greenvile Collins (1693); Alexis Hubert (1693); John Seller (1703); Hermann Moll (1714); Jacques Nicolas Bellin (1756); Lewis and William Morris (1802) and John William Norie (1835). The chart index of the NMM should also be consulted; Irish material will be found listed under 'Ireland: north coast'; Ireland: south coast' and so on. There are yet other headings under which the Irish coast will appear, as in the Robert Adams' charts depicting the series of

34 For example, C (Cork Hill, 1882 C12; City of Dublin Bride's Alley area, 1894 C24); I (Ireland, Local Government Board, Chancery Lane scheme, 1920 L1); N (North Dublin Street Tramways, 1880 N3; Nelson's Pillar, 1891 N1)); R (Royal Dublin Society, 1877 R 9); and S (St Stephen's Green, 1877 S2 and South Dublin Railway, 1877 S8).

engagements between the English fleet and the Spanish Armada in 1588.[35] In pursuit of maritime maps, Ireland must always be considered within its larger British Isles/North Atlantic/North European contexts. The NMM is well known as the principal repository for navigational charts, dating from the earliest surveys of the Hydrographic Office (established 1795). It holds a comprehensive collection of Admiralty charts; however, only the current sheets are available to readers on request as early editions are stored off-site, requiring prior notice. The PORT website, hosted by the NMM [http://www.port.nmm.ac.uk] is a major database for maritime studies. Its research guides are available to the public online, including a guide to charts and maps in the NMM, and links to other repositories of interest to scholars in this field. Access to the NMM is freely available to persons over 18 years; student status or college affiliation is not required. The museum itself has an outstanding collection of navigational instruments.

The Royal Geographical Society (London) with the Institute of British Geographers [http://www.rgs.org] is in possession of numbers of maps relating to Ireland, including some in the Michael C. Andrews collection, donated 1935. Its published catalogue titled *Catalogue of [the] Map Room of the Royal Geographical Society, March 1881* (London, 1882) is supplemented by an unpublished card catalogue; as with most other major map repositories, an information leaflet outlining its collections can be requested directly from the Society, while it is also developing its online cataloguing system.

Most universities and third level colleges in Ireland have at least some maps in their collections, but the range, quality and accessibility will vary enormously. TCD Map Library, already introduced, provides a generous and exceptional service to students and staffs of other colleges. Other college map libraries are on a much smaller scale and largely restricted to currently registered students and members of their own staffs. The School of Architecture, University College Dublin, has a dedicated map section at its library in Richview, Clonskeagh Road, and researchers are welcome, but as in the case of all map libraries outside of the major national repositories, the intending visitor should make advance contact and ascertain the conditions governing access. The UCD Architecture Library at Richview operates as an agent for the Ordnance Survey of Ireland for their PLACE mapping series; researchers can obtain up-dated computer-generated maps centred exactly on their area of study (such as a particular road intersection or premises), at the scales of 1: 10,000, 1:5,000, 1: 2,500 and 1:1,000, as A4 and A3 printouts, at a cost which is reasonable considering the value of the product.

Cartographic material will, by its very nature, feature in most archives; where a particular institution is physically present, such as a convent, hospital or charity school, it is likely that maps or plans are extant, which may at a minimum note the site boundaries and adjoining holdings. The *Directory of Irish archives* 4th edn (2004) is the standard entry point to such private holdings. Also valuable is the CD ROM

35 Robert Adams, *Expeditionis Hispanorum in Angliam vera descriptio* (London, 1588); a full set is held in the NMM, Greenwich.

searchable database of 420 public and private archives in the Republic and in Northern Ireland, titled *A dictionary of sources for women's history in Ireland,* Maria Luddy et al. (Dublin, 1999). Many major architectural practices, such as Ashlin and Coleman, were employed by church and educational bodies; where the institution no longer exists or does not hold maps, the Irish Architectural Archive (45 Merrion Square, Dublin 2) might be approached as it holds a large collection of building plans and engineers' drawings for every part of Ireland, and is freely open, without appointment, to researchers [http://www.iarc.ie]. Even when potentially useful material has been identified in private hands, the researcher should be aware that such institutions are unlikely to provide formal 'opening hours' and ready access in the way that major state-funded repositories can, so that advance contact, preferably by letter, is recommended. It is also worth reminding the researcher that in most cases they have no legal right to see the records, the archivist is often unpaid, and facilities such as photocopying may not be readily available, even to the archivist. That said, private archives in Ireland are on the whole extraordinarily obliging to the genuine and well-prepared researcher.

NEW ACQUISITIONS

As with other fields of historical research, the map enthusiast needs to be alert to the channels through which new collections or new cataloguing achievements are publicised. By definition, by the time a printed or CD-ROM catalogue is available it is already dated; online computer cataloguing overcomes this obstacle, but currently there are relatively few map catalogues fully and freely available on the Web. The annual *Report of the deputy keeper of the records* (Dublin, 1869–1964) makes occasional reference to map accessions; its equivalent for Northern Ireland, *Report of the deputy keeper of the records* (Belfast, 1925–) similarly describes records deposited with PRONI. Even though these earlier 'new' acquisitions are usually integrated into the catalogue, the extra detail provided in the reports may still prove useful. Recent annual reports for the National Archives of Ireland and the National Library of Ireland should also be scrutinised. In addition, the NLI publishes its newsletter, *Nuacht Leabharlann Náisiúnta na hÉireann*, giving more frequent updates. The *Irish Economic and Social History* journal publishes an annual report (since 1983) on business records deposited, in most cases, in a national or regional archive; maps may feature among these collections. The journal *Irish Geography* traditionally devotes space to a review of maps and mapping. The section entitled 'new maps of Ireland', prepared by Paul Ferguson, is the best point of information to go for an update on recent map publications (including facsimile reprints) and informed commentary. Updates on new holdings, which may include maps, are also to be found in *Irish Historical Studies; Irish Archives,* and *The Society of Archivists' Journal. Analecta Hibernica*, which is devoted to providing transcripts and calendars of primary source material, can occasionally yield map-related gems, as in the transcripts of a petition to the privy council, and expenses for a mapping commission in the south of Ireland, 1568–71.[36] These latter publications can all be consulted in the NLI and in most university libraries.

IDENTIFY THE CARTOGRAPHER

To 'know' a map one must really get to know its creator. In the pursuit of biographical details the researcher can delve deeper into the making of the document, and identify links with co-workers and competitors, patrons and customers, and of course the places which he mapped (and generally, though not always, traversed!). The identification of cartographers is fraught with difficulties as they were variously styled as 'plattmakers', surveyors, land surveyors, map-makers, engineers and architects, not to mention the fact that some were accomplished in a number of other occupations – Rocque, most famously, was an international *dessinateur de jardin*. On the changing state of the profession of surveyor the classic study is *Plantation acres* by J.H. Andrews (Omagh, 1985). In pursuit of any individual cartographer, the essential reference is the new edition, by Sarah Bendall, of Peter Eden's *Dictionary of land surveyors and local map-makers of Great Britain and Ireland, 1530–1850* 2nd edn (London, 1997), a two-volume British Library publication. Tooley's dictionary of mapmakers, first published in 1979 (supplement 1985), aims to cover all cartographers world-wide; vol. i of the revised edition, edited by Josephine French, covers A–D (Tring, 1999). Also useful is *Who's who in the history of cartography: the international guide to the subject*, edited by Mary Alice Lowenthal (Tring, 1998). In Bendall's British Isles text, the indexes proper are organised by date, by areas in which surveyors practised (on a county basis), by types of maps which were produced (for example, tramway, urban estate, valuation, waterworks), and by places where surveyors are known to have lived. In the latter case the addresses are given in great detail, including street name for urban areas. There is also a general index; entries under 'Irish' include Board of Works, boundary survey, forfeited estates trust, general Valuation Office, plantation surveys and post-roads surveys. The dictionary entries themselves are listed in alphabetical order, with the cartographer's dates of birth and of death, residences, occupations, maps produced and dated, and where original copies are currently held. Although a large production – 13,744 cartographers have been listed – it is thoroughly cross-referenced throughout. One particularly useful feature is the cross-referencing between cartographers, allowing the reader to make connections that would otherwise stay unnoticed. From the local history perspective, this allows the names, dates and 'connections' of surveyors known to be working in any given area to be identified, a very good start in any cartographic study. This is a reference text which county libraries should hold, or be encouraged to purchase.

The Oxford dictionary of national biography [http://www.oup.com/oxforddnb/info], major ongoing project (to be printed online and in sixty print volumes September

36 J.H. Andrews 'Robert Lythe's petitions, 1571', pp 232–41.

2004), also includes cartographers of note, several of whom had experience of map-making in Ireland. A specifically Irish production is Rolf Loeber's *Biographical dictionary of architects in Ireland, 1600–1720* (London, 1981), while he has also published a biographical dictionary of engineers in Ireland (*Irish Sword*, four parts, 1977–79, xiii, nos.. 50–53). Biographies of engravers and printers will be found in Walter Strickland's *A dictionary of Irish artists* (Dublin, 1913, reprinted Shannon, 1969). The effort of D.G. Lockhart at identifying surveyors at work in Belfast city by screening the advertisements in the *Belfast Newsletter* 1738–1840 (see *Irish Geography*, xi (1978), pp 102–9) is an example of what the local historian needs to do to ensure that all leads are followed up. Most library and archive catalogues will file maps under the surveyor's name as well as the map title and date. PRONI has an extensive catalogue section devoted to surveys and surveyors, which along with the maps and associated materials, identifies correspondence to, from or about individual map-makers, making it an invaluable starting point for any historian of cartography.

GUIDE TO SECONDARY READING

Paul Ferguson's manuscript 'Irish map history, a bibliography and guide to secondary works, 1850-present, on the history of cartography in Ireland', provides the most direct routeway into the literature on map history in Ireland. As already noted, this can be consulted in the Glucksman Map Library, TCD, and is both comprehensive and up-to-date. Headings which could usefully be consulted include county name, name of cartographer, and foci of inquiry such as placenames, railway commissioners, soils, military plans. An early published edition of this bibliography (Dublin, 1983) can be consulted in most major libraries. Perusal of Ferguson's bibliography will at the very least assure the local historian that he or she has massive scholarly support. There is indeed an extensive literature even if it is initially difficult to locate what might be fruitful.

Journals dedicated to the history of cartography and/or contemporary developments in the field include *Imago Mundi, The Map Collector, Mercator's World, Cartographica*, and *The Cartographic Journal*, while *Irish Geography* and *The Geographical Journal* also include substantial if occasional papers on historic maps. The British Cartographic Society produces a newsletter *Maplines*, whiles Map Curators' Group of the BCS produces its own newsletter, *Cartographiti* [http://www.cartography.org.uk/]. *Mapforum* is an online journal [http://www.mapforum.com]. Careful perusal of the *Journal of the Royal Society of Antiquaries of Ireland* and both *Transactions* and *Proceedings of the Royal Irish Academy* are guaranteed to be fruitful. The *Irish Sword*, dedicated to military history, not surprisingly has many essays based on map evidence: sieges, battleplans, forts, artillery fortifications and castles all feature. The *Royal Engineers' Journal* includes some articles of Irish interest in a similar vein. Geology, drainage, reclamation of bogs and the potential of mineral deposits all lend themselves to mapping; the *Scientific Proceedings of the Royal Dublin Society* and the *Journal of the Royal Geological Society of Ireland* will also repay careful examination. The standard history or historical geography journals similarly include map studies; journals devoted to specific historical periods (early modern, eighteenth, nineteenth century and so on) should also be scoured. The *Journal of Transport History,* and the *Journal of Canal and Railway History* deal with cartographic matters, as maps are such an intrinsic part of the records with which they are concerned; although largely dealing with UK case studies, road, canal, tramway and rail developments in Ireland are all particularly well suited to consideration on a British Isles basis. Unlikely sources may pleasantly surprise the local historian; the *Irish Ecclesiastical Record* and the *Irish Georgian Society Quarterly Bulletin* have both yielded excellent material. *Long Room*, the journal of Trinity College Library, includes essays on selections from its vast map collection, and the comparable journal for the British Library map collection, *The British Library Journal* (formerly the *British Museum Quarterly*), also makes occasional reference to maps or individual cartographers of importance to Ireland. The Charles Close society (London) is devoted to the study of the Ordnance Survey, and its remit covers Northern Ireland and the Republic; its journal *Sheetlines* has published some two dozen papers relating to Irish OS maps since 1981, while vol. xxx (1991) was a special issue devoted to Ireland.

CHAPTER 4

Map-making in local history

CASE STUDIES OF WORKING WITH MAPS IN LOCAL HISTORY

The location of maps is but the first stage; more critically, how does one move into meaningful interpretation? How can a really critical stance (Appendix 1) be taken with maps? As with other historical sources, merely to reproduce the 'raw data' is usually inadequate. Much time has been spent in this guide on the routes that might be followed to access maps, but what can the researcher do when this has been achieved, other than reproduce a section? A few straightforward examples of ways in which maps have been used in local history and historical geography might prove more instructive than lengthy discussion. Researchers generally rely on existing published maps as base maps for their own cartographic endeavours; the matter of copyright therefore also needs to be investigated. The 'guidelines on map-making' with which this chapter concludes will deal with the technical aspects of moving from base map or 'raw material' to the presentation of your own data in cartographic form.

Examples of imaginative and professional use of maps in local history are to be found in a huge variety of publications, some of which have already been introduced, such as the local history journals. For a rural perspective, *Townlands in Ulster, local history studies* (Belfast, 1998), edited by W.H. Crawford and R.H. Foy illustrates the way in which mapping can be used to better understand the nineteenth-century record. *Dublin through space and time* (Dublin, 2001), edited by Joseph Brady and Anngret Simms, provides many examples, at city-wide and more intimate scales, of how historical data might be mapped to great effect. Each fascicle in the *Irish Historic Towns Atlas* series has a selection of text maps and can act as a model of how a researcher might go about 'using' maps in local history.

Perhaps the primary reason for using maps in local history is for location, as a way of introducing an area, to situate and structure a local history study, or as an 'anchor' source around which other information is gathered. Assembling a succession of different maps, of different types, at different scales and from different authorities, can act as a launching pad for a local history study, as is illustrated in this guide in the cases of Baltimore (fig. 1), the Four Courts, Dublin (fig. 3) and Waterford (figs. 31 and 32). Comparative study reveals changes over time, and will help to identify key questions for exploration, while all the time remembering that the map itself is a sophisticated human construct with its own agenda, and must be treated in a critical way (Appendix 1).

Contemporary maps can be used as the basis upon which early archaeological patterns can be reconstructed. Circular monastic enclosures can be identified by

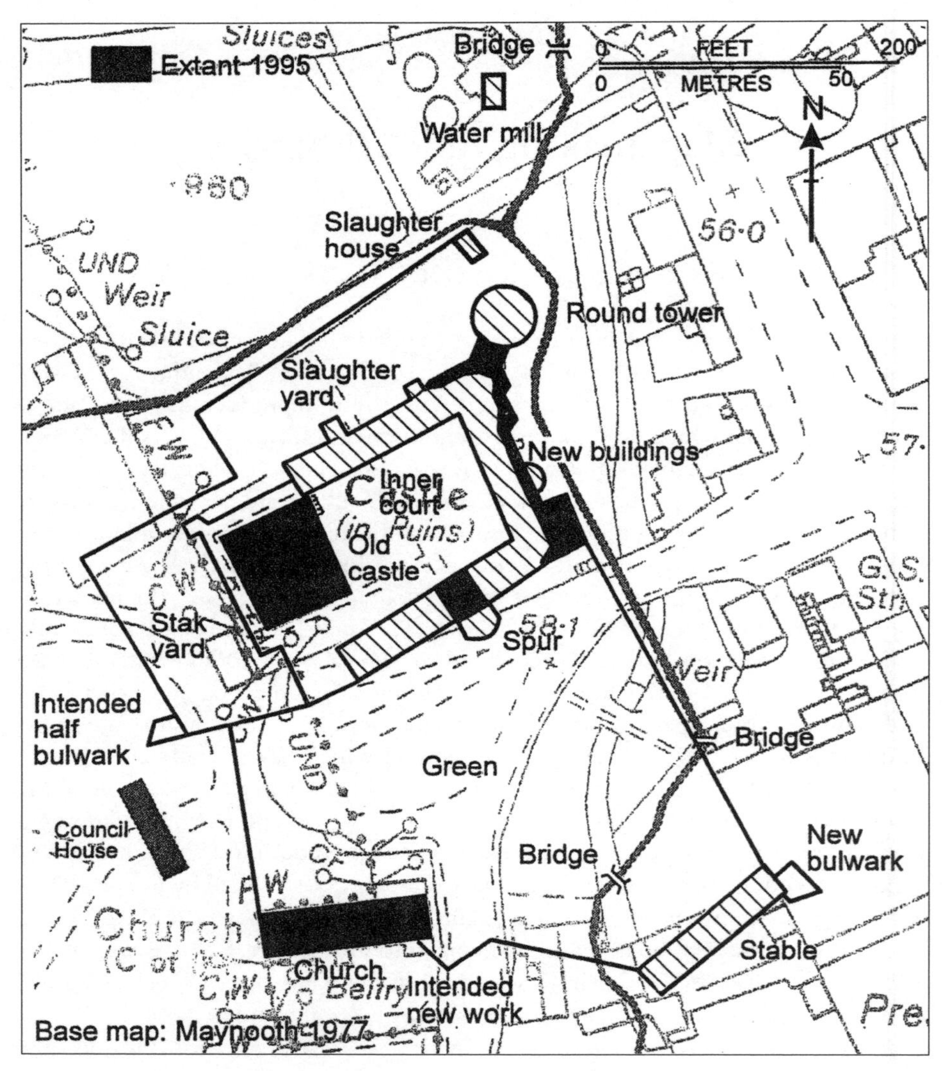

59 Maynooth Castle, *c.*1634, superimposed on OS 2,500, edition of 1977, Arnold Horner, *IHTA*

careful map-based detective work, noting the alignment of routeways and property boundaries, the location of round towers, crosses, burial plots and other ecclesiastical remains. Examples will be found for Kells (A. Simms), Kildare (J.H. Andrews), Downpatrick (R.H. Buchanan and A. Wilson) and Dublin (H.B. Clarke) in the *Irish Historic Towns Atlas* series (Appendix II), produced by the Royal Irish Academy (Dublin, 1986 onwards). Similarly, the line of the medieval or early modern town wall, gates and towers, or the ground plan of an early castle, can be traced on a recent large-scale OS map, following the clues offered by the map itself in conjunction with field and documentary evidence. In the case of Maynooth, for example, Arnold Horner (fig. 59) has superimposed a 1634 plan of the castle on the OS 1:2,500 revision of 1977. This map work allows the visitor to understand the buildings which survive today as part of an impressive complex, and to gain an insight into the ambitious scale of the redevelopment undertaken by Richard Boyle earl of Cork in the 1630s, when he ordered a local stonecutter 'to pull down all the rotten, decayed, disproportioned and unuseful old buildings [at Maynooth] … and agreed … to rebuild three ranges of the square court in a fair and uniform manner, according to a model and articles now in making, and to re-edify the decayed church'.[1] Home scanners and even the most basic computer graphics programmes (such as MS Paint) allow background images to be overdrawn with ease.

There is a case to be made for redrawing of early maps in many instances, especially where the quality of the print and the size of the original do not allow reproduction as it stands. Figure 60 from Irishtown, Dublin 4, illustrates this direct approach. Estate maps which showed this Dublin village before and after landlord remodelling have been redrawn, to the same scale and oriented in the same direction, to enable the researcher to recreate the stages in the transformation of this village. As always, maps are used in conjunction with documentary, graphic, archaeological and oral evidence. In this case the estate agent of the 1790s, Mrs Barbara Verschoyle, has left impressive evidence of her personal ambitions to 'straighten out' and tidy up this fishing village on the edge of the Fitzwilliam (later Pembroke) estate.[2] While Irishtown was only a short distance from the estate's highly profitable developments of Merrion Square and Fitzwilliam Square in Dublin city, its thatched cabins and fishing women were a world away from it socially.

A useful practice in local history is mapping land ownership. The single-sheet county index maps to the six-inch survey can be used as base maps on a county basis (fig. 49a). Alternatively the townland index maps to the 1:2,500 survey, which also refer to the six-inch sheet number (fig. 49c) may prove useful either alone or in combination. In either case, the townland boundaries are legible, and the indexes lead the researcher directly to the larger-sale OS and Valuation Office maps. The practice of using the six-inch county index to map landholding has a long history; in 1852–3 a map of estates in County Roscommon was constructed on this basis.[3] In figure 61, for example, all immediate lessors with land valuations above £4,000 in the 1850s have been identified by Martina O'Donnell, using the OS townland index maps for the county (edition of 1956), and the general valuation record (Griffith's valuation) which was completed for Donegal 1856–8. Six major landlords were identified, allowing the researcher to move into the records of individual estates and places.[4] The process of mapping, when compared to merely listing in tabular form, opens up the investigation. How did individual holdings arise? Why

1 Richard Boyle to Edward Tingham, Apr. 1632, from the Lismore papers, quoted in Arnold Horner, Maynooth, *IHTA*, p. 2. **2** Prunty, 'Estate records', p. 123. **3** Andrews, *Plantation acres*, p. 138. **4** Martina O'Donnell, 'Settlement and society in the barony of East Inishowen *c.*1850' in W. Nolan, L. Ronayne and M. Dunleavy (eds), *Donegal history and society* (Dublin, 1995), pp 624–48.

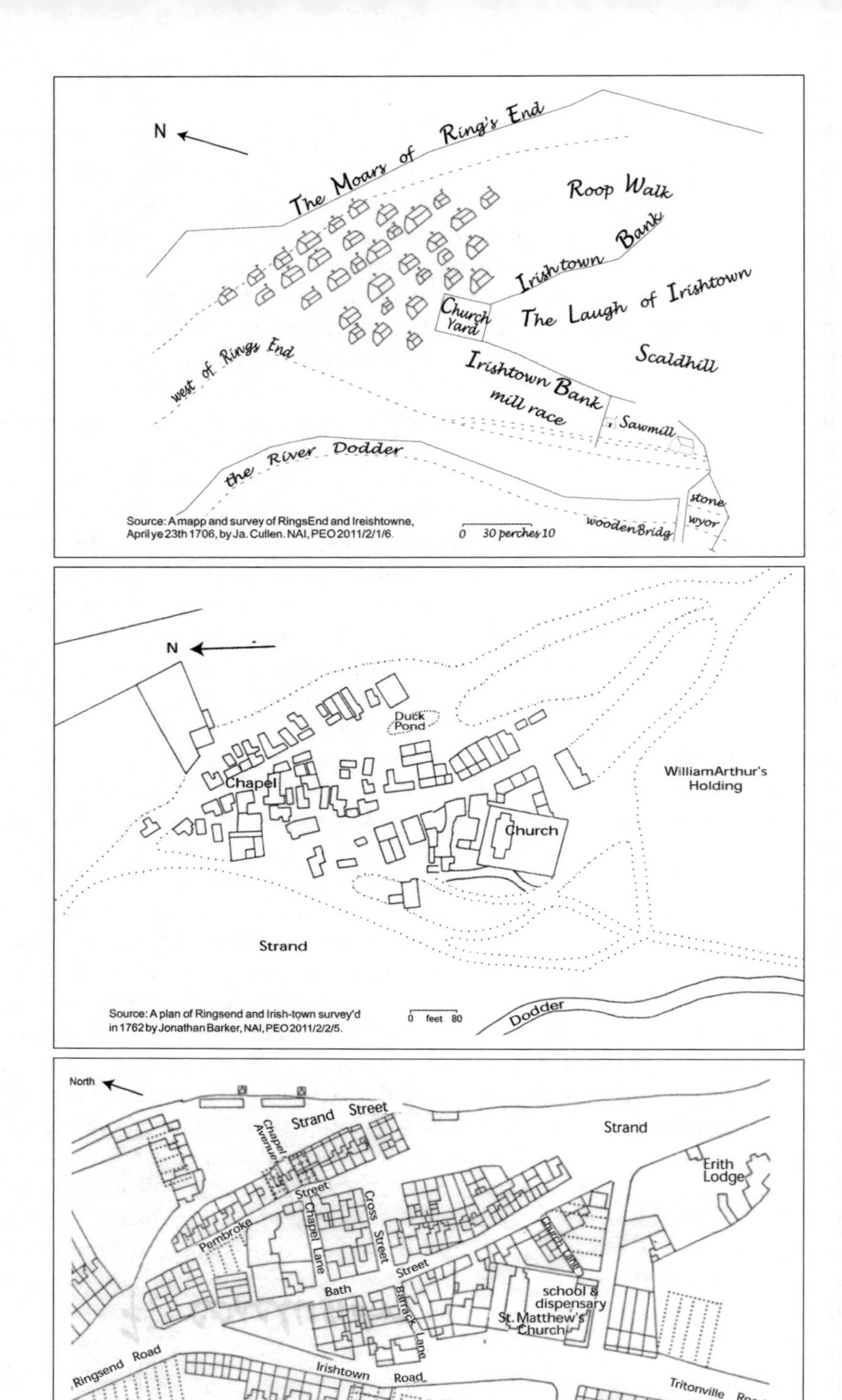

60 Irishtown, Dublin redrawn by J. Prunty from Pembroke estate and OS maps, 1706–1865

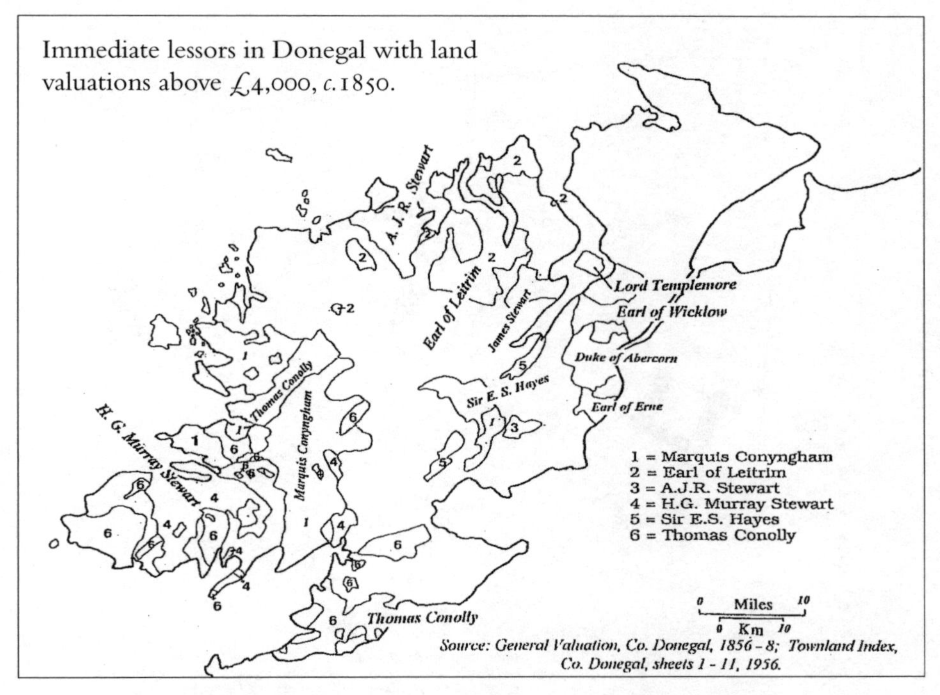

61 Landholding in County Donegal *c.*1854, by Martina O'Donnell

are some holdings so fragmented? How were these managed? What difference did it make to be a tenant on one estate rather than another, or did it matter? How did the Congested Districts Board and the later Land Commission operate in Donegal? Mapping the disposition of the holdings of the major estates is merely the first step; who were the middle-ranking landlords and where did they hold land? The six-inch maps and the cancellation books as well as the published valuation, ensure that land-holding at the intimate local level can be determined; the index maps provide the townland network within which information can be entered. The final drawing (fig. 61) is greatly simplified, but conveys an immense amount of information economically and effectively.

The valuation record is intrinsically 'mappable'. There are many other data sets which do not include maps, but are nevertheless better examined through the process of mapping. Some listed information, such as the Civil Survey of 1654, can be mapped, as was done by Anna Byrne for Kilkenny city (fig. 62).[5] There are inherent difficulties in the Civil Survey as a source. Only the property of a selected few, namely Catholics whose property was valued over £10, is recorded. The data-collection was carried out by 'inquisition' or courts of inquiry not by mapped

5 Anastasia Byrne, 'The use of the Civil Survey for the reconstruction of the socio-economic topography of Kilkenny city', Geography Department UCD, BA dissertation, 1985; see also the *IHTA* fascicle by John Bradley, Kilkenny (Dublin, 2000).

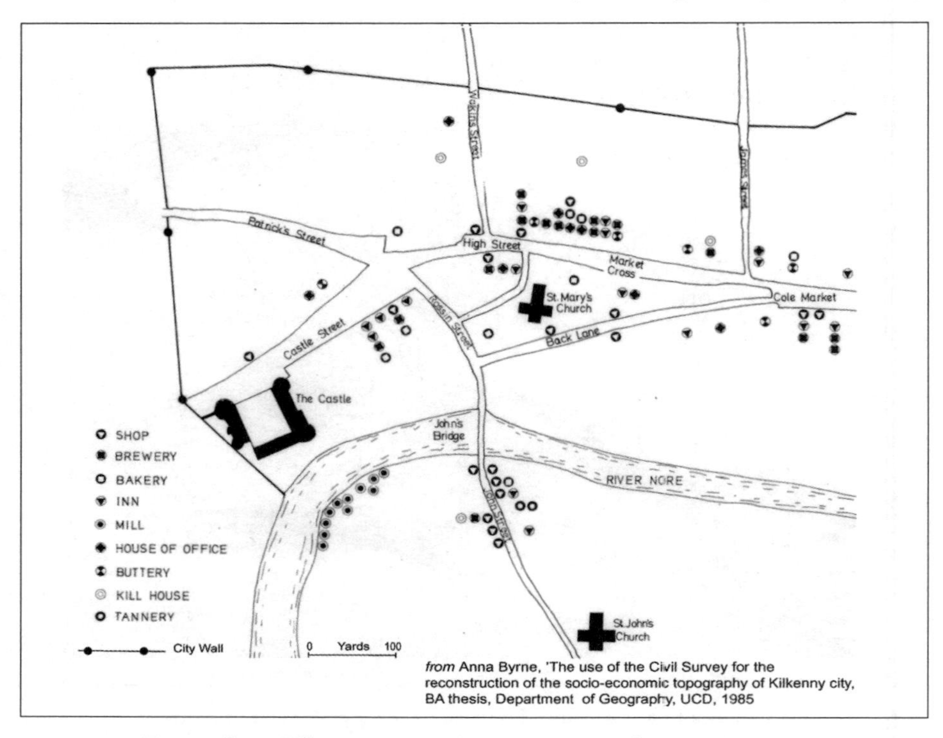

62 Extract from Kilkenny, economic activities 1654, by Anna Byrne, 1985

measurements; the reliability of the evidence therefore depends on the capabilities of the jurors, and indeed whatever interests they themselves promoted or protected at the time. Their familiarity with the local scene resulted in vague descriptions of location – 'on ye East side neere Market Cross', 'on ye North side' – which complicates the cartographer's job. While length, breadth, number of rooms and value in 1640 is given, it is not always clear on to which street or lane the premises face. And even where the exact location can be pinned down, the descriptions of premises need some deconstruction. At a time of massive upheaval following on Cromwell's destruction of the city in 1650, how should terms such as 'fitt for brewing' or 'this house is suitable for an inn-keeper', be treated? As recent uses to which they would soon revert, or as new proposals by jurors with an eye to their own futures? In some cases a brewery was sited behind an inn and held by the same owner, in other cases the brewery was associated with a shop, and in still others it was probably for home consumption as it was connected with a substantial town house. While allowing for these and other difficulties, mapping is still an immensely potent tool in local history. In the Kilkenny case, a succession of maps based on the Civil Survey allows this rich but frustrating source to be exploited more fully. Mapping often confirms the expected, such as tanneries on the river bank beyond the city wall (fig. 62). In addition, mapping arouses curiosity as to what is associated with whom, where and why. The clustering of the most important RC families in

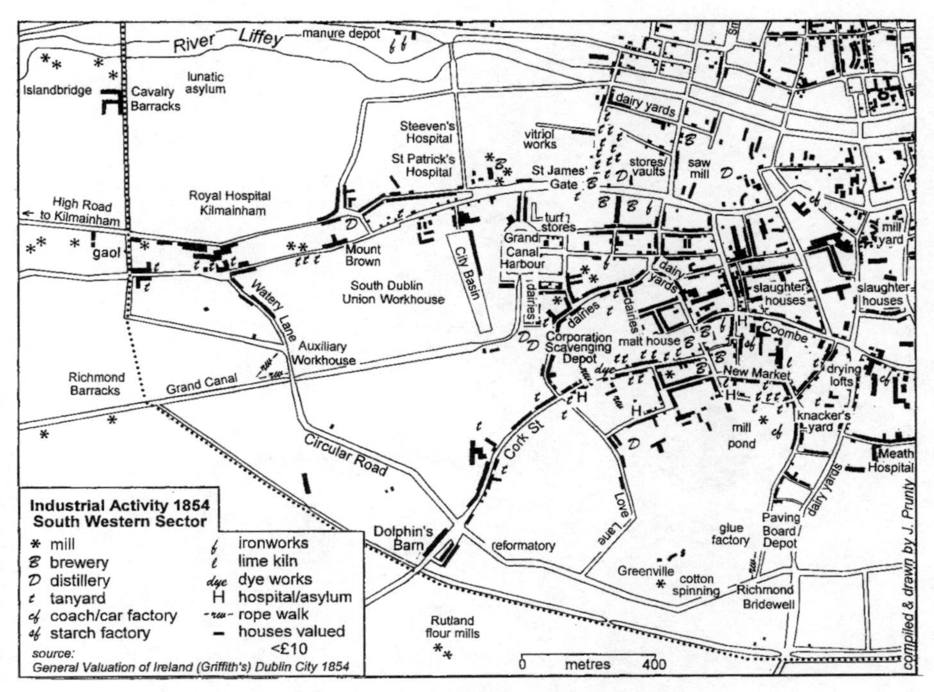

63 Dublin city 1854, SW industrial sector

1650s Kilkenny (fig. 62) can be found from High Street to Colemarket, the area with most extensive outhouses, where each property has a stable, most have a brewery, and gardens and orchards are most numerous. Valuation alone is a crude guide to status, as inns, mills and other industrial properties are rated highly, but by bringing together the scraps of information provided in the Civil Survey some insight can be gained into the socio-economic topography of the city. The primary data in historical research is most often piecemeal and unsatisfactory; mapping as a tool can maximise the returns from that source.

Rather than specialist study of one data source, no matter how valuable in itself, local history typically builds on a wide variety of sources, which can be reflected in the mapping of smaller areas. Figure 63 is an examination of the south-western sector of Dublin city in the 1850s based on street directory and valuation information, and using the OS six-inch proof edition (1838) both as a primary source and as a base map.[6] The five-foot (1:1056) valuation sheets, with their individual house or plot numbering, were used in conjunction with the published

6 This map was not published until 1844 (with information up to 1843); the delay was due at least in part to the reform of the barony boundaries. The term 'first edition' therefore is reserved for the published version of 1844. However, there were proof copies of sheet 18 in circulation in 1838, while the OS made copies from this 'abandoned' 1838 plate intermittently in the 1960s through to at least 1990, whenever a researcher asked for a copy of the 'original' Dublin six-inch survey sheet 18.

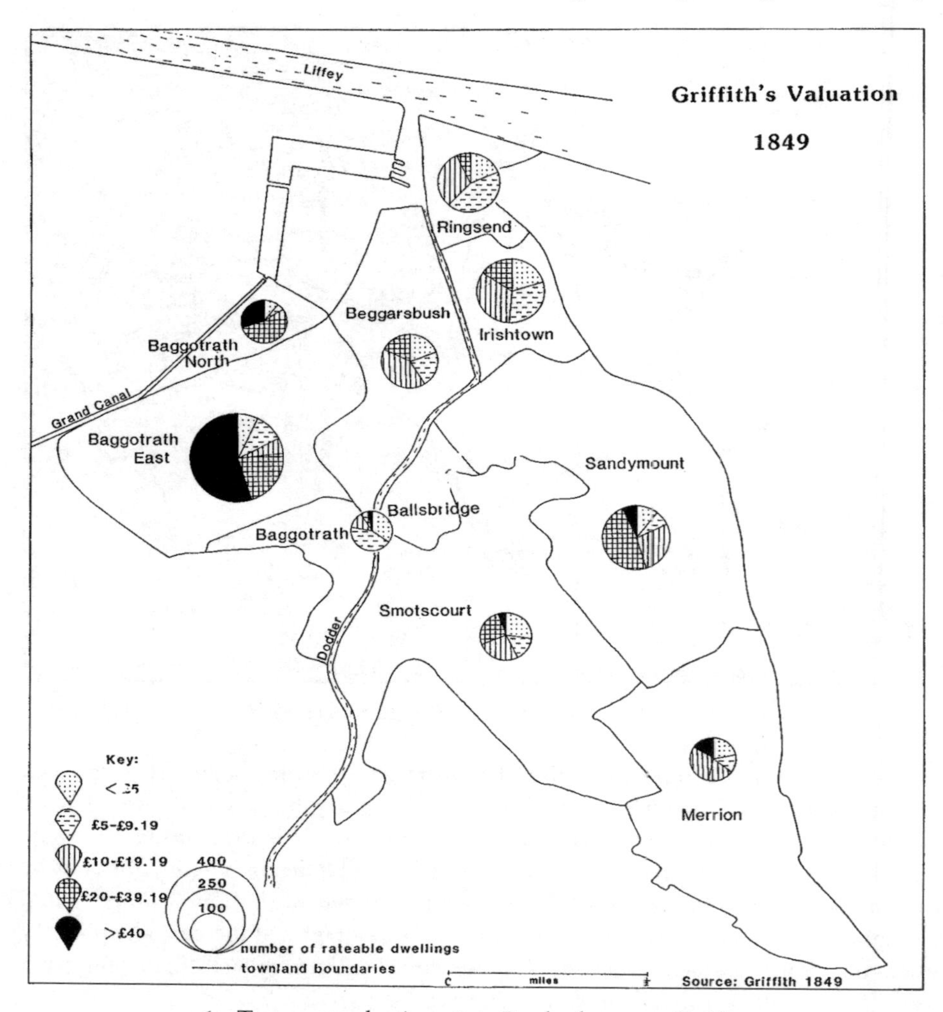

64 Tenement valuation 1849, Pembroke estate, Dublin

valuation record to enable exact location. Redrawing was essential to enable such a compendium of information to be legible; information on dairies and slaughter houses had to be excluded, as that would have made the final picture impossible to decipher. This mapping exercise (fig. 63) displays how tanneries, breweries and mills clustered together, and how they were related to the city watercourses. The extraordinary mixture of activities in this heavily industrialised and densely populated districts is evident, while it also explains the basis for the Guinness expansion of the 1890s, taking over a number of smaller breweries and distilleries in this district.

The process of overlaying maps with statistical information is long-established; the Census of Ireland, in conjunction with the OS, was a pioneer in this field, overlying maps with statistical data from 1841 onwards. Symbols can provide

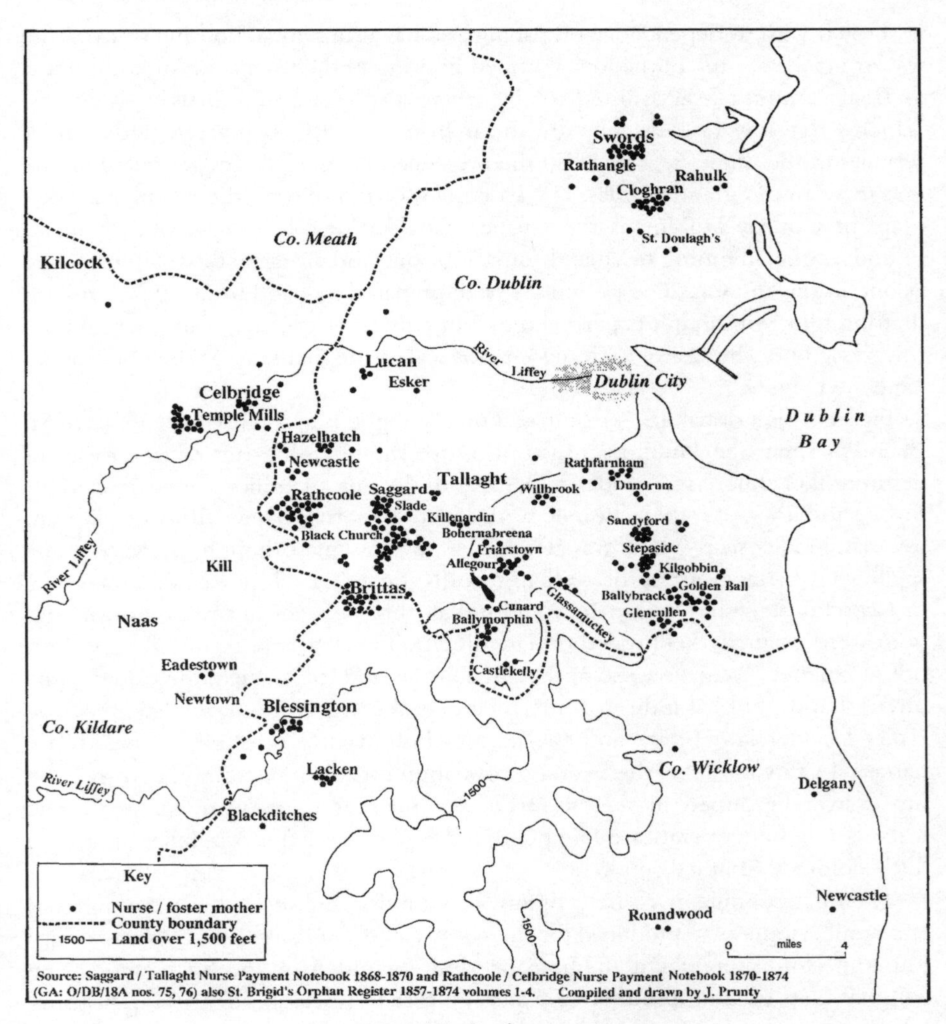

65 St Brigid's Orphans, 1868–74

quantitative as well as locational information if they are drawn to scale; bar charts, pie charts, scaled symbols (such as circles depicting town population) have all been widely used. Figure 64 is a simple example showing how pie charts were used to display the General Valuation figures for selected townlands in south Dublin. Griffith's valuation, as a centralised state-run record, is based on the townland/ parish/barony/county hierarchy; it pays no regard to individual estate boundaries, so that figure 64 though covering the Fitzwilliam/Pembroke estate is not identical with it. Direct comparisons with estate rentals cannot be made; however, it does give a good indication of where the lower-value housing was concentrated. Closer examination reveals that the small industrial villages of Ballsbridge and Milltown have much in common economically with the village of Ringsend, while

Irishtown with its dependence on fishing also has a substantial housing stock in the lowest bracket. Some of the lower-valued housing in the townlands of Smotscourt and Sandymount is accounted for by terraces of estate-built artisan dwellings, which carry low rates as they are so far from the city (compared with similar terraces in the inner city), and also the presence of some gate lodges guarding the entrances to the grander villas. The base map for figure 64 is the townland index map for County Dublin; as the village of Ballsbridge, is split by the townland boundary, the adjoining townlands of Ballsbridge and Baggotrath were combined as one in this instance. The pie-charts were prepared by hand in situ (black ink on drafting film, with graph paper behind), but computer packages can now simplify this procedure. The size of each pie-chart is scaled; the mathematics behind this are explained below.

Figure 65 is a distribution map based on the registers and nurse notebooks of St. Brigid's orphanage, Dublin, an institution for the boarding-out or foster care of destitute RC children.[7] Townland addresses of the nurse families were extracted for the period 1868–74; these homes were located approximately from the Dublin townland index map, which was also used as the base map, though the place-name spellings were taken from the orphan records (fig. 65). Each nurse could take one or two children; rarely were there more than three, except in the case of siblings who were generally kept together. The distribution therefore is of nurse-mothers not of children. To make sense of the distribution, and to orientate the reader, some physical and political indicators are included: the OS contour at 1,500 feet, the rivers Dodder and Liffey, and the county boundaries. Selected townlands are named. In this instance, the exercise of mapping the addresses revealed that the nurses were grouped in neighbourhoods, and that there were relatively few instances of 'solo' operations. The published reports bear this out, referring to the 'little colonies' of nurse children.[8]

The final example given here, figure 66, is a redrawing of an overly complicated and cumbersome map produced for the 1913–14 Local Government Board inquiry into the condition of Dublin housing, and appended to the final report. The original map was the product of painstaking field research, identifying housing which was unfit for human habitation and otherwise condemned as dangerous and locating new housing schemes, either public or private, completed or in planning. From the rather crude line work on the original it is obvious that it was completed by a draftsman or engineer tracing from an OS map rather than by the OS itself. Yet exact location is central to the purpose of this map, and so this base is redrawn (at A3 size) rather than transferring the information onto a more exact OS street map. Only a fraction of the information in the crowded original is reproduced in this case; with rub-down *Lettratone*, explained below, used to mark individual housing estates. A second map was drawn (using the same base) to carry the

7 Jacinta Prunty, *Margaret Aylward, lady of charity, sister of faith* (Dublin, 1999), pp 56–78. 8 *St Brigid's Orphanage sixth annual report* (Dublin, 1862), p. 11; *St Brigid's Orphanage sixteenth annual report* (Dublin, 1873), p. 10.

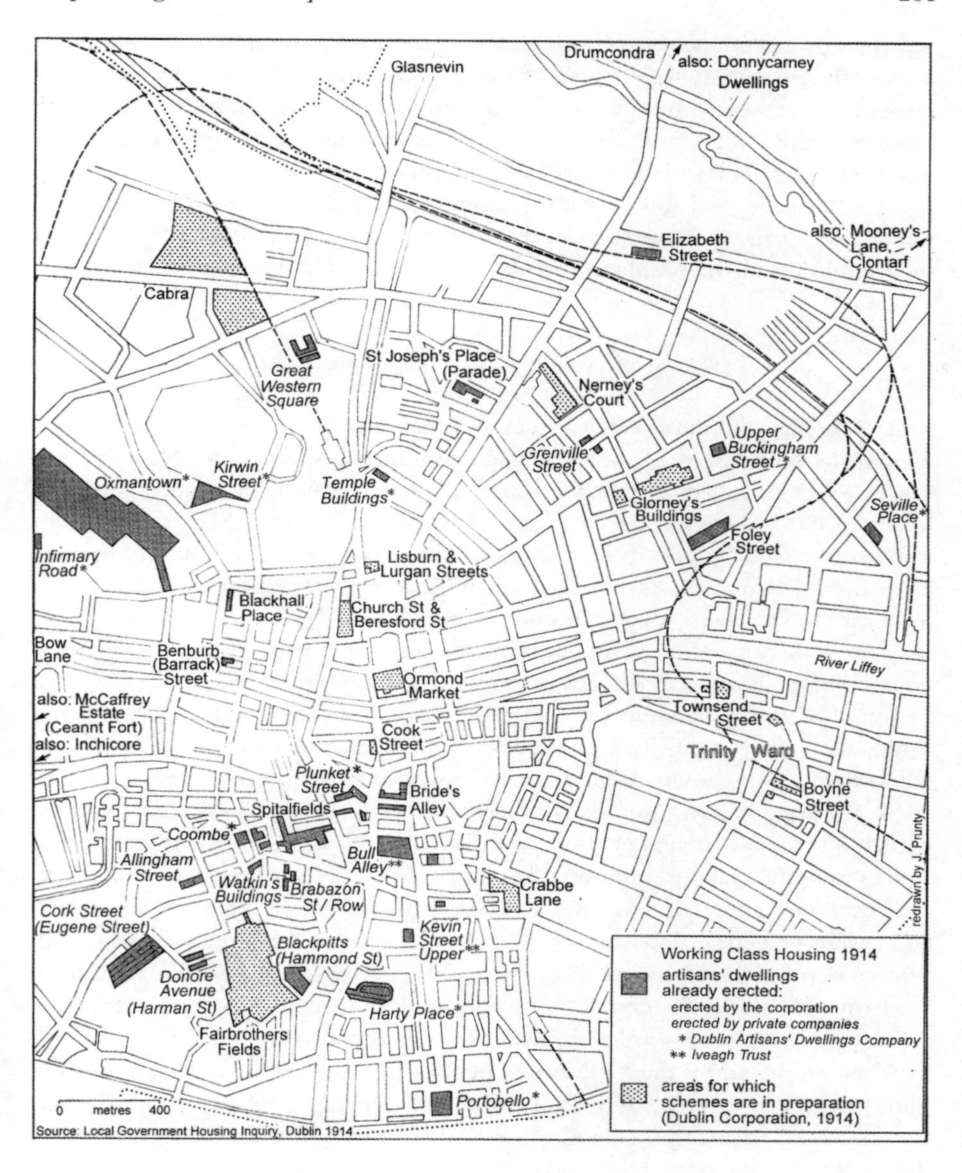

Drumcondra
also: Donnycarney Dwellings
Glasnevin
Elizabeth Street
also: Mooney's Lane, Clontarf
Cabra
Great Western Square
St Joseph's Place (Parade)
Nerney's Court
Upper Buckingham Street*
Grenville Street
Glorney's Buildings
Oxmantown*
Kirwin Street*
Temple Buildings*
Seville Place*
Foley Street
Infirmary Road*
Lisburn & Lurgan Streets
Blackhall Place
Church St & Beresford St
Bow Lane
Benburb (Barrack) Street
Ormond Market
River Liffey
also: McCaffrey Estate (Ceannt Fort)
also: Inchicore
Cook Street
Townsend Street
Trinity Ward
Plunket* Street
Bride's Alley
Boyne Street
Spitalfields
Coombe*
Bull Alley**
Allingham Street
Watkin's Buildings
Brabazon St / Row
Crabbe Lane
Cork Street (Eugene Street)
Kevin Street Upper**
Blackpitts (Hammond St)
Donore Avenue (Harman St)
Harty Place*
Fairbrothers Fields
Portobello*
0 metres 400
Source: Local Government Housing Inquiry, Dublin 1914
redrawn by J. Prunty
Working Class Housing 1914
artisans' dwellings already erected:
erected by the corporation
erected by private companies
* Dublin Artisans' Dwellings Company
** Iveagh Trust
areas for which schemes are in preparation (Dublin Corporation, 1914)

66 Artisans' dwellings, Dublin 1914

information on condemned housing. Italics are used in figure 66 to distinguish private developments from those of the municipality, with asterisks to further separate the schemes built by the Dublin Artisans' Dwellings Company from the projects of the Iveagh Trust. In the final map, everywhere east of Seville Place was deleted as not relevant to the topic (city centre slum clearance schemes), and arrows point to areas beyond this core area, rather than extending the map too far and forcing the scale to be reduced even further. The tighter frame, exclusion of superfluous detail and simple key results in a more readily understood map.

MAPS AND COPYRIGHT OWNERSHIP

One of the first and most practical questions that must be put to any map relates to ownership. With low-cost photocopying and scanning widely available copyright can be easily infringed, knowingly or unknowingly. Both the Ordnance Survey of Ireland (OS*i*) and the Ordnance Survey of Northern Ireland (OSNI) provide very full written explanations of the rules governing copyright, and should be applied to in the first instance. The written permission of the legal owners of any map material should also be sought; in some cases there will be no payment, but a copy of your book or article will be requested for their own records. Under no circumstances should a researcher presume to include OS maps (extracts, tracings) in any publication, handout or website without first contacting the appropriate Ordnance Survey authority. This applies to 'out of date' maps as well as to current editions, to OS maps which the researcher overwrites with his/her own notes, to all 'sketch maps' redrawn (manually or electronically) from OS base maps, and to all digital information. It applies when material, such as leaflets, are distributed free, as well as to profit-making ventures. While this is primarily a matter of ethics, writers who transgress copyright laws may also be leaving themselves open to prosecution and severe penalties. Publishers will, as a rule, require the author to secure all written permissions. Therefore the author is held personally responsible, along with the firm which does the copying or printing. Both Ordnance Survey authorities actively protect Crown or Government of Ireland copyright interests.

Maps produced by the Ordnance Survey of Northern Ireland are protected under Crown Copyright, and all unauthorised copying is prohibited; full details may be obtained at http://www.osni.gov.uk. In the Republic, OS Ireland maps, data, and publications are protected under the terms of the Copyright Acts of 1963 (amended 1987). Under the heading 'artistic works' any diagram, map, chart, plan or photograph is protected for seventy years from the end of the calendar year in which it was first made available.[9] Under the heading of 'literary works', tables and compilations, including computer programmes and databases (and therefore all spatial data, whether in printed or electronic format) are also protected. Whether

9 The following notes are based on the OS*i* copyright leaflet 1: general, 1 Jan. 2001 and Leaflet CP1, 1 Mar. 2004.

viewed on a VDU or on paper, the product is protected. Under EU legislation the duration of copyright (Directive 93/98/EC), and the legal protection of databases (Directive 96/9/EC) have been harmonised. The main exception is that Government of Ireland copyright is protected for fifty rather than seventy years from the date it was first made available. It is under this clause that selections from 'old' six inch and twenty-five inch sheets have been widely reproduced by local historians, and dangerous assumptions made about the legality of using later extracts. There is also an exemption for the making of a single copy for personal (strictly non-commercial) use, without which no serious study could be undertaken. However, as the OS*i* makes very clear in its current *Catalogue of products and services*, 'anyone wishing to reproduce OS*i* material, or use it as a basis for their own publications, must obtain a licence from OS*i*, for which a fee may be payable'. Before issuing a permit (for once-off reproductions) or a licence, OS*i* needs to be provided with the following details in writing: the type (scale) of mapping, the area of the map being reproduced and its paper size, the scale of reproduction, the number of copies, and the purpose of reproduction. Other information may also be sought. The licence will expire when the number of copies allowed is reached, and/or on the date specified on the contract. In those cases where the publication is for academic research on a non-commercial basis, the OS*i* currently does not charge a fee; however, this is entirely at the discretion of the OS*i* and a permit must still be sought well in advance of publication. The unique licence or permit number provided by the OS*i* must be included prominently on the face of all reproductions. The wording of the acknowledgement is specified as follows: 'Mapping reproduced from Ordnance Survey Ireland by permission of the Government. Permit no. [*xxxx*] © Government of Ireland [*year*].' Reprints will require a new permit. Usually the OS*i* requires a copy of the publication to be forwarded to them within four weeks, which demonstrates exactly how the permit or licence conditions were fulfilled. When the map extract is to be displayed on an Internet site, a licence is always required; in addition to the general information about type (scale) of map to be displayed, technical information on the type of site (Internet, Intranet or Extranet), form of the display (fixed extract or user-definable window) and usage permitted at the user end (viewing, printing, downloading etc.), must be provided. Digitising is perhaps the most difficult field to police. The OS*i* will not allow its own conventional paper products to be digitised except in instances when they cannot supply the digital data themselves in the form required. Where permission is given for the user to create the raster digital data themselves, there are very strict conditions governing the use and final destruction of the converted data. This is an area in which policy is all the time trying to catch up with technology, and researchers should keep themselves fully updated on the current regulations. In short, it is the responsibility of the researcher to ensure that copyright regulations are complied with in full; the procedures to follow are well established and the charges (where there are charges) are modest relative to the value of the product. If in any doubt at all, ask.

GUIDELINES IN MAP MAKING

The encouragement of map-making among local historians is a primary aim of this research guide. The basic trigonometrical framework within which survey details can be plotted has been long established by the Ordnance Survey on land and by the British admiralty by sea. While it is possible to undertake an entirely new survey of a small area (hand-held global positioning plotters and advanced theodolites make this feasible), for most history purposes this is unnecessary. The skills required for basic map-making today are therefore within the grasp of most local historians and geographers: imagination, patience, perseverance, an appreciation of the language of maps (both mathematical and graphic), an eye for detail. With some investment of time and resources, a very fair return can be made. There are several compelling reasons for scholars to take responsibility for the research, design and final production of their maps. Most obviously, this will minimise the errors of transcription and comprehension that can occur when persons unfamiliar with the research undertake the processing and presentation of findings. The person who has collected and or manipulated the data being mapped, and – ideally – has an intimate knowledge of the district under study, is surely the best placed to follow through with the information display and image processing, that is, with the map-making. The following guidelines are intended to provide practical direction to ensure that the final product is indeed of a high quality.

DATA SOURCES FOR MAPPING

The information to be mapped must be approached as critically as any other source used in historical research: who generated it, for what original purpose, the status of the data at the time of its creation, its completeness, the manipulation already undertaken, and the technical and other limits under which its creators laboured are all questions to be explored. How up-to-date was the information by the time it was first assembled and published? Some of the data (boundaries, placenames) will be taken directly from an OS or other map, and so the questions in Appendix 1 are applicable. Try to work out the data collection techniques which were used at the time; this will assist in evaluating how accurate, or otherwise, the base map and others sources are likely to be.

New sources of data pose new questions as well as opening up possibilities; the wealth of satellite images that can be downloaded even for smaller areas brings its own challenges. The Eurimage site, and links [http://www.eurimage.com/] provides an entry point to this new world. When mapping data that is indeed spatial but was not produced with mapping in mind – such as the Civil Survey 1654, charity records, or commercial directory listings of 'tenement dwellings'– the local historian must tread lightly. Similarly, great care must be taken where integrating material from different databases, as with a map showing use of premises, combining valuation, street directory, oral and census sources. The reader needs to know in full the sources upon which any map relies, the mathematical framework around which it is constructed (base map(s), scale, orientation), the ways in which the data now presented was manipulated, the name of the author of the map, and the date (year) of its creation. Acknowledging databases leads directly to the important question of dataset ownership. Copyright for Ordnance Survey maps is vested in the Government of Ireland (OS*i*) and Crown (OSNI). As already explained, within the EU government-owned information is copyrighted for fifty years, and privately-owned information for seventy-five years. Permits or licences must be sought in accordance with current regulations from the relevant Ordnance Survey. Both the base maps and the information therein are protected by copyright, and prosecution may follow illegal use. Ignorance of the regulations, educational, 'charitable' or otherwise non-commercial use does not absolve the researcher from complying with the law, which, as already stated, may not (in some instances) even require payment. The generation of digital data is expensive, and commercially most valuable, hence it is jealously guarded by its rightful owners. GEO–ID (*Geospatial Information Directory*) maintained on-line by the GIS laboratory, Department of Geomatics, DIT Bolton Street, Dublin 1, is a listing of available digital databases and their legal ownership; at the time of writing this is an inventory only, providing a starting point for researchers interested in purchasing databases. Developments in this field move on at breakneck speed, but the GEO–ID website [http://www.gis-ireland.com], c/o DIT Bolton Street, will update researchers.

COMPUTERS AND CARTOGRAPHY

Most professional cartography is now computerised. Both Ordnance Survey Ireland and the Northern Ireland Ordnance Survey office are at the forefront of computerised mapping, with all of their new map products digitised since 1987. User-friendly explanations of digital mapping are available on the websites of the three Ordnance Survey bodies in Ireland, Northern Ireland and Great Britain; each body has links to the other authorities, as well as a link to a common site titled http://www.osmaps.org/. Virtually all maps world-wide are in fact now digital files of numbers representing the type of data and its location. GPS (global positioning systems) ensure that all contemporary mapping is truly international. The researcher who wishes to embark directly on desktop cartography will find a range of graphics, GIS (geographical information systems) and other map-making programmes that are user-friendly and increasingly transferable, allowing even the beginner to speedily produce quite a professional-looking map. Maps may be digitised directly on screen, or by using a digitising tablet; in either case the computer process involves defining each geometric element (points, lines and polygons or closed areas) by a grid co-ordinate, in what is termed a 'link and node' structure. Each line or vector is related to the next, and information is assigned in layers (one layer could be the river network, another third class roads, another the railway lines) which can be combined or separated out.

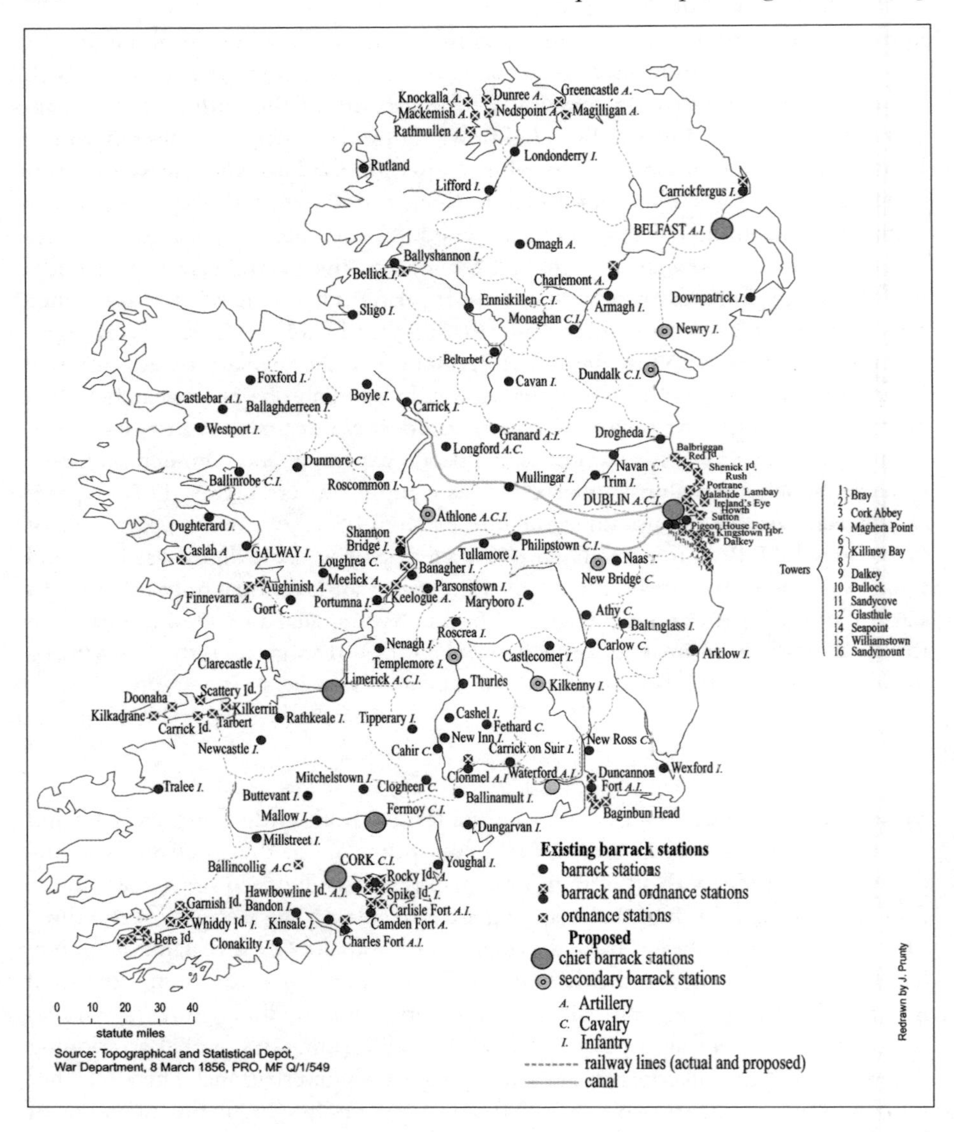

67 Barracks in Ireland, 1854

Figure 67 is an example of computer mapping using Adobe Illustrator 9.0. It is a direct redrawing of a map produced to accompany a statutory report in 1855 advising the rationalisation of barracks in Ireland. The published map is seriously skewed NNE, while the quality of the print and the size of the original would not allow reproduction as it stands. In this case, the invisible bottom 'layer' consists of the original map (a poor quality photocopy) acting as a non-printing template, which was dimmed to 50% and overdrawn. Each element in the map was added in a separate layer: coast, county boundaries, internal waterways, symbols of barrack stations, station names, with decisions about layer ordering made at the last stage (to decide what elements would come to the fore or 'overwrite' other elements). Working in distinct layers (each of which can be 'locked') ensures that elements do not run into each other, making detailed maps possible and legible at this small scale. It also allows formats or styles to be applied to a complete layer, as in (for example), the light-weight, broken style applied to the lines of the county boundaries.

Figure 68 is produced using ArcView GIS (by James Keenan). What distinguishes a map generated by a GIS programme is that information about individual features, represented as lines, points and areas or polygons, can be stored in a complementary database. A unique ID code is given to each feature, but countless different types of information can be stored about it. In this case, the RDs (rural districts), each with a unique ID code, are linked to a large database holding population information from the CSO (Central Statistics Office). This 'attribute information' is stored in rows and columns, as in any standard database. The map-maker selects which attributes to display in any particular map, as he or she explores the massive volume of information available to them. Percentage population change (fig. 68) is a good example of where data needs to be displayed spatially before in-depth analysis can proceed; tabular data alone is clearly insufficient in this case. In a dynamic on-line or electronic environment, pointing on any feature on the map displayed on the computer screen will bring up, in table form, the attributes belonging to that feature. Conversely, pointing on any row in an attribute table brings up the feature it is linked to on the map. The connection between database and map is dynamic, namely, if a change is made to an attribute in the table, this will be displayed also in the map. If, for example, an error had been made in the database attached to figure 68, then correcting the error would automatically update the map on the computer screen. And there are many other situations in which the RD divisions could be utilised, that is, other attributes could be listed, justifying the initial investment of time and energy in drawing or, more accurately, digitising these polygons and giving each a unique ID code. GIS links sets of features, with a set of common attributes, and calls them 'themes'. Roads (lines), counties (polygons) and town locations (points) could each be grouped as a theme. Spatial overlay is the tool by which different themes or layers of information are combined. The earlier discussion of the 1:50,000 *Discovery* maps (fig. 47) explains how different themes (water courses, relief, settlement, road networks and so on) can be combined to make a topographic map. Similarly, in a map showing population change in rural districts (fig. 68) the map-maker could choose to show settlement or other features as an aid to the reader

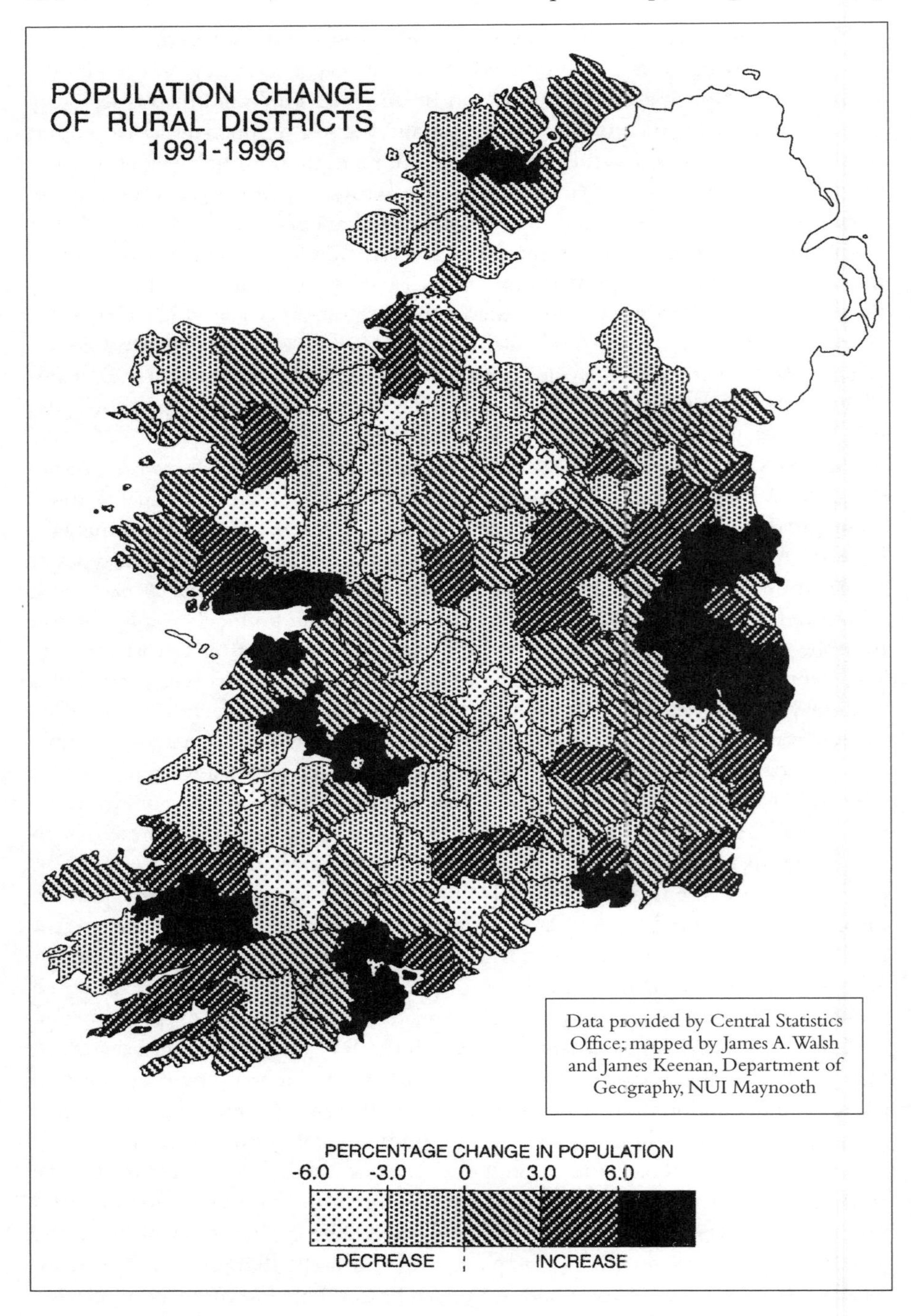

68 Example of choropleth mapping (James Keenan)

(where these attributes were already entered in the GIS database). The flexibility offered by GIS packages makes them extremely attractive to the historian as well as to geographers, planners, engineers, property developers and others.

Steps in computer mapping are clearly explained in Kevin Mooney's 'Skerries town map, an application of desktop cartography' (*Survey Ireland*, no. 14 (1997), pp 11–17). But as Mooney and other expert practitioners in computer mapping insist, these welcome technological advances make an understanding of basic cartographic principles even more essential. While map-making can be used to enliven a newspaper story – pinpointing the site of an armed robbery and the ensuing car chase, roadblocks, cross-country pursuit with helicopter back-up – it is lamentable to see it reduced to this alone. Regardless of the time or the skill levels now required for their production, maps continue to be powerful if understated cultural creations, and the language of maps still demands intelligent and sustained examination. The local historian who has been persuaded to engage in map-making, and who possesses the computer skills to manage typical map-making or graphics packages such as Adobe Illustrator, Macromedia Freehand, Coreldraw, Arcview or Microsoft Map Maker, still needs to be alert to the basic principles of cartography. Faced with overwhelming choice, at the click of a mouse – what information is to be mapped and how, font size and style, design and size of symbol, scale, layout – it is even more urgent that the map-maker devote time and thought to what their map is to communicate, and how that might most fairly and effectively be achieved.

The Dublin Institute of Technology, Bolton Street, offers part-time and full-time professional courses in surveying and map-making, including courses in desktop cartography and GIS; its library holds an excellent collection of relevant publications including practical manuals. The universities also offer courses in this area, including an MA in Geographical Information Systems at NUI Maynooth (Geography department). The International Cartographic Association has its own three-volume publication titled *Basic cartography*,[10] targeted at students and technicians, which covers both theoretical and practical issues in a most accessible way. It is generously illustrated, with the minimum amount of text, and highly recommended to those embarking on traditional map-work, those wishing to upgrade their skills, and (vol. 3) those moving into computer cartography. Online courses in computer mapping and map design are also available, for example at the ArcView site http://campus.esri.com. The journal *Survey Ireland*, produced by the Irish Society of Surveying, Photogrammetry and Remote Sensing, covers recent advances in computer mapping and GIS, and although aimed at professionals is accessible and lavishly illustrated. *Introducing remote sensing: digital image processing and applications* (London, 2000) by Paul J. Gibson and Clare H. Power is recommended as a comprehensible student text, and also heavily illustrated. Rob Kitchin (with Martin Dodge) introduces very different perspectives on map-making in *Mapping cyberspace* (London, 2000). Geographers at the University of Newfoundland have produced their own guide

10 Anson and Ormeling (eds), *Basic cartography*, ii (London and New York, 1993); R.W. Anson (ed.) vol. 2 (1988, reprinted 1996).

to computerised map design and production (using Coreldraw software).[11] And as with all computer developments, there is much to be gained from time spent on the Internet, noting in particular how interactive mapping has become. Where, for example, a paper map may indicate a number of towns, a digital map or GIS product may link those sites (points or nodes) with layers of data (attributes), including pictures, population numbers, notes on historic sites, website links, email addresses, video clips and sound. One click of the mouse, and endless possibilities open up.

The following guidelines apply regardless of whether the map is to be entirely hand-produced in the traditional manner, or computer generated. No effort is made to cover the details of computerised mapping, as each product comes with online and tutorial help, while there is also a wide range of commercial 'home tuition' manuals. The speed of programme updates would rapidly render obsolete any instructions that might be proffered. It is presumed that the map will be produced in black and white, as this is still the preference for publication. Computer mapping in colour needs special technical support, not least to ensure that the final product is indeed the shade(s) intended. One note only is offered here: when mapping in colour, very light pastel shades are best for most purposes.

In terms of map design, the local historian may embark on anything from the simplest construction – the minimum linework and place-naming needed to locate a village mentioned in the text – to a major 'informative map' which is really complete in itself, telling the story in a more compelling and absorbing way than any amount of prose could manage. There is also enormous potential in the combination of other visual images (such as photos, line-drawings or aerial views) with maps. But whatever the level of sophistication, there are several distinct steps along the way to final publication. The stages in map production are illustrated by reference to a North Dublin city map produced by the author. And the reader should be alert to the fact that the notes which follow are but one way of embarking on the task. Each practitioner brings their own favoured ways of working, and there are many possible routeways to the same destination.

1. Selecting a source map

Selecting a source map to serve as one's template or framework within which information can be plotted, is the first requirement. The importance of choosing well cannot be overstated; where a series of maps are to be created, the same base map can be used again and again, and at different scales; the full map or subsections within it can be used; and where carefully drawn, will allow the direct addition and subtraction of detail, without the tiresome job of redrawing (either by hand or with a computer package). Ensure that you are operating within the law, both with the choice of base map, and the information you are inserting; refer back to the notes on copyright protection already offered. When submitting maps to a publisher, you are likely to be asked to provide the following information, in three columns: i)

11 Charles M. Conway, Gary E. McManus, Clifford H. Wood, *A guide to map design and production using Coreldraw7 and Coreldraw8* (Newfoundland, 1998).

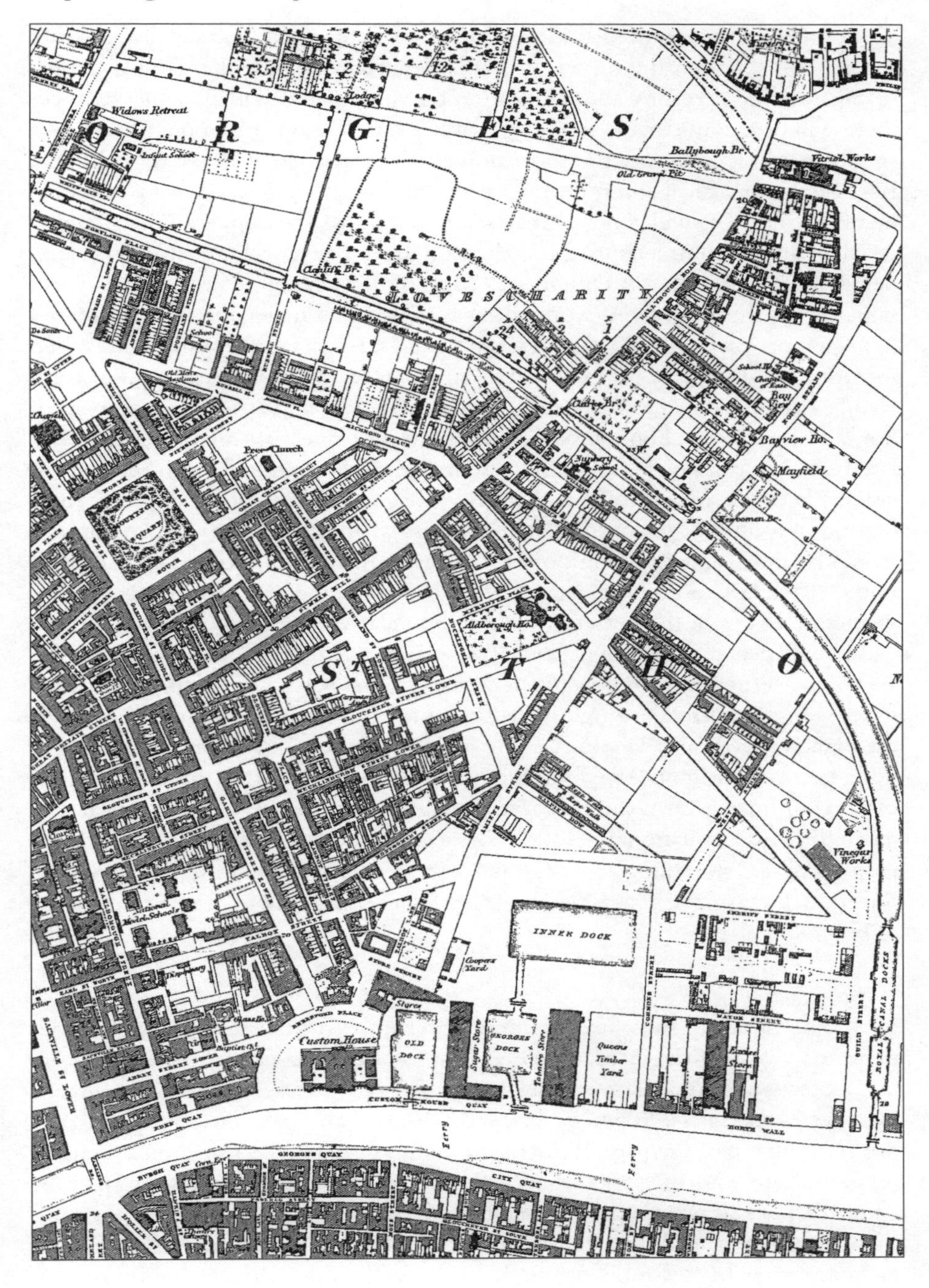

69 North Dublin city, source map, OS six-inch, County Dublin, sheet 18, proof edition, 1838

caption (title or subject), ii) source, iii) owner of copyright. Full details therefore should be kept from the outset. It is also important to note the dates given for original survey, revisions, and publication on your source map; in some instances there will be a considerable time lag between survey and final production, so you may be working with survey information much earlier than you first thought. As a rule, the map-maker works from a detailed source map, and selects what is required; he or she therefore will choose a map series that shows far more than they require, moving always from detail to generalisation, from larger scale to smaller scale. A detailed six-inch sheet of Dublin city (fig. 69), for example, was the basis upon which maps showing street layout, the downgrading of tenements, and the homes of recipients of relief in 1850 and 1900 (figs. 92–94) were constructed. When intending to show change, the source map selected can be an early edition, with changes (such as a new road, realignment of field boundaries, new terrace of houses) superimposed on the skeletal framework of an earlier map. Alternatively, a later OS map can be chosen as the base map, and distinctions drawn (perhaps in colour) between what existed at the early date and later additions. Coverage must expand beyond the area of interest; in the case of an urban centre, for example, it would be short-sighted to limit oneself to the immediate built-up area as mapped in the 1830s, but 'room for expansion' to for example the situation in 1900, should be factored in from the outset. At county boundaries something of the adjoining county will generally also be needed. It is a simple matter to cut or crop to the core area of concern; it is an entirely different story to find you have cut off your options by confining yourself too tightly at step one. This is also the point to question cartographic tradition; do you really want to use the county, barony or municipal boundary as your limit? Are you using it unthinkingly, following the silent dictates of previous maps? Or is there another boundary that might be more true to the research? The circulation area of a provincial newspaper? The Roman Catholic parish limits? Commuting zone? Coastal settlement that cuts across other divisions? River catchment area? This is the time to think creatively about the whole concept of bounding place.

2. Preliminary layout

It is important from the outset to consider what the final page layout will be, the preferred 'shape' of the map, and where best to place each element – legend, scale bar, source(s) and so on. For published maps 'portrait' page set-up is to be preferred over 'landscape', as far as practicable, and OS maps are traditionally aligned with north to the top of the page. If the map is not to be aligned north, then a compass point indicating north is essential. Whether producing in thesis format (A4), or for a journal, book, brochure, poster, wall display, website or power-point presentation, the cartographer needs to take careful note of the final area dimensions from the planning stage. In electronic terms, the speed with which the map will arrive on screen needs to be considered, and the more limited 'web safe' palette of colours employed. Where, for example, it is intended as a map that visitors will take on a walking tour, the author might consider whether perhaps it should be laminated. Where printing on both sides of the page, should there be some overlapping coverage? Perhaps it could be structured as a strip map? If the map is to be folded, exactly where the folds will cross needs to be considered. Unfortunately, poor planning will probably result in a product that 'fits' badly in the space allowed, being too wide and short, too long and narrow, with wasted blank space, needing to be reduced further to allow for margins and page binding, or the fold coming in exactly the place that suits least, all of which could be avoided. To that end, the preparation of a preliminary layout, based on the final print area, is strongly recommended; this can be a rough pencil sketch (or a preliminary mock-up done on-screen), but crucially it will be within the bounds of the final 'true' printing area. Consultations with the publisher or printer should be held as early as possible. Many important maps are created for thesis research; while the format here is A4, even at this early stage the author should consider the way in which his or her map might eventually be published. What percentage reduction will be needed to fit the A4 map to a particular book page size? How legible will it then be? How will the page layout appear? If factored in from the outset there may be no need to redraw thesis maps.

Typically, the first stage of production of a base map involves the pasting together, by hand or by scanning, of OS extracts to cover the area of interest and its immediate environs. It is possible to purchase base map information in vector format directly from the OS; the OS also sells six-inch map information in raster format, while the purchase of 'seamless' maps centred on the area of interest and at a variety of scales (as PLACE maps) has already been introduced. The researcher is reminded that they must comply fully with copyright regulations at every stage, whether using digital or paper data, and acquire the correct permit or licence as their project dictates. Where early OS paper maps are concerned, the map(s) will be identified by consulting the six-inch county index maps (see figs. 49 and 54) and noting the sheet numbers, or (in the case of the six-inch sheets) by referring to the printed *Townland index* 1851 (facsimile reprint 1984) for the number of the sheet which includes the townland or village under discussion. Where the area extends over several sheets, the grid can be used as a 'best fit' guideline. In the case of the original six-inch and 1:2,500 maps, and any maps derived from these, neighbouring counties may not fit together exactly. This may be scarcely visible, but is a reminder that each county was originally plotted relative to its own point of origin (although all were produced on the same type of projection).[12] This problem arises also where maps produced before and after the 1850s adjustment are used together: when a one-inch townland index map is superimposed on to the standard one-inch sheet (derived from the 1850s adjustment), very slight discrepancies will be noted. The map-maker should also note that there is usually some small element of distortion at the edges of any photocopied image. The scale bar of the original should be included in the

12 See also Andrews, *History in the Ordnance map* (1974) p. 18.

area to be copied or scanned; this can be cut and pasted closer to (but still on the edge) of the area of interest.

When photocopying or scanning maps it is important to include the scale bar, wherever possible, and always at exactly the same reduction or enlargement. In practically all cases at least one 'map edge' should also be included, as this will help later with orientation (whether by hand or on screen). The size of the base map is a very practical issue: on paper, the base can be an unwieldy 'larger than life' creation, or may need to be scanned electronically in pieces and then fitted together. Where maps are complex, the file can be too large in size to save on a floppy diskette. As a general point, scanning in true colour takes up much more memory than scanning in greyscale or black and white; the default can be changed under 'set output type'. Scanning to a very high resolution when not needed (for example, scanning a black and white image as true colour) simply results in a large file size without contributing anything to the quality of the final document. The format in which the file is saved will also affect its size. As already introduced, bitmap (.bmp) images are raster files, that is, made up of numerous individual pixels or tiny tiles, and can be very large. GIF (.gif) is a raster file format which compresses solid areas of colour and yet preserves sharp detail, such as linework; it is well suited to scanning base maps, yielding significantly smaller files. In deciding on the type of file format to use, the cartographer needs to consider matters such as portability and readability by other programmes, whether the file type can carry both raster images and vector drawings, and whether the file is to be manipulated further (by an editor) or if it is to be delivered ready-to-print. File options (Save As) are provided at each stage of the process from scanning through to final editing. For example, an image scanned as a bitmap or GIF file and overdrawn with line work and text in a programme such as Adobe Illustrator, may be saved initially as an Illustrator file (.ai) and then when the work is complete, saved as a .pdf file and forwarded to the publisher. PDF files (portable document format) are highly compressed files which can represent both vector and bitmap (raster) graphics, and is the industry standard for delivery of documents to be printed. PDF files can be opened (with Adobe Acrobat Reader) and viewed or pasted into documents; they cannot be edited, so this format represents the final stage of a computer-generated map.

Where the cartographer continues working in the traditional mode with a rather unwieldy paper base map, the matter of size must be considered before any tracing or digitising is undertaken. There is little point in reducing the base map too drastically, as the precious detail will be rendered indecipherable also. This new source map (reduced) should also include a note of the original scale (in words: 'six-inch'), edition and date (proof copy, 1838), sheet number (Dublin, sheet 18), as well as the scale-bar from the original (now reduced). By adding one or two inconspicuous N/S and E/W crosses, it will be easier to preserve exact orientation. Any company that copies architectural drawings or building plans will have the facility to reduce large sheets like this. If several paper copies are made of the base map, they can be used for preparatory work, using coloured pencils.

3. Calculating scale

If there is no scale bar on the extract(s) in hand, it is advisable to insert one of your own at an early stage, before scanning or copying is undertaken. When you start reducing or enlarging maps it is the scale bar that you need; the scale in words (six inches to the mile) or as a representative fraction (1:2,500) tells you the scale of the original map, and should be noted, but is now useless for calculating 'real' distances. One simple system is to take a map of the area which includes a scale bar (such as an OS six-inch sheet), that is, where you are certain of the map scale, and from this to construct a scale bar for your own extract. Firstly, work out the 'real world' distance between two points on the map whose scale is known. Using a ruler (or the edge of a sheet of paper), measure the straight line distance between two points A (for example, a church) and B (for example, a crossroads) on this map; draw a line marked line A to B on a sheet of paper, representing this length. Now calculate (using the scale bar which is on that map) what is the 'real world' distance of this line. You now know that a certain straight line distance A to B (at 18 cm) is (for example) three miles in the real world. Now go to your own map, the one without a scale bar. Measure the same straight line distance A and B (the church to the crossroads) and draw a line representing this distance on your sheet of paper. You know that on this second map the church and crossroads will still be three miles distant, nothing has moved in the real world! Taking it that this second line is 11 cm in length:

Map 1 (scale known)	distance A to B	18 cm representing 3 miles
Map 2 (scale not known)	distance A to B	11 cm representing 3 miles

What percentage reduction is map 2?

$^{11}/_{18} \times 100 = 61.1$

Therefore map 2 is 61% reduction of map 1

Your own map is therefore a reduction to 61% of map 1. You can now reduce the scale-bar of the original (map 1) to 61%, and paste it onto map 2. This is very simply done by reducing it (to 61%) on a photocopier or scanner, or by calculating by hand and redrawing the line. It is, however, strongly advised that you do not rely on a single measurement, but calculate the straight line distance several times, each time choosing two points in different parts of the map going in different directions, so that you have the same criss-cross of lines covering both maps. You calculate the percentage difference in each case as outlined above; you should get the same answer repeatedly, an assurance that your measurements and calculations are indeed accurate. Where your own scale-less map is an enlargement of the map for which you have a scale, the process is identical, except that in the final step map 2 will be an enlargement of map 1, and the scale bar will be enlarged proportionately. The first option, where the scale is taken from a map drawn at a larger scale (that is, where the scale bar is reduced), is always to be preferred over the second option, where enlargement is required. The scale-bar thus created will be accurate, but possibly unwieldy in size. A simplified or neater version can be made now or at a later stage (when designing the final page layout). Miles or feet may be maintained,

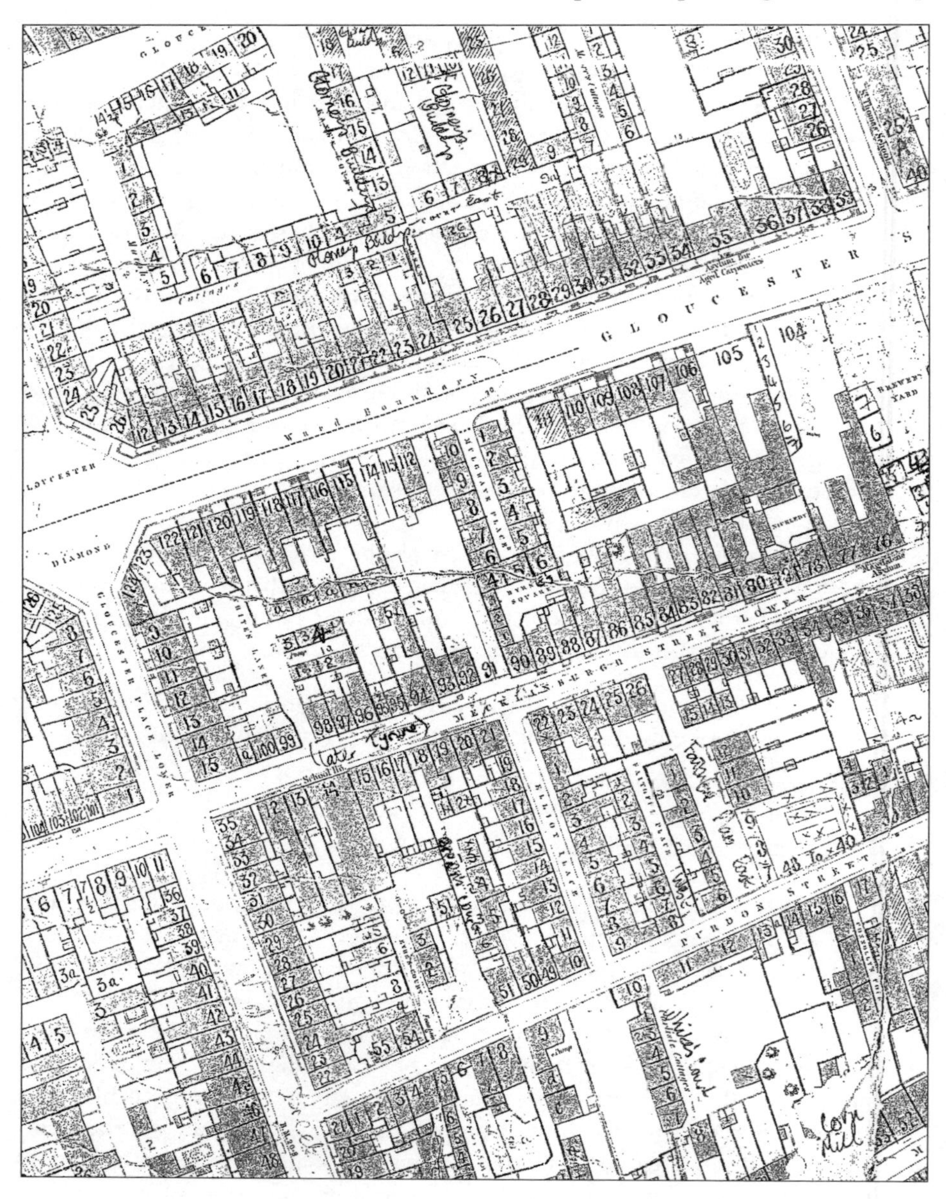

70 North Dublin city, Valuation Office map, 1:1056, to locate houses, *c.*1854

but it is preferable to work in metric, especially where there is to be a series of maps. Appendix 7 will assist in making the conversion.

4. Creating your own base map: the process of generalisation

The detailed source map, and its scalebar (either taken from this map, or constructed from another map as explained above), may be scanned into the computer, if necessary in sections, into a single base layer. It can then be used as a locked, non-printing template, whereby selected items are overwritten electronically (working in a new layer) and a new map created. If a full GIS approach is to be taken, this is the stage to ensure that the map is overdrawn or digitised in such a way that unique codes can be ascribed to each feature. The more traditional hand tracing can be equally effective and in some instances may even be quicker. When tracing by hand from the single-page master copy, polyester film (not 'tracing paper') is recommended, as it is less likely to shrink or buckle. Whether tracing electronically or by hand, it is first necessary to decide what features are to be selected, and what details may be left aside for the purpose now in hand. This process of generalisation is fundamental to map-making. If everything was copied and merely reduced photographically, the human eye would be unable to distinguish one feature from the next. What to retain? What needs to be simplified? What to combine? Exaggerate? Alter? In the case of the North Dublin map (figs 68–71), both the main streets and the network of alleys and courts are retained from the original six-inch sheet, along with some street names; there are additional placenames (identified from the more detailed Valuation sheets and from field research), one church only is selected, and this is represented by a simple cross. The exact location of each tenement dwelling was decided from the street directory numbering combined with the house numbers on the larger scale (1: 1,056) Valuation Office maps (fig. 70), but plotted (with small red stickers) on large photocopies of the six-inch paper master map. All this preliminary work was accomplished before embarking on any tracing.

5. Tracing

In your own map-making, it may help first to highlight the features to be retained (such as the street or road pattern) in colour on the paper master, while also noting the preferred place for inserting symbols, or other additional information. If working on a very large sheet by hand, it is best to trace a few major lines first (such as the railway, road or lengthy river) and then return to fill in the details, on the principle that cartographers work from the whole to the part so that systematic errors are not magnified. The early OS fieldworkers, you will recall, first established their country-wide framework and then filled in the details, for the same reason. Richard Griffith's orders to his assistant boundary surveyors (1832) still ring true. In making tracings of maps he recommends 'a finely pointed pen, so that the lines of the map shall be distinctly and neatly represented on the tracing, and he is to take care that the tracing paper does not shift during the operation'.[13] Hand tracing

13 Richard Griffith, *General instructions* ... (1832), p. 53.

is always done with a quality black ink drafting pen (brand names include *Rotring* and *Edding*). Size 0.35mm is a good standard where (as is usually the case) there will be reduction before final reproduction; line width of 0.25mm is recommended for finer lines, and where there is to be no further reduction. Each layer is taped down securely in turn, with masking tape: the paper copy, then the tracing film, not to be lifted until all tracing is complete. Ideally a light table (that lights the map from below)[14] or drafting table will be used, but where this is not possible, ensure that you can work on the tracing from all sides. Taping your work onto a large board may be one way of preventing it from being disturbed, as you can lift off the entire board and tracing film as a unit when the table is required. 'DIY' cartographers have also been known to use a sheet of thick glass or perspex as their tracing board, lighting from below with a regular desk lamp. Hand tracing is done systematically from top to bottom, and from all four sides, to ensure an even rhythm and to allow the ink to dry. Use of a small bevelled ('raised') ruler is highly recommended; do not trust to your eye, or the steadiness of your hand, but consider each piece, even arcs and curves, as so many short but carefully ruled lines. The pen needs to be held upright, not sloped as with biros, and each line should be made with a single, steady and decisive stroke. If an even pressure is maintained, and the pen held at the correct angle, the line will be perfect. The directive to work systematically downwards is to ensure you do not now smudge such well-executed work. Corners should meet exactly, and no more, at the first attempt. Errors may be scraped away (once only) with a scalpel or covered blade. A small cloth should be used to wipe the pen nib occasionally, and as with any large drafting project, the hands may need to be washed during as well as after the drawing process. Professional drafting tools include plastic templates for curves and circles. Flexible spline curves (with and without weights) can also assist in drawing lines, but for most local history cartographic purposes two bevelled rulers – approximately six inches and eighteen inches long – will suffice.

Ensuring that your base map does not move also applies when tracing electronically; the base map layer should be 'locked' immediately, so that it cannot be stretched, rotated or otherwise interfered with during the process of overwriting. When drawing curved and other irregular lines, numerous small straight segments can be easier and more accurate than freehand drawing, with most computer programmes allowing endpoints to be joined up exactly and 'smoothed' as necessary. Working in layers is strongly recommended, even for maps with little detail. By separating out each layer (for example, into road linework, placenames, scalebar and source) it is possible to format each layer independently, applying line weights, font

14 A white light source of between 200 to 1,000 lumens intensity as compared to ambient lighting in a ratio of 10 to 6 or 9 is advised; a light source of 20 W fluorescent or daylight tubes is recommended, with protection from the light and heat provided by sheets of 5mm thick glass, matt on one side or a sheet of milky-white plastic or opal glass. Anson and Ormeling (eds), *Basic cartography*, i, p. 104. Before constructing a light table, readers should consult an experienced electrician.

style and other attributes in an efficient manner. It also ensures that reworking of one element does not interfere with another, as may be the case if all the map detail is entered on a single layer. When all is completed, it is of course possible to combine several layers into one.

As mentioned at the outset, it is common in historical geography and local history studies to produce a single 'master' map from which a series of other maps will be constructed. In the case of the Dublin map (fig. 69), the north city section is a portion of a much larger project, covering the street detail within the canals, but extending along the major arteries to cover the suburbs as far as Donnybrook and Harold's Cross (south) to Inchicore and Kilmainham (west), and Phibsboro and Drumcondra (north). Whether constructing by hand or by computer, one carefully-drawn 'master map', can act as the base for countless others. The secret is to include all the linework that is likely to be common to all of the maps: municipal boundary, river(s), principal roads or major street patterns, the type of feature which will enable the reader to locate him or herself immediately, but will not clutter the map. Map-makers using a computer package such as Adobe Illustrator or ArcView have the advantage of being able to decide later which 'themes' or 'layers' of information to combine on their map, and which to leave unused. Minimal lettering could also be introduced: in the Dublin case, *River Liffey* was inserted on this master map (fig. 71), and the river very lightly shaded, as it was certain to feature in each of the resulting maps. Ensure that a scale bar is included (its placement is not crucial, as it can be relocated later), and most importantly (when working by hand or electronically), again include a N/S and E/W cross, to enable exact alignment once you remove the tracing from the paper base. The finished map at this stage is solely crisp linework, and larger in size than planned for your final map; there is no lettering, shading, border or additional graphics (fig. 71). When multiple maps using this base are planned, you need to make a sufficient number of copies and work with each map in turn, putting the 'master tracing' into safe keeping; if you are working by hand, copies can be made on acetate film. If your map is extra large and unwieldy, this is the point at which you could reduce it to a manageable (but still larger-than-final) size. If you have digitised the map (overdrawn it using a computer programme), you are now at the same stage, but with the advantage that you can save your 'master copy' on disk, correct errors, change size and orientation, make revisions, updates and multiple copies, and follow through the next steps with great flexibility and ease. Alternatively, you could combine both traditional handwork and computer resources: trace by hand to this stage (including scale bar, and N/S/E/W cross), scan in the 'empty' map (once you have reduced it to a 'scannable' size, such as A4), ensure it is 'locked' into shape, save as your 'master' map; and then continue, on a copy, with the insertion of text, symbols and other matter. If you have digitised your map, you will be working with points, lines and areas; the individual lines can be manipulated in terms of weight and pattern, and closed areas (polygons) infilled or shaded. Whether working by hand or computer, the cartographer needs to remember that in the final printing process, 'dot gain' causes ink to spread, and colours to darken. This effect may be counteracted by ensuring that

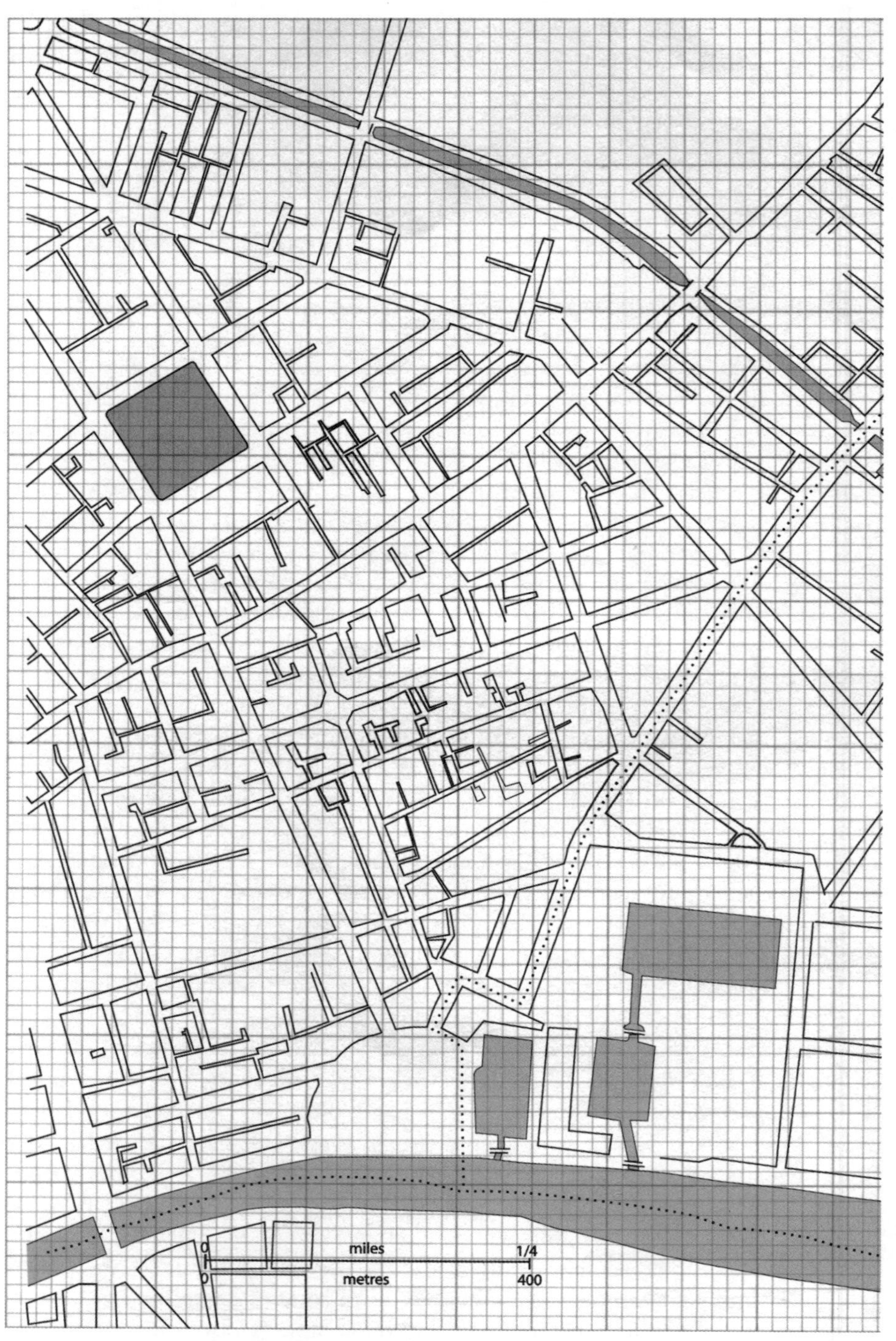

71 North Dublin city, *c.*1850, linework only

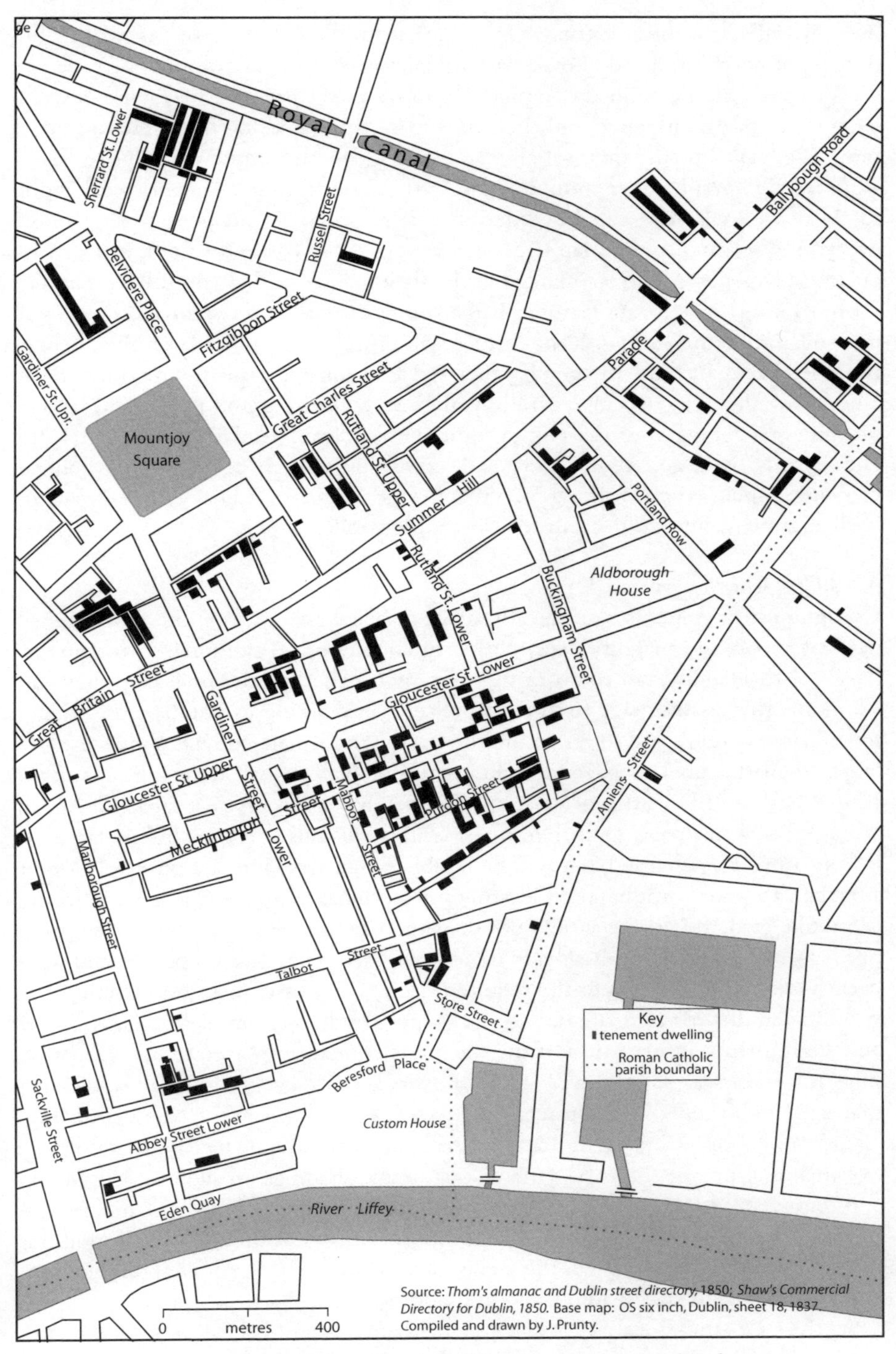

72 Tenement dwellings, North Dublin city, *c.*1850, completed map

there is sufficient contrast (or white space) around each piece of ink work, and (when colour printing) to choose pastel shades over stronger colours.

When tracing by hand is completed, remove the paper base map, tape down a large sheet of graph paper, and then on top of this, tape down the tracing once more (fig. 71), ensuring that it is aligned exactly (the N/S and E/W cross serves as a guide). The graph paper should be larger than the tracing, to enable work right to the edge. A border can be added at this stage. As its purpose is to enclose everything relating to the map (legend, source(s), scale bar, and so on), it is best to enclose a larger area than is intended in the final map(s), so that you still have ample room to insert or relocate features. It is a simple matter to move in and cut down to size later. Taking care with the border will prevent difficulties later; the corners must be exactly 90° (use the graph paper and a set square). Tape down both ends of a long bevelled (raised) ruler, carefully aligned with the graph paper, and in one movement, keeping your ink pen upright, draw a clean, steady line, that starts and ends exactly on target. Allow to dry, and repeat this on each side. The same general principle applies when working on screen: work against a grid background to assist with alignment, and do not 'cut to size' too drastically at first.

6. *Adding lettering, symbols, key*

Computer cartographers have the advantage of being able to zoom in on individual parts of a map, and magnify their work area enormously. Traditional cartographers have not this facility, but by working at a larger scale than the final map they can overcome this disadvantage to some degree. When working by hand, before adding in lettering, symbols and so on ensure the tracing (or acetate copy) is about one size larger than the final map. If intending to present as A4, all lettering and other elements should be introduced at approximately an A3 size; if preparing for an average 'book size page' (A5) then make your inclusions at A4 or slightly larger. It is impossible to visualise what will be legible when working at a massively larger size than the final product, and legibility is a crucial matter. If the final lettering cannot be read, the feature should not be named, or at least some alternative system needs to be explored. If not already done, your own tracing (with border and scale bar) can be reduced down to the required size on to acetate or to transparent film; when taped down, yet again, onto graph paper, it will be a simple matter to insert and align further information. If the tracing is copied onto acetate or transparent film as a reverse image it will make the final work even easier; all additional material can be added to the front, and excess line work (from the original tracing) scraped away from the back. At this point computer cartography has all the advantages over manuscript work, as text, symbols, colour and shading can all be added, altered and removed with ease. This author has produced figure 72 both by hand and on screen; it is at this final stage that the computer outscores the traditional methods most significantly, especially in the facility it allows for enlargement and reduction, and changing font size, shading, and line widths.

In the case-study local history maps already discussed (figs. 59 to 65), a range of lettering styles, symbols and boundary lines are featured, with each map having its own content, purpose and effect. Some of the principles governing the design of symbols and lettering are discussed below. Computer cartographers can generate their own symbols on screen, but the computer can also be harnessed to serve those using more traditional methods. Symbols and lettering can now be readily created, in a range of sizes, using any standard word processing package such as MS Word (under symbols, autoshapes and clip-art), or downloaded from an Internet site, and printed out on paper or (even simpler) on a standard A4 acetate sheet. These can be cut and placed onto the map using a scalpel (to hold the tiny cut-out pieces), and non-reflective adhesive tape, with the graph paper as a guide. If the computer-generated symbols are printed out directly onto adhesive-backed acetate the process of 'cut and place' is even simpler. Alternatively, rub-down symbols (marketed as *Lettraset* and under other brand names) are available in shops which sell stationery or art supplies, and are worth the investment when a considerable amount of work is in hand; in this case it is important to use a proper rounded 'rub-down' instrument (not a sharp pencil). It is also possible to purchase ready-made lines of different dimensions and styles, both rub-down and flexi-line (under the brand name *Lettraline*), which are valuable for borders, rivers and boundaries. Perfectly good symbols can also be drawn using a plastic stencil and drafting pen; with a little practice it will be impossible to distinguish these from the more 'professional' (and expensive) rub-down version. Regardless of the method used to generate point symbols, in the positioning of each, it is the middle of the base of the symbol that serves as the 'exact point'. Where using bar symbols, for example, it is the base of each column that the reader takes to be the exact spot referred to. The key or legend panel, with or without a border, can be prepared on computer and simply cut and pasted in as a single entity – quite literally, with paper adhesive ('Pritt') in the case of hand-drawn maps (figs. 59 to 65). Where a graphics package is being used, the legend is simply a combination of text and symbols, and may be created as a separate layer. In the case of GIS and other computer-mapping programmes, the legend panel is created automatically, though with the option of formatting, altering or deleting. Entirely freehand drawing (by hand or on screen) is not recommended; few people can manage to reproduce the same symbol exactly again and again.

In terms of lettering, whatever typeface is chosen should be distinct, without serifs, and taking up the minimum of space with the maximum effect. Both Arial and Helvetica fonts fit this bill, but there are several others also. Size matters – reflecting the importance or insignificance of the named feature – but so too does style, as the nature of the feature can be expressed by the style of the text. In figure 56 the ingenuity of the Ordnance Survey is clear in the variety of fonts it used in the six-inch and 1:2,500 series. The location of the feature is indicated by the placement of the text, and its extent by the spacing: the OS stretches townland names (in a Latin font) across the entire townland, or mountain names across the heart of the range. To name a particular feature, such as a village indicated by a black circle, the placement of the label depends on the space available, and what might, or might not, be obscured. The preferred place for the lettering is to the right and slightly above the symbol. In English and Irish we read from left to right;

the symbol therefore is followed by the lettering, where space permits. Where this is not possible (due to overlap with other linework, or the closeness of other lettering), the text label goes to the right and slightly below the symbol; the next choice for the lettering is to the left but above the symbol, and finally (if all else fails), to the left and slightly below the symbol.[15] In no cases should the final (reduced) font size be less than six point; indeed, the smallest size in general use should be eight or nine point. Both upper and lower case should be used; block capitals make individual letters more difficult to distinguish, especially when the font is of a small size. Italics are also unnecessary for 'regular' text, and should be reserved for special effect.

Practically all maps will need placenames. These may be handwritten, when the compiler is confident of his or her skills in calligraphy, or more usually, printed, cut and pasted down (literally, when working in the traditional mode) or as text inserted on screen; their exact placement therefore needs some consideration. Lettering is generally positioned to minimise the rotation of the paper necessary for easy reading. In most cases therefore lettering will be aligned horizontally (graph paper will assist in this) 'with the exception of the names of rivers, canals, chains of mountains etc. which require to be adapted to their natural sinuosities', as Major Colby of the OS spelled out to his underlings in the 1824 *Instructions*. Where curved lettering is used (perhaps for a mountain range) it should always be arranged along a regular base line. This can be done very easily on screen with most computer graphics packages by drawing a text path first (of any shape) and then inserting the text. 'Uneven' lettering (where the eye alone was trusted) is immediately obvious to the reader. Place the lettering adjacent to the feature being named as noted above, and ensure that each piece of lettering has a small area of white around it, so that it does not 'run' into other linework or obscure the symbols (Appendix X). Lettering takes precedence over the line work; if necessary, obstructive linework can be scraped away from the reverse of the acetate (where a reverse image was made), or (on screen) masked by ensuring that each letter has a small white margin or that each piece of black text is set against white infill. On the final all-Ireland 1854 barrack map, for example (fig. 67), the coast appears 'broken' to ensure that the placenames and symbols stand out. On a crowded map (as in fig. 69) the question needs to be asked: what needs to take precedence in each case? The network of roads and streets? Placenames? Heights? Long names should not be hyphenated, or broken up over two lines; the 1:50,000 extracts (fig. 47) show how this is studiously avoided. Even where the name consists of several words (*Cill na Manach*) the preference is to keep it together for ease of reading. When labelling a river or boundary the lettering where possible goes on the 'appropriate' side, that is, on top of the river, or on the side of the boundary facing into the unit it describes; this can

15 'Names of cities, towns, villages, &c. are to be commenced on the east side of the places to which they refer, when no special obstacle occurs to interfere with this arrangement', Major Colby, 'Instructions for the interior survey of Ireland, 1824', reprinted in Andrews, *A paper landscape*, p. 319.

be done easily on screen as already explained, by inserting a text path first and then ascribing style, thickness and colour (attributes) to the line, shifting the text itself (the name of the river) slightly above the line (adjusting kerning). Inserting street names can be a little more contentious, but in general the lettering follows the line of the street, is kept horizontal as far as the street line allows, and faces into the centre of the map. Any large-scale OS map can be trusted as a guide. This again makes the case for having a rough 'scribble copy' alongside the final map.

7. *Map design and page layout*

A fussy map, packed with florid lettering and exotic symbols will distract rather than enlighten. There are a few major considerations to bear in mind: what is the map intending to show? How can it communicate this most directly? And what minimal ancillary information is needed to ensure that the central purpose is indeed clear to the reader? In the map showing the downgrading of houses to tenement status on a North Dublin street (figs 68–71), for example, it was decided to represent each premises by a simple rub-down rectangle, rather than drawing the exact shape of each unit. In this case also, it was decided to name two or three major places only (the Custom House, River Liffey, Mountjoy Square and Gardiner Street), sufficient to allow the reader to locate this district relative to the larger city. In a rural settlement map (fig. 65) the 1,500 feet contour line is a sensible addition, implying why settlement is limited beyond this point. This is one of the advantages of maintaining control of the map-making process from design through to production; from your own wider researches you will be well equipped to decide which 'landmark' features should be added, and which can be ignored.

On the page itself, each element of the map should be positioned in a logical, unambiguous and graphically-balanced way. Items must be placed in order of importance with respect to the rules of visual perception.[16] As noted already, we read from left to right and from the top of the page downwards. When scanning a page we therefore expect the most important information to be in the top left hand corner, the least significant material to the bottom right hand corner. Where the title is contained within the map border (that is, where there will not be a caption), it should be placed in a visually dominant position, such as upper left corner, lower left corner, or centred across the top, and in a larger font size (fig. 64). The legend should be in such a position that it neither dominates the page nor tries to excuse itself. It should be a well-designed, compact unit in its own right, carefully aligned, and may have its own light border. If it is in a 'long and narrow' format, or in two columns, it is easier to read; think of how newspapers prefer multiple narrow columns over the wide format, as more immediately engaging, and taking less effort to read. The legend must not obscure the information being mapped, but can of course take over a 'redundant' part of the map. In all cases 'empty' areas should be used to optimum effect, rather than simply expanding the border to take in more white space. Visually it is best to place the scale bar where it can be consulted while

16 *Basic cartography*, ii, p. 31.

not commanding undue attention. It must be exactly parallel to the page border. Similarly there is no need for the information on the base map and data sources to be excessively obvious; it is best located at the bottom of the page, aligned to the right (or left) margin, and in a small font. It is important that neither this nor any other text is positioned too tightly to the map border, as this will reduce legibility. There should be nothing arbitrary or automatic about where to place the various elements that make up a map; some thought and experimentation will result in a visually pleasing and helpful layout.

In the final map (fig. 72), ensure you have included the following (within the frame): the map upon which your outline is based (publisher [OS], series, sheet number, year of publication and/or edition), the source(s) of the information being mapped, and where appropriate (in small font) your own name and the year – for example, 'Compiled and drawn by J. Prunty, 2004. A north-point is essential wherever your map is not oriented due north (that is, as on OS maps); a scale bar is also needed, as discussed above. The title may be inside the frame or alternatively given as a caption below (there is no need for both); there should always be an indication of the date or dates to which the map refers, usually incorporated into the title or caption. Finally, there is always need for third-party proof-reading, examining the map for completeness, and also checking against draft materials, before final submission for publication. At each stage of the cartographic process there is room for error – the OS itself has immortalised the occasional mistake – while the graphic nature of the map ensures that a misspelling or misplacement is even more public than in a passage of prose.

8. Choice of lettering and symbols

Returning to the question of symbols in more detail, the map-maker needs to think about the types that are available, the dimensions to which they will finally be reduced, and their relative spacing. There are certain 'thresholds' to be considered. At what stage can the reader discern the presence of a symbol? What scaling is required to ensure that the different categories indicated by identical symbols (such as graduated circles showing town populations) are clear to the reader? What is the minimum point at which the shift to a new class or size category is perceptible? There are three ways in which information is symbolised on any map: point, line and area. Many of the principles, such as the question of visibility, apply under each heading.

The visibility and legibility of a point symbol depends partly on its shape (Appendix X). A solid black circle of diameter .4mm is visible. If the circle is 'open', a slightly larger size (.5mm diameter) is required for equal visibility. Because of its shape and bulk, a solid triangle of base .6mm is about as 'visible' as a solid square of base .5mm, or the circles already mentioned. These variations are indeed tiny, but they illustrate how the human eye perceives different shapes in different ways. Once the 'threshold of perception' is crossed, that is, the symbol is visible, what is the minimum size at which the reader can clearly differentiate between the shape of different symbols (squares, circles, triangles and diamonds) all on the one map? Once this 'threshold of differentiation' is passed, at what point can the reader be absolutely certain in the identification and separating out of each symbol, i.e. has crossed the 'threshold of separation'? The cartographer also needs to consider the spacing between symbols, relative to their visual perception. A line of small dots too close together becomes, to the eye, just a black line; place the same dots too far apart and you lose the sense of a continuous dotted line. Differences which are scarcely noticeable defeat the purpose of the map, while excessively large, 'loud' and overlapping symbols will aggravate the reader.

Each of these 'thresholds' must be considered empirically, with regard to individual symbols; good sense, and experience, will guide the aspiring cartographer. By critiquing published maps, in journals, books and theses, the trainee cartographer will develop an appreciation for quality map-making, and at the very least an awareness of what can go desperately wrong, based, for the most part, on the lack of training in the fundamentals of map production and graphics. While experience is an unrivalled teacher, there are still some general points to take on board from the outset.

In the creation of point symbols, clear, simple shapes are more effective, keeping in mind always what these will look like in the final reduction. Basic shapes – square, circle, triangle, diamond – in solid and in open format – can be recombined to produce countless new symbols; the Kilkenny map by Anna Byrne (fig. 62) illustrates this point exactly. Where there is an immediate association between the symbol and the feature this is welcome – a small flag for the boat club, cross for the church, aeroplane for the airport, handset for the telephone exchange, skull and crossbones for the toxic dump – but beware of having excessively detailed pictorial symbols. Computer-generated graphics have made this most tempting. Within the total map environment, picture symbols only succeed if they are greatly simplified: strong, distinct, flat shapes, requiring minimum line-work (figs 62 and 72). The top edge of any symbol, and of lettering, has been proven to be the piece that the eye most rapidly identifies, so special care should be taken here. A cross placed on top of a small circle will, for example, catch the eye more effectively than a cross in the centre of a circle. As with all point symbols, solid shapes are easier to distinguish than open. They will always capture the attention of the reader more rapidly than open shapes, and this may not be what is intended. Very large point symbols may imply that the information is associated with the extensive area covered by the symbol rather than with a discrete place; again this may not be what is intended.

Point symbols are commonly used to show settlement. While the compiler might wish to draw a black square for each house, the scale limits what is possible. One solution may be to further generalise or group the information, for example, a black square for ten houses, a single dot for less than ten. The number of 'classes' or 'categories' is always a matter of debate; in most cases five will be sufficient, as the reader can only cope with so many distinctions at the one sitting. However, it should be emphasised that it is not merely a matter of limiting the number of classes, but of ensuring that each category can be easily distinguished. As already mentioned, point symbols require to be surrounded by a small area of white, otherwise they will merge into the surrounding ink; this may mean breaking linework, deleting another feature, or slightly relocating the symbol (fig. 67 and Appendix X). The reluctance to overwrite what is there must be overcome by reminding yourself of the main purpose of the map. What is it intended to show? And how best to display this? Arrows imply movement, or a point beyond this place; it is therefore problematic, and unnecessary, to use them for labelling.

Features can also be symbolised by lines; figure 57 shows the ingenuity of the OS in the creation of different lines to distinguish different boundaries, some of which will coincide. To make the distinction between different line symbols the width can be varied, but these must be well-differentiated (for example, widths of .25mm, .35mm, .7mm). Double lines can be used in place of single, and lines can be 'pecked', that is, dashed, dotted or combinations of both (Appendix X). In the case of pecked lines the space between dashes or dots must be worked out carefully, so that the line does not 'break up' and get lost in the other map detail, or alternately become just another single line, albeit less sharp. When tracing lines that will be significantly reduced in the final map (such as the course of a river, or the outline of a coast) there is no advantage to be gained from including every tiny deviation; where this is done the final line work will simply appear to be badly executed or indecisive. Significant bends and changes in direction need to be included, but only beyond a certain threshold, as there is no way that every minor zig-zag can be discerned in the final map. This 'smoothing' of lines (very easily accomplished on screen), is similar to replacing picture-like symbols with simpler geometrical shapes; the final size is kept in view at every stage of the process.

Symbols that are adjacent need to contrast sufficiently to ensure that each is readily distinguished. And there will always be the problem of overcrowding in some part of the map, where there is so much to show and so little space to show it. In the 1610 Speed map of Dublin, for example, St Patrick's cathedral outside the walls is depicted handsomely, while Christ Church cathedral, of equal standing, is treated quite meanly under the pressure of space. Where the problem is one of crowding, those symbols which need to be most precisely located take precedence: spot heights, the cross for a church, the road junction. Symbols are 'read' from the bottom up; the base of each therefore is taken as the exact location of the feature. Moving them 'aside' is not always an option, so alternatives need to be explored. Symbols can be 'tiled' or overlapped to good effect (fig. 67 and Appendix X), but as always each needs a small 'clear space' (or 'halo') around it, to ensure that it continues to stand alone. When overlapping symbols, the smallest symbol should come to the fore, the largest becomes the background, the overlap in each case being edged with white. And all symbols, as with lettering, legend, scale bar, borders and so on, should be aligned exactly using the graph paper and a ruler as necessary.

Choropleths (shading) can be effective in conveying information (fig. 68 and Appendix X) and excellent quality is quite easily produced by hand (computer 'fill' is instantaneous). Sheets of 'film screen' or commercially-produced stick-down shading can also be used (brand names include 'Lettratone'). The variations in perceived lightness are expressed as percentages, the 'value variable' or relationship

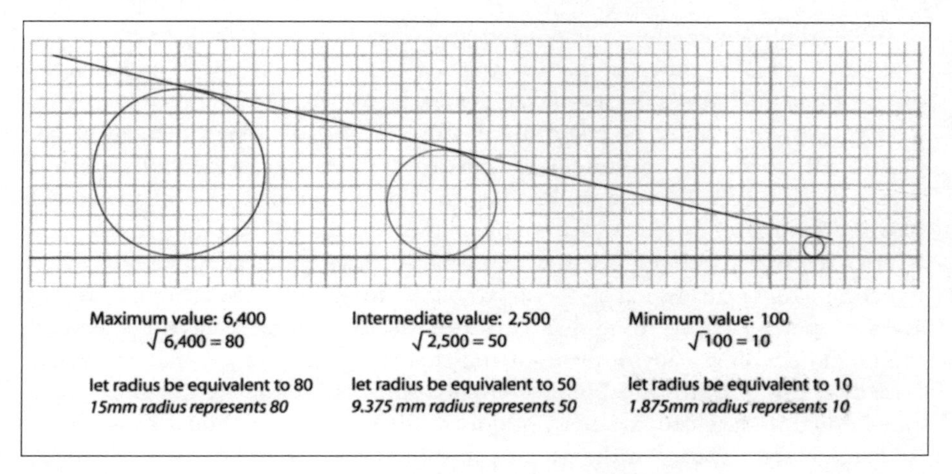
Maximum value: 6,400
√6,400 = 80
let radius be equivalent to 80
15mm radius represents 80
Intermediate value: 2,500
√2,500 = 50
let radius be equivalent to 50
9.375 mm radius represents 50
Minimum value: 100
√100 = 10
let radius be equivalent to 10
1.875mm radius represents 10

73 Graduated or proportional circles

between the amount of the sheet covered by black ink (whether lines or dots) and the amount of white space in between. Thus a sheet marketed with a value variable of 10% will be visually very light (lots of white space relative to the size of the black dots), another at 50% will appear darker and more crowded, with less white space relative to the size of the dots; 100% would indicate a completely black sheet. This technical digression is useful in highlighting one of the key matters in the construction of choropleth patterns by hand (or on screen): firstly consider the thickness of each line or diameter of each dot and how close the spacing. The same size dots, each time brought more closely together, can create a graded series; a pattern of lines moving from dashed horizontal to straight horizontal to lines criss-crossing at right angles is another clear sequence. Horizontal and vertical line shading combined can be graduated from a very loose criss-cross to a dense mesh of lines (Appendix X). Other basic points to consider are line direction: at angles of 90° (horizontal or vertical) or at 45° (diagonal)? All going in one direction or criss-crossed? Are the lines broken (dashed) or continuous? If dashed, in what way are the dashes arranged? Similarly, the arrangement of dots can be investigated; they too may be aligned horizontally or diagonally, to different effects. When deciding on the 'fill' to create for each unit, a few questions need to be posed. Are the classes or groups scaled in some way, for example, population ranging from high to low density? If so, the shading should follow this sequence, going (visually) from 'high' to 'low', from darker to light (fig. 68 and Appendix X). Alternately, does each category carry the same weight, such as a map showing the baronies in a particular county, or the land owned by different landlords? If so, the choice of shading should imply 'equally important but distinct'. There is always the temptation to leave one category as white or unshaded; this should be avoided as visually the message conveyed is that there is 'nothing' here. Similarly, care should be taken with an all-black section, as this in turn implies 'completely full', which may not be the intended message.

In choropleth mapping it is possible to create innumerable distinct patterns through combining different line spacing and orientation, and (less importantly) different line thicknesses (fig. 68 and Appendix X). Where lettering is to overlie the shading, extra care must be taken that it can still be read. Experimentation will prove well worth the investment of time. In all cases the borders of each unit should be defined before the shading begins. Where adjoining units happen to fall into the same category, the cartographer needs to decide whether to keep the internal divisions, or to remove these lines and shade as one large unit. Unless there is a compelling reason to keep these extra lines, they should be excluded, as they are now merely additional noise. The pattern decided on for each unit should be indicated on a photocopy or printout and kept to hand, to ensure that no mistake is made at this critical stage. Using a bevelled ruler and following the graph paper taped down underneath, a pattern of dots, dashed lines or full lines can be drawn by hand, based exactly on the graph paper divisions. As with all traditional ink work, it is best to work systematically from the top to the bottom, then to turn the table (or oneself) around and work systematically from top to bottom again. To achieve diagonal patterns, the graph paper is turned around at 45° and the drawing continues as before. Each dot should be made in an assured manner, not just a tip. This is another area of cartography which the computer has simplified immensely, whereby polygons are ascribed a pattern and colour as 'fill'. The map-maker need not rely entirely on the 'automatic' choice offered by the computer programme, but should consider the above points as they relate to traditional mapping.

Symbols can provide quantitative as well as locational information if they are drawn to scale. Circles, squares, columns, triangles and other shapes can be employed. What matters is that the areas are made in proportion to the quantities they represent. As well as showing total quantities they can be divided proportionately to display further information, as with pie charts (fig. 64). While most map-makers are likely to rely on computer programmes to generate their pie charts (either using GIS software or simply cutting and pasting pie-charts made elsewhere onto the map), those relying entirely on traditional methods may find the following explanation helpful. It will suffice to calculate proportional sizes for three entries (maximum, minimum, and roughly midway); from this it is very simple to read off what sizes the circles 'in between' should be (see fig. 73). In terms of graphics, the results are well worth the initial effort. The area of any circle is πr^2, but as π is a constant it is r^2 (the radius squared) that needs to be calculated. On a piece of graph paper write down the quantity to be represented by the largest circle ('round up' a little to an even number, say, 6,400). Find its square root ($\sqrt{}$ of 6,400 is 80). Now draw a circle (using the graph paper divisions as a guide) with a radius equivalent to 80. For example, a circle with a radius of 15 mm might be decided upon, that is, 80 is represented by 15 mm radius. That circle now represents the largest quantity (6,400). Take a second quantity, say 2,500. As before, find its square root ($\sqrt{}$ of 2,500 is 50). If a radius of 15 mm represents 80, then a radius of 9.375 mm will represent 50. A third quantity of 100 ($\sqrt{}$ of 100 is 10), will be represented by a radius of just 1.875 mm.

Once a scale is established (as in fig. 73), the size of circles in between can be 'read' off directly.[17] Some experimentation will be required before the final scale is determined upon – the first choice may result in sizes ridiculously small or large, but it is merely a matter of deciding what distance (in mm) 'on the paper' should represent a given quantity (based on its square root).

Where the size of symbols has a mathematical basis – for example, graduated circles or bars (fig. 64) the final design of the key panel needs special care. The reader should be able to work out from the legend the approximate quantities represented by the proportional circles. Alignment within the panel is generally to the right, with (for example) due regard made for the correct placing of numerals.

When all the map work is completed satisfactorily, the cartographer is advised to store intermediate work, whether as paper copies or as computer files, as well as the completed master copy, as future map-making projects may allow some of this material to be re-used. Layers may be extracted from individual computer files (for example, where saved in Illustrator format) and used again separately and in combination. Many of the territorial divisions used in GIS mapping by historians and geographers alike – townlands, electoral divisions, counties – can be utilised in very many different mapping projects, covering land valuation, demographics and agriculture, as these were the units for so many major statistical ventures. And finally, specific reference to the map(s) created with such care should be part of the writing of local history: 'see fig. x' or more usually (fig. x). While any map will 'speak for itself' in some senses, the author still needs to ensure that attention is drawn to it as appropriate.

17 For a fuller treatment see F.J. Monkhouse and H.R. Wilkinson, *Maps and diagrams* (London, 1971), pp 29–33; see also *Basic cartography*, ii, p. 36.

Conclusion

In *The compleat geographer* (London, 1723), Herman Moll celebrates the art of map-making as perhaps the most valuable of all the projects to which 'ingenious men' might direct their labours:

> For by the help of them, Geography is made plain and easie, the mariners are directed in fetching us the commodities of the most distant parts. And by the help of them, we may at home, with pleasure, survey the different countries of the world and be informed of the situation, district, provinces, cities, and remarkable places of every nation.[1]

Maps have continued to fascinate the public as well as proving their usefulness in countless fields, from archaeology to land transfer. There is an especial enthusiasm for map use among local historians, who recognise in the map not alone a valuable primary source but an entry point to the complex relationships which mark any community, and an ideal vehicle for the communication of their own research findings. However, despite this undoubted interest, there is a real dearth of background in cartographic matters. The general hesitancy to go beyond using maps merely as illustrations is regrettable but understandable. 'Carto-literacy', as it might be termed, is a neglected field. NUI Maynooth offers the country's first, and still only, full university course titled 'Maps in history', offered jointly to both History and Geography undergraduates, while maps and map-making feature under both sources and skills in the MA (Local History) programme. Geography departments continue to offer practical classes in map-making. This writer benefited from the inspired teaching of Dr Arnold Horner as an undergraduate student in UCD. However, the overwhelming predominance of computerised map-making and GIS data processing, and the strain to keep abreast of the technology, may lead to the neglect of the more philosophical questions about the purposes of map-making and the aesthetics of map design. In local history publications, the quality of the maps can be less than impressive. This is often for very simple reasons, such as lettering poorly aligned or too small, matters which could so easily be rectified. And yet, as I hope this guide has indicated, it is an area of rich potential for the local scholar. To know your own place, visually, in plan and from a bird's eye perspective, to trace its lineaments over time, the ways in which it has been remade and stretched, and yet holds in its very shape the story of its past, brings great personal satisfaction. Maps as sources can open windows on the past; they are dangerous, in that they can expose enormous vistas beyond! The reception afforded to each fascicle in the *Irish Historic Towns Atlas* project bears direct testimony to the enduring fascination with

1 H. Moll, *The compleat geographer*, p. xiv.

'old' maps; the way in which these maps are actively used in the reconstruction of the town's past, especially through its text maps, can serve as 'models of best practice' for the local historian. A well-researched, designed and produced map is one of the best media for the storage, communication and analysis of spatial information.

There have been massive map losses in Ireland, but there are very hopeful signs of improving archival and cataloguing situations. The production of good quality map facsmiles, in the *Irish Historic Towns Atlas* project, has been especially helpful. Advances in archival science have led to more map collections becoming available to the local researcher. The computerisation of catalogues in the UK and Ireland, along with the ongoing deposit and purchase of records, mean that many existing county 'guides' are outdated or grossly incomplete. To undertake a proper 'search' of UK, Irish and other archives, and to compile updated lists (including accession numbers) of relevant maps would be a very worthwhile project for any local history society. The results could be made available through the local journal and updated more frequently through the society's website. Another key difficulty is the expense and difficulty of getting good-quality copies of historic maps, including early OS sheets at six-inch and larger scales. In this regard, serious investment in county library map resources would be money well spent. The example of South County Dublin's principal library at Tallaght has been highlighted in this guide, and others also have given a lead. Local collections, original prints or manuscript and/or later copies, of a wide range of map types, over different periods and scales, immediately accessible, could be posited as the starting point. But there is also the need to foster critical study. Very often maps become available but analysis can go no further; researchers are without the intellectual and research tools to truly engage with the product. In conjunction with any county library collection, there is ample opportunity for lectures, workshops, fieldtrips, training courses and the full range of activities that characterise local history at the grassroots. But having a 'specialised' map collection in any county library brings its own problems, arising from practical storage needs. The separation of map material from the essential accompanying material continues, to the detriment of sound historical research; supporting the vision of the map as part of a larger historical collection is a continuing challenge.

Maps are composite documents, works of art and of science, which need to be explored on their own terms, to be examined, evaluated and appreciated in context. To that end, the greater the understanding of the world within which a particular map or series was created, the political, commercial, military and other pressures behind the undertaking, and the resources devoted to its production, the sounder will be the historical judgement. Recent technological advances – remote sensing and the development of GIS – have moved data capture and the process of drawing into new spheres. However, the deeper questions about what we choose to map and why, why we choose to do it in a particular way, for whom, with what funding, to serve what larger purposes (articulated or not) persists. Map making is still a powerful but understated tool. As historical sources, maps are among the most intriguing materials that any local scholar could work with. The end of local history and of historical geography is to people the past, to truly enter into this earlier world; maps with their record of places and land use, of routeways and landmarks, of lands improved, enclosed, planted, bought, sold and confiscated, continue to be one of the most fascinating entry points, and means of making sense of the past. The map is not the end of the study in itself (that would be cartographic history), but it opens doors to earlier communities.

List of Appendices

Appendix I

Questions to be asked of maps

Who was the creator of this map? Inspiration? Sources (new survey? other maps? travellers' reports?) Created for what purpose? When? Over what time span?

Who funded its creation? Why? On what basis? Patron? Client? Customers? Promotion/advertising undertaken? Who actually used this map and how? What was the process of commissioning?

What is included and why? About what is the map silent? How has the content been manipulated? Does it map the future? Did that future happen? Why or why not?

What is represented symbolically? What is named? Orthography? Representation of settlement? Who is 'present' in the map? In what ways has it desocialised the territory it represents?

What sort of power structures did this map serve/reinforce/threaten? How?

What can we tell of the individual cartographer/surveyor? Of the profession of map-making/surveying at that time, in that place? What was his background, skills, and personal agenda? His other occupations? Family/business connections? Previous experience?

Produced for what audience? Who in fact was the map user? What expectations did the user bring to this 'picture'? Changing readership over time? Contemporary/later responses to this source?

Was this a stand-alone, once-off production? Or was it part of a series? Where are its companion maps?

What was its original physical appearance? Wall display? Pocket map? For an atlas? Part of a published report? Folded as an enclosure with a letter or memorandum? Where was it likely to be stored: estate agent's office? Church? Military headquarters? What other maps were kept with it and why? How did it come to be in its present repository? Are there other copies of this map? Where are they to be found?

How did it enter the public arena? Or was it kept from public scrutiny? Why? Who had access to it? And where? Was it offered for sale? If so, to whom and where? If not, why not? Why did it survive: as a special collector's item? lady luck? never left its first home?

How did this compare with previous output in that geographical area or in that branch of cartography? Is there original thinking in evidence? New sources?

New perspectives? Or reworking of existing (even outdated) material? Inertia? Censorship (which may be covert or overt). Did it have an impact on current or later cartographic work?

When was it fossilised? Of antiquarian, not current geographical, interest? What was it compared with? And why?

Publication history: subsequent editions? Corrections/adjustments/additions? Made by whom? Why?

Method of data collection? Carried out by whom? Level of experience? Support back-up? Technology available? Utilised? Where was the production centre?

'Hidden rules of cartographic imagery' – geometry, projection, ideological filtering out (of the poor, of the native, of the itinerant & the temporary, of that which has no economic value to the outsider); hierarchy of symbols? Size and design? Shading? Decoration (title pages, cartouches, scrolls, lettering, vignettes, compass roses, borders, N point). Aesthetic fashions? Printed or MS? Lithograph? Handcoloured? Who held 'editorial' role? Firm or individual production? What do we know of the creator's/firm's other work? Recruitment and training of cartographers? The 'profession' of map-making at that time? Expertise gathered elsewhere?

Why was a particular feature emphasised? Whose interest did that emphasis serve? Who might have paid for it?

Vision of originator or of the compiler? His formative influences? To whom was he accountable? Accomplishments in other areas?

Context: local, national and international trends? Religious? Philosophical? Political? Military? Commercial? Immediate political climate? Public concerns? Personal/private agenda?

Associated/complementary sources? Was it part of a larger memoir, or is there ancillary written information to be consulted? Interaction with other sources? Shared vision? Part of a larger movement (British Isles, international?) Competition? Contradiction? Hostility?

In the overall process of creating and using this map, what sort of interests did it serve? With which social group(s) is it associated? Beyond the cartographer alone, who controlled its production and dissemination?

Appendix II

Irish Historic Towns Atlas, Royal Irish Academy, Dublin.

Editors: J.H. Andrews (1986–92), Anngret Simms (1986–2004), H.B. Clarke (1992–2004), Raymond Gillespie (1995–2004); Cartographic editors: Mary Davies (1986–99), Sarah Gearty (2000–2004)

FASCICLES PRODUCED TO DATE:

1. Kildare, J.H Andrews, 1986
2. Carrickfergus, Philip Robinson, 1986
3. Bandon, Patrick O'Flanagan, 1988
4. Kells, Anngret Simms, 1990
5. Mullingar, J.H. Andrews with K.M. Davies, 1992
6. Athlone, Harman Murtagh, 1994
7. Maynooth, Arnold Horner, 1995
8. Downpatrick, R.H. Buchanan & Anthony Wilson, 1997
9. Bray, K.M. Davies, 1999
10. Kilkenny, John Bradley, 2000
11. Dublin, part 1 to 1610, H.B. Clarke, 2002
12. Belfast, part 1 to 1840, Raymond Gillespie and Stephen Royle, 2003
13. Fethard, Tadhg O'Keefe, 2003

For forthcoming towns and towns in preparation see http://ria.ie/projects/ihta2/forth.htm.

Appendix III

Ordnance Survey series list, on-line version as of March 2004, reproduced with the kind permission of the National Archives of Ireland

- OS Parish List: Index of over 2500 parish names
- OS 1: Progress reports and returns
- OS 2: Correspondence registers and indexes
- OS 3: Registered correspondence 1824–46 (not listed separately, see OS 2)
- OS 5: Registered correspondence 1847–90 (not listed separately, see OS 2)
- OS 6: Registered correspondence 1891–1935 (not listed separately, see OS 2)
- OS 7: Registered correspondence 1936–1952 (not listed separately, see OS 2)
- OS 12: Engraving journals
- OS 13: Orders, circulars and memoranda
- OS 43: Parish observation books
- OS 51: Registers of documents
- OS 54: Descriptions of trigonometrical stations
- OS 55: Boundary remark books
- OS 58: Content field books
- OS 59: Road field books
- OS 60 Levelling field books
- OS 62 Content registers originals
- OS 63: Content registers duplicates
- OS 65: Levelling registers originals
- OS 66: Levelling registers duplicates
- OS 95: Templemore memoir
- OS 96: Memoir, places other than Templemore
- OS 102: Plots of rivers and lakes
- OS 103: Common plots
- OS 104: Plots
- OS 105: Fair plans
- OS 107: Proof impressions
- OS 140: Manuscript town plans
- Town Plans: Index of town plans (OS 138 to OS 146)

The following OS series have been listed by parish number: OS 54, 55, 58, 59, 60, 62, 63, 65, 66, 104, 105. For more information about the archives of the Ordnance Survey including a note on the place names of parishes and the five Districts (A to E) employed, see the online guide to the archives of the Ordnance Survey.

http://www.nationalarchives.ie

Appendix IV

Ordnance Survey Memoirs: Heads of Inquiry, *c.*1834

Thomas A. Larcom, NLI, MS 7550

SECTION I: GEOGRAPHY OR NATURAL STATE

1 NAME

2 LOCALITY

3 NATURAL STATE: *divided into*

a) Natural Features
Hills, lakes, rivers, bogs, woods, coast, climate

b) Natural History
Botany, zoology, geology

SECTION II: TOPOGRAPHY OR ARTIFICIAL STATE

divided into

1 MODERN

Towns, public buildings, gentlemen's seats, bleach greens, manufactures, mills, communications

2 ANCIENT

The history of the parish, as shown by objects of antiquity (pagan, ecclesiastical, military), and ancient buildings which remain

3 GENERAL APPEARANCE AND SCENERY

SECTION III: THE PEOPLE, OR PRESENT STATE

divided into

1 SOCIAL ECONOMY

Early improvements, local government, dispensaries, schools, poor, religion, habits of the people

2 PRODUCTIVE ECONOMY

Occupations, manufactures and agriculture: rural, commercial and manufacturing, possibilities for improvement

SECTION IV

1 DIVISIONS OF LAND (other than counties, baronies, parishes and townlands)

2 TOWNLANDS

Appendix: tables of schools, occupations, and the supporting statistical information

Appendix V

Valuation Office classifications, Richard Griffith, *Instructions to the valuators and surveyors* (Dublin, 1853)

Classification of buildings, with reference to their solidity

		House or office (1st class)	Built with stone or brick and lime mortar
		Basements to ditto (4th class)	
Buildings	slated	House or office (2nd class) 5 to 12½ feet high	
		House or office (3nd class) 5 to 8¾ feet high	Stone walls with mud mortar
			Dry stone walls pointed
	thatched		Good mud walls
		Offices (5th class)	Dry stone walls

Classification of buildings, with reference to their age and repair

Quality		Description/Signification of the letters
	A +	Built or ornamented with cut stone, or of superior solidity and finish
First Class	A	Very substantial building, and finished without cut stone or ornament
	A –	Ordinary building and finish, or either of the above, when built 25 or 30 years
	B+	Medium, in sound order and good repair
Second Class	B	Medium, slightly decayed, but in repair
	B –	Medium, deteriorated by age, and not in good repair
	C+	Old, but in repair
Third Class	C	Old, out of repair
	C –	Old, dilapidated, scarcely habitable

Appendix VI

Popular map scales commonly used in local history work

Scale as a representative fraction or ratio	*Scale in words (all miles are statute miles)*	*Typical map series*
1:360	10 yards to one inch	
1:540	15 yards to one inch	The War Office used a variety of natural scales such as these in its maps of barracks and fortifications
1:600	50 feet to one inch	
1:720	20 yards to one inch	
1:1,080	30 yards to one inch	
1:1,440	40 yards to one inch	
1:1,800	50 yards to one inch	
1:500	Approx. one mile to ten feet	Manuscript town plans produced for the Valuation Office, 1830–48, especially of smaller towns and villages
1:600	50 feet to one inch	
1:1,000	1m to 1km	OS town plans: detailed sheets of urban areas (from 1970s)
1:1,056	Five feet to one mile	Manuscript town plans produced for Valuation Office, 1830–48 (some also at smaller scales); manuscript Valuation Office maps (urban areas)
1:1,250	50.688 inches to one mile	OSNI 1:1,250 Irish Grid series, similar to the OSNI 1:2500 maps but the larger scale makes it possible to show topographical features in more detail and with greater precision

Scale as a representative fraction or ratio	*Scale in words (all miles are statute miles)*	*Typical map series*
1:2,500	25.344 inches to one mile; 1mm to 2.5m	Northern Ireland 'county' series, six separate series, one for each county in Northern Ireland; these are being replaced by the 1:2,500 Irish Grid system (*c*.23,000 sheets to cover the Republic)
1:2,500	25.344 inches to one mile; 1mm to 2.5m	Coverage is almost countrywide; large uncultivated areas and some islands omitted. Sixteen sheets cover the area of one 6 inch sheet
1:5,000	20cm to 1km	The limit for cadastral mapping; town plan of Athlone
1: 5,280	12 inches to one mile	
1:7,000	1cm to 70m	Ennis street map, OS*i* city & town series
1:9,000	1cm to 90m	Street maps, OS*i* popular series, eg: Limerick, Waterford
1:10,000	1cm to 100m	Replacing the 'six inch' series; Northern Ireland Irish Grid series
1: 10,080	One inch to forty Irish perches	Down Survey, Plantation and early estate surveys
1:10,560	Six inches to one mile	Six-inch 'townland' series, first major OS undertaking in Ireland; Valuation Office (rural); OSNI changed to 1:10,000 series in 1968

Scale as a representative fraction or ratio	*Scale in words (all miles are statute miles)*	*Typical map series*
1:12,000	1cm to 120m	Belfast street map OSNI; Galway street map, OS*i* city & town series
1:13,000	1cm to 130m	Limerick street map, OS*i* city & town series
1:15,000	1cm to 150m	Cork City street map, OS*i* city & town series
1:20,000	1cm to 200m	Dublin district map; OS*i* Dublin street map
1:21,120	Three inches to one mile	Some Down Survey parish maps; some DS barony maps
1:25,000	4cm to 1km	'Wayfarer' OS*i* series and other detailed OS maps for specific areas, including the Aran islands, the Wicklow Way.
1:50,000	2cm to 1km	*Discovery/Discoverer* series, all island of Ireland, OS*i* and OSNI; each covers an area 40km x 30km. 89 sheets in the series; 71 by OS*i*, 18 by OSNI
1:63,360	One inch to one mile	205 one-inch sheets cover the island (1899–1900 ed.); this scale used also for 'district' maps of Dublin, Cork, Killarney, Wicklow
1:70,000	1cm to 700m	OSNI, the Gateway to Ulster heritage and activity map, one sheet covers SE Ulster
1:100,000	1cm to 1km	
1:126,720	Half inch to one mile	Most of OS county index sheets. Coloured OS*i* series; takes twenty-five sheets to cover all Ireland; takes four sheets to cover N. Ireland

Scale as a representative fraction or ratio	*Scale in words (all miles are statute miles)*	*Typical map series*
1:190,080	One inch to three miles	County index sheets, larger counties
1:210,000	1cm to 2.1km	OS*i* 'Complete Road atlas of Ireland'
1:250,000	1cm to 2.5km	Four sheets to cover Ireland, the OS*i*/OSNI 'Holiday' series, titled *Ireland West, Ireland South, Ireland East, Ireland North*
1:253,440	Quarter inch to one mile	Six sheets required to cover Ireland; Griffith's railway map (1839); *Map to accompany the general scheme for the defence of Ireland* (War Office, 1885–6)
1:380,160	One inch to six miles	
1:500,000	Approx. one inch to eight miles	GSGS maps of Ireland; OSNI map of Northern Ireland (one sheet covers all N. Ireland)
1:600,000	1cm to 6km	Road map of Ireland/ Bord Fáilte tourist map of Ireland; one sheet covers Ireland
1:625,000	1cm to 6.25km	Monastic map of Ireland (OS*i*)
1: 633,600	One inch to ten miles	Railway map of 1838 (published version); OS peat bogs of Ireland map. One sheet covers Ireland
1:750,000	Approx. one inch to 12 miles	One sheet geological map of Ireland, in colour

Scale as a representative fraction or ratio	*Scale in words (all miles are statute miles)*	*Typical map series*
1:760, 320	One inch to twelve miles	
1:1,000,000	1cm to 10km	
1:2,000,000	1cm to 20km; 1¼ inches to 40 miles	One-page map of Ireland, school atlas
1:10,000,000	1cm to 100km	Western Europe map, school atlas
1:100,000,000	1cm to 1000km	World map, school atlas

Appendix VII

Conversion table: 'Irish' /Imperial/Metric units

LENGTH

1 inch = 2.54 cm

1 cm = 0.3937 inches

1 foot (12 inches) = 30.48 cm

1 metre = 1.094 yards

1 foot of the Imperial Standard Yard = 0.304 79947 International Metres

1 yard (3 feet) = 0.9144 metres

1km = 0.6214 miles
1000 m in 1km

1 statute mile (63,360 ins) = 1.6093 km or 8 furlongs or 1760 yards

2 Irish miles = 2.55 statute miles

1 pace = 30 inches

1 English or statute perch (or pole) = 16½ feet

1 Irish perch (or plantation perch) = 21 feet

One link = one hundredth part of the surveyor's chain

One statute chain = 22 yards

One furlong = 10 chains or 40 poles or one eighth of a mile

1 nautical mile = one sixtieth of a mean degree of longitude, or one minute of latitude at the place; *c.*6080 feet; a cable is one tenth of a sea mile

AREA

40 perches in a rood; 160 perches in one acre; one rood is ¼ part of an acre

1 square yard = 0.8361 m^2
1 imperial or statute acre = 0.4047 hectares
1 square mile = 2.590 km^2
1 acre Cunningham = 1.291322 statute acres

1 m^2 = 1.196 square yards
1 ha = 2.471 acres
1 km^2 = 0.3861 square miles

1 acre statute = .7744 Cunningham acres

1 perch or pole = 272¼ square feet or 30¼ square yards

1 imperial or statute acre = 4840 square yards

50 Irish acres = 81 imperial acres

A R P
100 Irish acres statute = 161 3 37.4 statute

A R P
100 acres statute = 61 2 37.6 Irish

1 Irish acre = 7840 square yards

A R P
10 Irish acres = statute 16 0 31.7 statute

1 acre statute = .617347 acre Irish

1 perch statute = .617 perch Irish

1 rood statute = .617 rood Irish

10 square chains make a statute acre

Appendix VIII

Questions to ask of your place, in the search for pre-OS and other non-OS maps:

Where are we? Which townland, street, parish (civil, RC), ward, rural district, poor law union, registrar's district, electoral district, barony, county, diocese, province?

How is the area named on maps? Spelling(s)? Font style and size employed? When and why is it named? Associated symbol?

Topographical headings under which it might be mapped? – coastal (marine/admiralty charts), inland navigation (canal mapping); geology, bog, mining resource (geological survey); scenic lakelands (tour guides); floods frequently (drainage schemes); rugged mountain (military road). What are the various names under which local forts, railway lines, rivers, bogs, bays and inlets are known? Alternative spellings?

Who held administrative responsibility for this place? city corporation? county borough, Poor Law guardians? county council? congested districts board? harbour board? Who was in charge of what and when? What maps might they have produced or used? Abandoned or failed development plans?

Who were their administrative predecessors? manorial lord? grand jury? ballast board? commissioners with defined powers? What maps might they have produced? What other maps might they have acquired or held in their own records?

History of settlement here: was it among the escheated or planted counties? Who owned local property? Was it part of a landed estate? Who might have had an interest in mapping this place or at least a part of it? Harbour developers? The congested districts board? A railway company? Canal company? Board of Ordnance? Lighthouse services? What improvements or changes have been made? By whom?

What key features exist that might have generated special maps? Railway, mines, bogs, forestry, barracks, forts, bridges, roads, reservoirs, piers, harbour, anchorage? Major events? Battles or rebellions? Inland navigation potential? Botanical gardens?

What organisations were based or operated here? Congested Districts Board? Evangelical missionary society? Religious order? Barracks board?

Who visited here? What are its scenic or visitor attractions? Who might have commented on your place?

Are there likely 'antiquarian' or archaeological attractions? Churches, cemeteries, round towers, forts, raths, lighthouses? Are any the subject of special reports?

Appendix IX

Blank Valuation Office sheet, to enlarge and enter details from manuscript 'cancellation books'

General Valuation of Ireland

15 and 16 Vic. cap. 69.

City of Dublin

________ *Ward.*

Number		Situation of Property	Names of Occupiers	Names of Immediate Lessors	Description of Property	Net Annual Value		Length in Yards	Breadth in Yards	Item	Height and Quality	Area in Square Yards	Price	Amount by Tables	Multiplier for Street	Relative Value	Observations and Rents
On Map	On Street					1853	185										

Appendix X

Drawing symbols, using graph paper as a guide

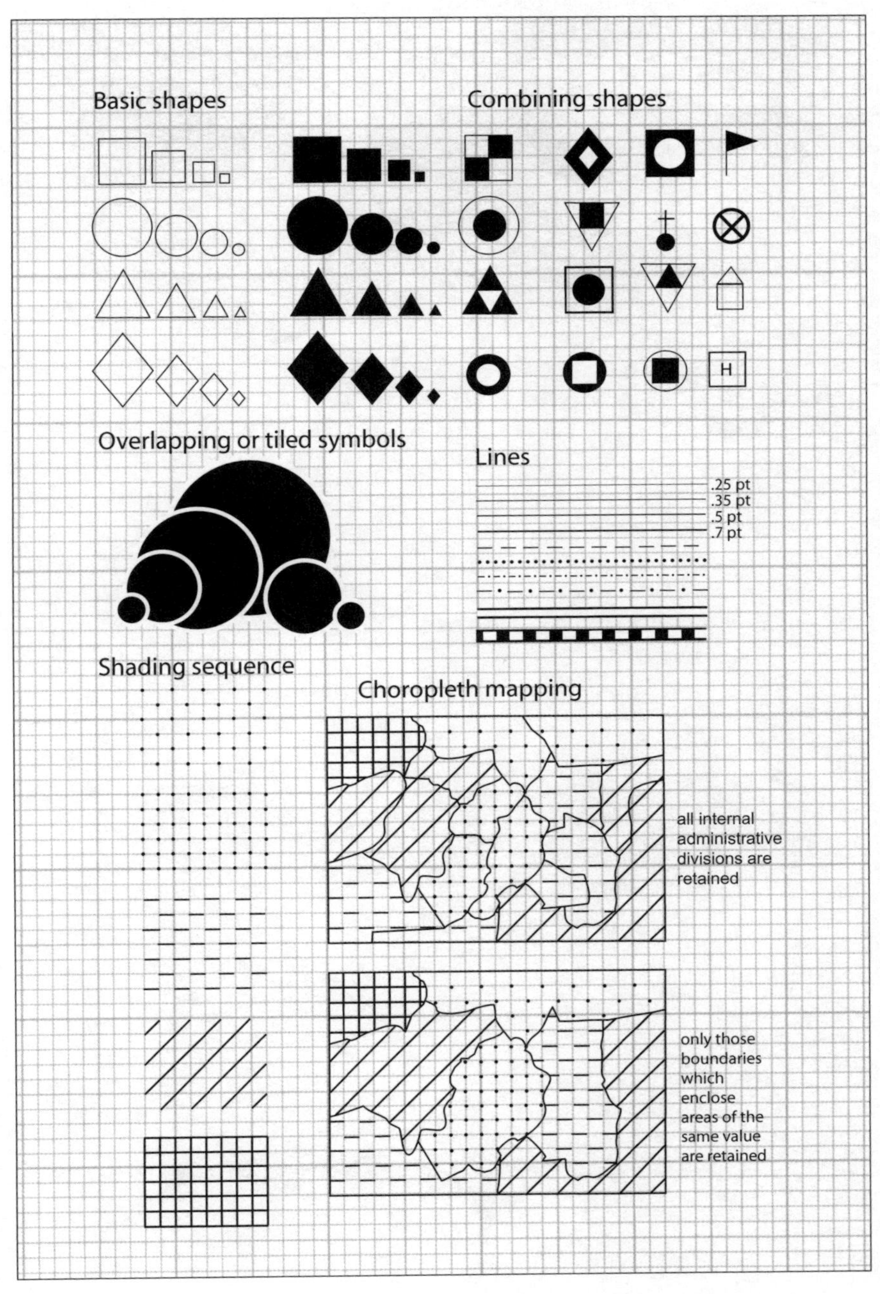

Bibliography

Admiralty manual of hydrographic surveying, i, 3rd edn (London, 1965)

Andrews, J.H., *A paper landscape: the Ordnance Survey in nineteenth-century Ireland* (Oxford, 1975; 2nd edn, Dublin, 2002)

—— 'Colonial cartography in a European setting: the case of Tudor Ireland' in *The history of cartography*, iii, Cartography in the age of renaissance and discovery (Chicago and London, 2001)

—— 'Maps, prints and drawings' in William Nolan and Anngret Simms (eds), *Irish towns: a guide to sources* (Dublin, 1998), pp 27–40

—— *Shapes of Ireland, maps and their makers 1564–1839* (Dublin, 1997)

—— *History in the Ordnance map, an introduction for Irish readers*, 1st edn (Dublin, 1974; 2nd edn, Kerry, Montgomeryshire, 1993)

—— 'Mapping the past in the past: the cartographer as antiquarian in pre-Ordnance Survey Ireland' in Colin J. Thomas (ed.), *Rural landscapes and communities, essays presented to Desmond McCourt* (Dublin, 1986)

—— 'Map and mapmakers' in W. Nolan (ed.), *The shaping of Ireland: the geographical perspective* (Cork, 1986)

—— *Plantation acres* (Omagh, 1985)

—— 'Maps and the Irish local historian' in *Bulletin of the Group for the study of Irish Historic Settlement*, 6 (1979), pp 22–31

—— *Irish maps* (Dublin, 1978)

Anson, R.W. and Ormeling, F.J. (eds), *Basic cartography*, ii (London and New York, 1993)

Anson, R.W. (ed.), *Basic cartography*, ii (London and New York, 1988, reprinted 1996)

Anson, R.W. and Ormeling, F.J. (eds), *Basic cartography for students and technicians*, i, 2nd edn (London and New York, 1993)

Black, Jeremy, *Maps and history, constructing images of the past* (New Haven and London, 1997)

Colby, Thomas, *Ordnance Survey of the City of Londonderry*, i, 1837 (facsimile reprint, Limavady, 1990), preface

Delano-Smith, Catherine and Kain, Roger J.P., *English maps: a history* (London, 1999)

Gearty, Sarah, 'Irish historical maps as a source for the archaeologist' in *Trowel*, viii (1997), pp 1–6

Harley, J.B. and Woodward, David (eds), *The history of cartography*, i, *Cartography in prehistoric, ancient and medieval Europe and the Mediterranean* (Chicago and London, 1987)

Harley, J.B. 'Maps, knowledge and power' in Cosgrove, Denis and Daniels, Stephen (eds), *The iconography of landscape* (Cambridge, 1988, 1994), pp 277–312

Hayes-McCoy, G.A., *Ulster and other Irish maps, c.1600* (Dublin, 1964)

Herity, Michael, *Ordnance Survey letters: Donegal* (Dublin, 2000)
—— *Ordnance Survey letters: Dublin* (Dublin, 2000)
—— *Ordnance Survey letters: Kildare* (Dublin, 2000)
—— *Ordnance Survey letters: Kilkenny* (Dublin, 2003)
—— *Ordnance Survey letters: Meath* (Dublin, 2000)

Herries Davies, Gordon and Mollan, R. Charles (eds), *Richard Griffith, 1784–1878* (Dublin, 1980)

Kissane, Noel (ed.), *Treasures of the National Library of Ireland* (Dublin, 1995)

Larcom, Thomas A., *The history of the survey of Ireland commonly called the Down Survey, by Sir William Petty* (Dublin, 1851; reprinted by Augustus M. Kelley: New York, 1967)

Lohan, Rena, *Guide to the archives of the Office of Public Works* (Dublin, 1994)

Monmonier, Mark S., *How to lie with maps* (Chicago, 1991)

Nolan, William and Simms, Anngret (eds), *Irish towns: a guide to sources* (Dublin, 1998)

O'Flanagan, Patrick, 'Surveys, maps and the study of rural development' in D. Ó Corráin (ed.), *Irish antiquity* (Dublin, 1981), pp 320–7

Parkhill, Trevor, 'Valuation records in the Public Record Office of Northern Ireland' in *Ulster Local Studies*, xvi, no. 2 (1994), pp 45–58

Prunty, Jacinta, 'Estate records' in Nolan, William and Simms, Anngret (eds), *Irish towns: a guide to sources* (Dublin, 1998), pp 121–36
—— *Dublin slums, a study in urban geography* (Dublin, 1998)

Reeves-Smyth, Terence, 'Landscapes in paper, cartographic sources for Irish archaeology' in *British Archaeological Records*, cxvi (1983), pp 119–77

Refaussé, Raymond and Clark, Mary (eds), *A catalogue of the maps of the estates of the archbishops of Dublin, 1654–1850* (Dublin, 2000)

Seymour, W. A. (ed.), *A history of the Ordnance Survey* (Folkestone, 1980)

Skelton, R. A., *Maps, a historical survey of their study and collecting* (Chicago and London, 1972 H., 1975)

Slater, Terry, 'The European historic towns atlas project' in *Journal of Urban History*, xxii, no. 6 (1996), pp 739–49

Smith, David, *Maps and plans for the local historian and collector* (London, 1988)

Stetoff, Rebecca, *The British Library companion to maps and mapmaking* (London, 1995)

Index